6.80

Ritter

# HIGH ENERGY ASTROPHYSICS
# AND ITS RELATION TO
# ELEMENTARY PARTICLE PHYSICS

# HIGH ENERGY ASTROPHYSICS AND ITS RELATION TO ELEMENTARY PARTICLE PHYSICS

edited by

Kenneth Brecher

and

Giancarlo Setti

The MIT Press

Cambridge, Massachusetts, and London, England

PUBLISHER'S NOTE

This format is intended to reduce the cost of publishing certain works in book form and to shorten the gap between editorial preparation and final publication. The time and expense of detailed editing and composition in print have been avoided by photographing the text of this book directly from the author's typescript.

The MIT Press

Copyright © 1974 by
The Massachusetts Institute of Technology

All rights reserved. No part of this book may be reproduced in any form or by any means, electronic or mechnical, including photocopying, recording, or by any information storage and retrieval system, without permission in writing from the publisher.

Library of Congress catalog card number: 74-19794.

ISBN 0 262 02115 3 (hardcover)
ISBN 0 262 52035 4 (paperback)

Printed in the United States of America

# CONTENTS

Foreword

| | |
|---|---:|
| QSO's, Observations, and the Redshift Problem<br>    H. C. ARP | 1 |
| Big Bang Cosmology<br>    K. BRECHER | 77 |
| Observational Problems of High Energy Astrophysics<br>    E. M. BURBIDGE | 109 |
| Theoretical Problems of High Energy Astrophysics<br>    G. R. BURBIDGE | 173 |
| X-Ray Astronomy<br>    R. GIACCONI | 201 |
| Statistical Thermodynamics of Strong Interactions<br>    R. HAGEDORN | 255 |
| Microphysics, Cosmology, and High Energy Astrophysics<br>    F. HOYLE | 297 |
| Electrodynamics and Cosmology<br>    J. V. NARLIKAR | 345 |
| Weak Interactions and Cosmology<br>    Y. NE'EMAN | 387 |
| Structure and Dynamics of Galaxies<br>    K. H. PRENDERGAST | 415 |
| Strong Interactions, Gravitation and Cosmology<br>    A. SALAM | 441 |
| Extragalactic Observational Astronomy<br>    W. L. W. SARGENT | 453 |
| General Relativity, Collapse and Singularities<br>    J. A. WHEELER | 519 |
| Index | 573 |

# FOREWORD

This volume contains the lectures presented at the NATO Advanced Study Institute entitled "High Energy Astrophysics and Its Relation to Elementary Particle Physics" which was held at the "Ettore Majorana" Centre for Scientific Culture, Erice, Sicily, Italy from June 16 - July 6, 1972. It is the result of notes taken during the lectures by some of the more than 100 participants, amended by them after listening to tapes of the lectures and, in most cases, checked by the lecturers themselves. In editing this volume, we have decided not to change the informal style of the lectures so as to preserve the actual atmosphere in which they were delivered. What this loses in strict clarity of elucidation we feel sure is more than compensated for by spontaneity of thought: rigorously re-worked written notes would make the topics discussed appear much more settled than they did in presentation. By retaining this format, we hope we have captured some of the excitement and uncertainty inherent in the topics covered. Naturally, we accept full responsibility for any errors which have thereby found their way into the resulting book.

This Institute was organized as an interdisciplinary school by G. R. Burbidge, Director of the Course, and K. Brecher. As such, its aim was to acquaint students and young researchers in elementary particle physics with the observations, theories, and problems of high energy astrophysics (situations involving high energy particles, massive or dense objects, high effective temperatures, and large total energies) which might lend themselves to new physical interpretations. Also, it sought to suggest an astrophysical setting in which current elementary particle and field theoretical ideas might be tested or further extended. On the other hand, it gave particle physicists the opportunity to present experimental results and theoretical considerations which might be intimately connected with astrophysics, and which might help to solve some current astrophysical problems. There was also discussion of the constraints that microscopic physics might set on the macroscopic world as a whole. The simple application of known physical processes in elementary particle physics to astrophysics (e.g., gamma ray production in pion decay), which has been the subject of other summer schools, was de-emphasized in favor of extending known particle physics to new domains (e.g., quantum gravity, mass spectrum of elementary particles at high energies, $K^0$ decay and long range interactions, etc.).

Individual lectures deal with one or another of these problems. If there really do exist "anomalous" redshifts (as H. Arp claims in his lectures), its explanation will almost surely require some fundamentally new physical idea (rather than a contrived astrophysical model). The lectures by K. Brecher, K. H. Prendergast, and W. Sargent were offered to provide the

# Foreword

"rule," so to speak, to which Dr. Arp's examples are the exception. An understanding of the violent activity exhibited by quasars and galaxies, discussed by G. R. Burbidge, E. M. Burbidge and R. Giacconi, may well require a new approach to be found only at the frontiers of elementary particle physics. On the other hand, as discussed by R. Hagedorn, A. Salam, and Y. Ne'eman, particle physics may already have demonstrated its intimate relation to the Universe. The lectures of Hoyle, Narlikar and Wheeler all deal with even broader (more speculative?) connections between the microscopic and macroscopic world. If any of this collection of observation and theory helps to stimulate the search for such a relation, it will more than justify its production many times over.

We would like to thank the following people, without whom the present volume would never have been produced. First, Dr. Gunnar Randers, former NATO Assistant Secretary General for Scientific Affairs for providing the support for this NATO Advanced Study Institute. Second, to Professor A. Zichichi for his invitation to hold the Institute at the "Ettore Majorana" Centre. Many thanks to the secretaries who helped in both the organizational details of the meeting and who typed this volume: Paola Zanlungo, Brunella Arbizzani and Wanda Iasquier; to Giovanni Zamorani, whose help was fundamental in preparing the material for publication; to Luciano Baldeschi for making many of the drawings.

Above all, we would like to thank all the lecturers and students who together provided the substance of the book.

                                          Kenneth Brecher
                                          Massachusetts Institute
                                              of Technology
                                          Cambridge, Massachusetts

                                          Giancarlo Setti
                                          University of Bologna
                                          Bologna, Italy

QSO's, Observations, and the Redshift Problem

Lectures by

H. C. Arp

California Institute of Technology

Notes by

M. Shara, G. Cavallo, J. Kormendy, C. Bloch

Lecture 1

## EJECTION OF MATTER FROM GALAXIES

I want to start this first lecture with a new piece of observational, photographic evidence that is doubly interesting. First, it is new, and second, it embodies all the principles and important aspects of things we're going to be seeing a lot of in all the material I'll be showing.

Before I show this first photograph, I want to make a few remarks about Seyfert Galaxies. (The new photograph is of NGC 4151, a Seyfert Galaxy). What most people call a "Seyfert Galaxy" is often incorrect; the correct term would be "Seyfert spectrum". In 1943 Karl Seyfert wrote a little paper in the Astrophysical Journal in which he discussed six galaxies: NGC 1068, NGC 4151, NGC 1275 and three others. The feeling has since arisen that a spiral galaxy with a brilliant, semi-stellar nucleus defines a Seyfert Galaxy; this is incorrect. NGC 1275, for example, is an E galaxy ejecting a jet from its nucleus. The jet has a redshift of 3000 km/sec relative to the galaxy, and the entire object is completely different from a spiral galaxy. Thus technically, what we should say is that a Seyfert Galaxy is one which has a Seyfert spectrum. This is important because we're going to see a lot of examples of Seyfert Galaxies with very different morphologies. Our job will be to try to understand the connections and physical relationships between morphologically different Seyferts.

NGC 1068 and NGC 4151, for example, have brilliant, semi-stellar nuclei, high-brightness "shoulders" and fainter outer regions, as well as spiral structure. Seyfert spectra are also exhibited by E galaxies and by compact objects with jets.

The typical signature of a Seyfert spectrum is the presence of $H\beta$ and forbidden oxygen lines (OIII). The $H\beta$ lines can be up to 8500 km/sec wide, while the (OIII) lines are much narrower. In NGC 4151, about 8% of the nucleus' light is contributed by the hydrogen lines, and 14% of the nucleus' light arises in all the emission lines, including those of higher excitation. The Seyfert spectrum is interesting because it represents extreme physical conditions in very small regions of space. These spectra also indicate the existence of some relationship between Seyfert nuclei and QSO's or QSR's.

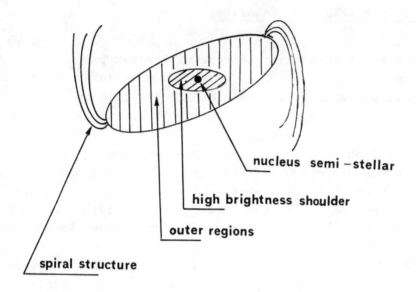

Figure 1.1  A spiral Seyfert

A thirty minute exposure of NGC 4151 with the 200-inch telescope shows a burned-out nucleus.  Shorter and shorter exposures seem to show a slightly fuzzy, semi-stellar nucleus that has a wider profile than an unresolved stellar image.  Accurate profile tracings are needed to decide for certain if a given Seyfert nucleus is resolved.  A maximum distance to a Seyfert Galaxy with a variable, resolved nucleus can be set by assuming that the nucleus varies coherently.

In figure 1.2, note the presence of a companion spiral galaxy with well-resolved arms.  The fact that the companion's arms are well resolved indicates that it may not be much more remote than NGC 4151.  We also see that a jet of material with a small galaxy on the end protrudes from one of the companion's spiral arms. A second companion galaxy is seen on the opposite side of NGC 4151.

A very deep plate of NGC 4151, taken by John Kormendy, shows spiral arms reaching out from the galaxy and ending on its two brightest companions.  On the original plate a "fountain" or

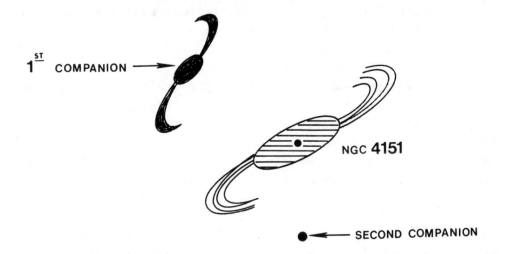

Figure 1.2  NGC 4151 and its two brightest companions

"splash" of material can be seen where the spiral arm meets one of the companions; this is a strong indication of some sort of physical connection. However, the redshift of NGC 4151 is about 900 km/sec while that of its brighter companion is 6700 km/sec! The other companions' redshifts haven't yet been measured but we can expect them to be about 20,000 km/sec, by analogy with other objects I'll be discussing.

The following four empirical rules of ejection were arrived at from observations like those above; they will be the main subject of my six talks.
1. Galaxies eject matter. This can be in the form of radio, luminous or compact matter, or in other states we don't understand.
2. Objects tend to be ejected in opposite directions. This is seen both in radio sources and in luminous matter.
3. Ejected objects tend to have higher surface brightnesses. Perhaps this is because very young, newly ejected material is highly compact with a high surface brightness.
4. The ejecta have generally higher redshifts than their parent bodies, ranging in velocity from a few hundred to hundreds of thousands of kilometers per second higher. This is the most controversial of my rules and many of my colleagues flatly disbelieve it.

We will try to establish that some galaxies have ejected very compact high redshift bodies, some of which are quasars.

I'll start discussing the first of these four ejection rules with something that almost everyone agrees about: double radio lobes have been ejected from galaxies. Figure 1.3 shows IC 4296 which is in the Centaurus Chain of galaxies. It's an E galaxy of about 11.9 apparent magnitude. There are two lobes on either side of the galaxy and the galaxy itself is seen to be a weak radio emitter. The polarization vectors of all three sources are parallel. This object is like the classical picture people have of a radio source: a central galaxy, and two extended side lobes not optically identified.

I want to make several points about this type of object which is probably a misleading stereotype. First, there aren't many radio sources of this type and, second, this sort of object skirts the question of what an E galaxy is, and what the relationship between an E galaxy and a spiral galaxy is.

Just roughly speaking the characteristics of the elliptical (or E) galaxies are the following:

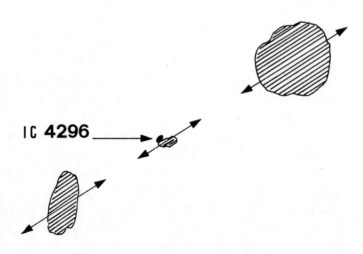

Figure 1.3  Sketch of  IC 4296 from Parkes measurements

a)  E galaxies are composed mainly of red stars and are free of gas and dust.
b)  They have low surface brightness.
c)  All the morphological forms of E galaxies we see are relaxed. That is, they are not in dynamically unstable forms.

The spirals, on the other hand, have almost exactly the opposite characteristics:
a)  They contain a lot of bright blue stars, gas and dust.
b)  They have high surface brightnesses due to the bright blue stars clumped in the spiral arms.
c)  The arms, at first sight, appear to be dynamically unstable, and should break up and become relaxed in $10^8$ or $10^9$ years.

People who haven't looked at a lot of galaxies tend to think that E and S galaxies are completely different. More attention should be paid to possible transition cases and relationships between E and S galaxies.

We'll keep this idea in mind as we examine the central galaxies producing ejection.

The first radio galaxies to be found were Virgo A and Centaurus A. Centaurus A (NGC 5128) is sketched in figure 1.4.

This strange object presents us with a contradiction: We have a lot of dark, spiral-galaxy type of material (type I material) superimposed on an elliptical galaxy. It was originally

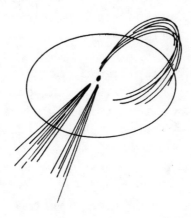

Figure 1.4   Sketch of Centaurus A (NGC 5128)

thought that NGC 5128 represented two galaxies (a spiral and elliptical) in collision, but after Baade revised the distance scale it was realized that collisions were very improbable. The dark material streaming away from the nucleus seems to originate in the center of the galaxy and is probably associated with the radio noise we detect.

A radio map of NGC 5128 by Frank Kerr shows a small inner pair of radio sources, and a larger radio pair lying roughly perpendicular to the absorbing material in NGC 5128.

The distribution of objects around NGC 5128 seems to be rather special. In fact, there are indications that the center of our local supercluster, extending out to about 100 Mpc, is near Centaurus A. By "supercluster" I mean to include not only galaxies, but also compact and peculiar objects, and quasars. Plotting the brightest radio sources around NGC 5128 we see one wide pair (about 18° apart). There also seem to be groups of bright galaxies on opposite sides of NGC 5128. (Including many fainter galaxies tends to clutter the field with unrelated background objects, destroying any significant pattern). The radio sources on either side of NGC 5128 define a "line of ejection". It is striking that the brightest galaxies around Centaurus A fall closely about this line, hinting that these objects, along with the double radio source, may have been ejected from NGC 5128. The redshifts of galaxies along this line are a few hundred km/sec greater than that of Centaurus A, but IC 4296 has a redshift of 3700 km/sec, compared with that of 400 km/sec for NGC 5128. Along conventional lines IC 4296 would be classed as a background object, but I think it, like other objects I'll show, represents ejection from the central galaxy.

The next radio source is Virgo A, or M87, a giant elliptical galaxy which dominates the northern ring of the Virgo cluster. It's been known for a long time that M87 has a jet leading away from the nucleus. Photographs of M87 in (OII) light show that the nucleus emits this line. The jet shows a blue continuum and is broken up into several small condensations, of the order of 1 arc second, which may not all be resolved. Two linear features which I call the counterjet can be seen pointing in the same direction as the jet but on the opposite side of the nucleus from the jet.

This is another strong radio galaxy which contradicts the stereotype and is highly peculiar. Most radio astronomers agree that the jet has been ejected from M87's nucleus. The compact condensations in the jet may be just resolved and they're seen to emit blue, featureless synchrotron radiation. If we could measure these objects' redshifts, and if they were high we certainly could call them quasars.

Another striking thing about the Virgo cluster is the chain aspect of galaxies in this region. The E galaxies in this region (classified by de Vaucouleurs) fall on a straight line which is an extension of the line defined by the M87 jet and counterjet. I'd like to suggest that this amazing coincidence is no coincidence at all, but rather that M87 has ejected these "second-generation" galaxies.

Another remarkable thing about the Virgo cluster was first pointed out by de Vaucouleurs. He noticed that the spiral galaxies in the northern wing of the cluster appeared to form a hollow oval about the linear chain of E galaxies. This observation is consistent with a series of ejections: M87 ejecting a chain of E galaxies, which in turn eject spiral galaxies. In fact, we would expect the third generation spirals to lie in a hollow oval about the center of the cluster. It's also important to note that the average redshift of the E galaxies is about 900 km/sec while that of the spirals is about 1800 km/sec in the Virgo Cluster, consistent with my fourth rule of ejection.

This chain of E galaxies is perplexing from a dynamical standpoint. Even with small peculiar velocities of say 100 km/sec relative to each other, the E galaxies should disrupt the chain in $10^8$ or $10^9$ years. We believe, however, that these galaxies are $10^{10}$ years old. Why, then, does the chain persist? Oort regards the galaxies on the chain as matter collapsing preferentially along one axis. I'll present a piece of evidence favoring ejection over condensation when I discuss NGC 383.

A spectrum taken along the line of the jet and the counterjet in M87 shows strong blue and weak red continuous emission from the jet, while the main galaxy is weak in the blue and strong in the red part of the spectrum. The counterjet shows weak hydrogen emission but no continuum, and this is a fact we have to live with: different ejecta from one object can show different spectra. We might picture the counterjet to be the track of some object that has passed out of M87, leaving a little wisp of glowing hydrogen in its wake. The passage probably occurred recently because the glowing hydrogen should decay rapidly. Radio maps of M87 show a source elongated along the direction of the jet,

Table 1.1
Radio-Source E Galaxies

| GALAXY | TYPE | $M_{pg}$ | RADIO SOURCE | $S_{1410}$ | REMARKS |
|---|---|---|---|---|---|
| NGC 5128 | E0p | 8.4 | CEN A | 1330 | CENTAURUS CLUSTER LINE |
| NGC 4486 | E1p | 10.3 | VIR A | 200 | VIR CLUSTER LINE |
| NGC 4374 | E0 | 10.8 | 3C272.1 | 6.0 | MARKARIAN CHAIN IN VIR |
| NGC 4261 | E2 | 11.8 | 3C270 | 17.9 | TWO VIRGO CLUSTER CHAINS |
| IC 4296 | E0 | 11.9 | PKS1333-33 | 7.0 | CENTAURUS CLUSTER LINE |
| NGC 1275 | E2p | 13.1 | 3C84 | 13.5 | PERSEUS CLUSTER LOOSE LINE |
| NGC 1265 | | (13.8) | 3C83.1B | | PERSEUS CLUSTER NO LINE |
| IC 310 | | (15.0) | | | PERSEUS CLUSTER NARROW LINE |
| NGC 547 | E1 | 13.4 | 3C40 | 5.2 | LINES IN CLUSTER |
| NGC 545 | E | 13.7 | | | |
| NGC 383 | E or S0 | 13.6 | 3C31 | 5.0 | MARKARIAN CHAIN |
| NGC 7720 | (D4)dbl | 13.9 | 3C465 | 7.7 | TERMINUS OF EIGHT NGC GALAXIES |

Table 1.1 (cont'd)
Radio-Source E Galaxies

| GALAXY | TYPE | $M_{pg}$ | RADIO SOURCE | $S_{1410}$ | REMARKS |
|---|---|---|---|---|---|
| NGC 3862 | DE1 | 14.0 | 3C264<br>PKS1142+19 | 5.5 | Zw. VOL.II,<br>6° LONG CHAIN<br>2° SHORT CHAIN |
| NGC 1218 | DE3 | 14.0 | 3C78 | 7.6 | Zw. VOL. V,<br>LOOSE LINE? |
|  | (D4) | 14.8 | 3C317 | 5.5 | Zw. VOL.I,<br>ELONGATED<br>CLUSTER |
|  | E0 | 15.0 | 3C66 | 9.7 | LINE OF<br>GALAXIES |

but no double lobe source.

Table 1.1 is a summary of all the radio source E galaxies' properties. This table was assembled in a systematic way and the results are very clear-cut. I just listed the 13 brightest radio-source E galaxies, down to 15.0 photographic magnitude. These could be examined in detail and no less than eleven out of the thirteen galaxies fall in distinct, clearly marked chains; two cases are indeterminate. This overwhelming result, that radio-source E galaxies tend to fall in lines, is a fact we have to accept and, if possible, explain. This hasn't been done yet and even my ejection hypothesis is unsatisfactory for several reasons.

Figure 1.5 is a sketch of a new Westerbork radio map of NGC 383 that Ron Ekers recently sent me. There's a chain of E galaxies with NGC 383 at the center and several fainter objects to either side. The radio source is clearly not double but extends along the line of E galaxies. The redshifts of all these E galaxies are roughly the same, but the outer, fainter objects' redshifts are systematically somewhat higher than that of NGC 383.

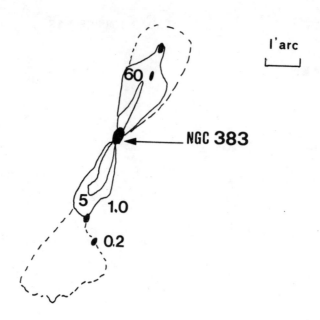

Figure 1.5  Sketch of a new Westerbork radio map of NGC 383

Let's come back to Oort's interpretation of preferential collapse along a line. Accepting this hypothesis, we are left with the paradox of the ages of the galaxies ($10^{10}$ years) conflicting with the line disruption time ($10^8$ to $10^9$ years). Also, rotation during collapse would curve the line of galaxies and the radio isophotes. Such curvature is not seen, raising a problem for that interpretation. It is also a problem as to how a direction of ejection can be locked in against rotation of the central galaxy.

Figure 1.6 is a radio map of the Perseus Cluster, containing NGC 1275, NGC 1265 and IC 310.

Ryle has hypothesized that NGC 1275 is the source of the radio ejection in the Perseus Cluster. He suggests that tubes of magnetic lines of force guide ejected particles from NGC 1275 past NGC 1265 and IC 310, producing the observed radio tails.

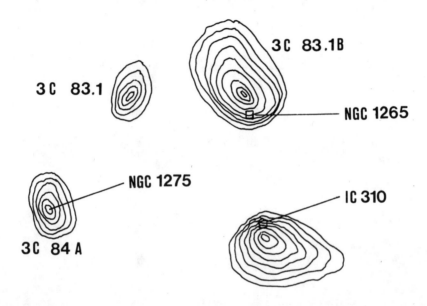

Figure 1.6  Radio map of the Perseus Cluster, containing NGC 1275, NGC 1265 and IC 310

I would like to suggest that these galaxies originated in some central explosion or fissioning process. Note that the radio tails point back towards the same point, approximately the centroid of the cluster, and I think that the luminous optical and radio matter has flowed outwards from this erstwhile center of the cluster. Another chain of galaxies associated with IC 310 is approximately along the line of the radio tail and points back towards the center of the Perseus Cluster. This again supports my theory of ejection from the centers of clusters.

Finally, the Westerbork observers have looked at the famous chain VV 172, and found no radio emission from it. All the galaxies have redshifts of about 15,000 km/sec except the distorted one which has a redshift of about 30,000 km/sec(fig.1.7).

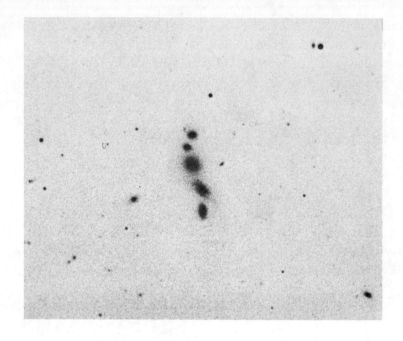

Figure 1.7

There's been a lot of argument about whether this latter galaxy is just a projected background galaxy. I don't think it is, because it's a peculiar bluish object which makes it unusual, and unlikely to occur just in the midst of a tight chain of galaxies.

I recently did some radio astronomical observations at 13 cm. with the 210 foot dish at Goldstone, California. Ken Kellermann tells me that one expects less than one-tenth of a source brighter than 0.1 flux units per square degree of sky at 13 cm.

I detected several radio sources brighter than 0.2 flux units around NGC 4151, in about four square degrees of sky, strongly indicating that there are radio sources associated with this Seyfert Galaxy. In my next lecture I'll show similar results about spirals and ellipticals. In particular, I'll talk more about ejection from E and S galaxies, and suggest that there are physical similarities between them.

An even more interesting configuration is exhibited by the radio source chain pointing to NGC 520. One additional feature is made up by the angular radio diameters as measured by Miley (1971). The diameters decrease in the succession 39, 9,

<2 seconds of arc, and also tend to point toward a progression of characteristics along the chain.

We can use the mechanism of synchrotron radiation as a loose working hypothesis for the explanation of these sources. The electrons responsible for the radio emission have lifetimes inversely proportional both to their energy and to the square root of the radiated frequency. Therefore, if they will not be regenerated, we shall expect that the older radio spectra will be steeper, while the younger spectra will be flatter. There are two kinds of chains of ejected objects: (a) curved, if the ejecting object was rotating, and (b) straight, if there is no rotation. In the latter case, we may suggest that the smallest objects are those which penetrated furthest. They are later in their development and appear younger to us. The latter is the suggested model for the chain of QSR's from NGC 520.

Lecture 2

REDSHIFT ANOMALIES IN ASSOCIATIONS OF LARGE GALAXIES AND THEIR COMPANIONS

During my first talk yesterday I think I failed to sufficiently emphasize a very important relationship between large galaxies and their companions.

As a first example of this we have M31, the large Sb spiral which dominates our own Local Group. Its very well known companions are NGC 185, M32, NGC 205 and M33. All four have redshifts more than 50 km/sec higher than that of M31 (See table 2.1). The same fact holds true for the companions of M81, and especially for M82. The residual redshift of this companion with respect to M81 is +234 km/sec, an unusually large redshift for a group that's supposed to be tightly held together. NGC 5128 and M51 also display companions with redshifts systematically higher than that of the parent galaxy.

Sixteen out of the nineteen companions in table 2.1 show positive differential redshifts, and the mean velocity of all the sample is about +70 km/sec, and this is a startling result. Regardless of whether the companions are in orbit around the main galaxy or are still receding from it as a result of ejection, we should expect to see as many negative as positive velocities, on the average, with respect to the central galaxy. If, in fact, we are seeing a significant predominance of positive redshifts, then we must be seeing the effects of some nonvelocity redshifts. Later I will show that this startling result was predicted from the evidence for decay or evolution to a larger redshift of older galaxies.

Let us return to NGC 4151 with a redshift of 900 km/sec and its companion with a redshift of 6700 km/sec. As far as we know spectroscopically, the companion to NGC 4151 is regular in that it is composed of gas, dust and stars, though it may have a more compact nucleus and possibly a higher surface brightness than normal galaxies.

People have suggested that anomalous galaxy redshifts might be explainable in terms of the Einstein gravitational redshift. This is not a workable solution, because the large gravitational field required would have a steep gradient across the galaxy. This would give rise to a varying redshift measure across the

Table 2.1
Galaxies Known To Be Companions Of Larger Galaxies

| DOMINANT GALAXIES | COMPANION GALAXIES | DIFFERENTIAL REDSHIFT (KM/SEC) |
|---|---|---|
| M31 | M32 | +85 |
|  | NGC 205 | +62 |
|  | NGC 185 | +58 |
|  | M33 | +57 |
| M81 | M82 | +234 |
|  | NGC 2976 | +81 |
|  | NGC 3077 | −104 |
|  | IC 2574 | +91 |
|  | HO II | +215 |
| NGC 5128 | NGC 5102 | +77 |
|  | NGC 5236 | +64 |
|  | NGC 5253 | −42 |
|  | NGC 5068 | +139 |
| M51 | NGC 5195 | +109 |
| NO. 48 | COMPANION | −120 |
| NO. 58 | COMPANION | +60 |
| NO. 82 | COMPANION | +90 |
| NO. 86 | COMPANION | +23 |
| NO. 87 | COMPANION | +180 |

galactic disc, with the central parts at the largest z, when the spectrographic slit was placed across the galaxy. Such a variation has never been observed, ruling out the Einstein redshift.

Another suggestion is that young, unstable galaxies blow out shells of particles which scatter the light we see. Because photons from the galaxy are somehow degraded by the shell, we see the galaxy redshifted. This theory can be ruled out both on the grounds of the exorbitant shell mass required and the lack of observations of the predicted redshift across the face of the galaxy.

## FEATURES OF RADIO SOURCES ASSOCIATED TO ELLIPTICAL AND SPIRAL GALAXIES

I want to go on now to further discussions of double radio sources. One well known double radio source is 3C 270, or NGC 4261 in the Virgo Cluster. It is a 12th magnitude E galaxy with a redshift of 2090 km/sec. At 178 MHz. it is a strong source producing 44 flux units at Earth. The two radio lobes are five minutes of arc apart. A similar, but somewhat fainter (24 flux units) object is 3C40. Virgo A, also known as M87, has radio components separated by fifty minutes of arc, and finally Centaurus A, a very bright, peculiar E galaxy, certainly has outer radio components thirty minutes of arc apart and possibly a pair eighteen degrees apart.

Some recent radio observations of spirals that I've made show some interesting properties, quite different from the ellipticals' characteristics outlined above. The radio companions of the spirals I measured were several times weaker (typically a few tenths of a flux unit at 13 cm) than those of the ellipticals (typically a few flux units). The spirals' radio companions were also typically five or six times further away from the central galaxy than those of the ellipticals. Finally, the spirals' companions are indicated to have flatter spectra than those of the ellipticals' companions. These characteristics, that the more distant sources are flatter and weaker, will be seen repeatedly in my lectures. But I want here to emphasize that not only the ellipticals have double radio sources. Spirals, too, seem to have double or multiple radio companions and looking at all the data one gets the impression that ejection is occurring in both types of galaxy. I would hypothesize that the ejection is connected with associated pairs and lines of galaxies. The mechanism and details of ejection may be slightly different in the two types of galaxy, accounting for the difference between the spirals' and ellipticals' companions.

One more point I'd like to make about radio sources concerns NGC 4258. This is one of the brightest apparent magnitude spirals in the North Galactic Hemisphere, with two well-defined spiral arms. Recent Westerbork radio observations of NGC 4258 by Vander Kruit and Oort clearly show two curved radio arms on opposite sides of the nucleus. These radio arms are distinct and separated from the optical arms. The conclusion of this observation is that new material is being ejected, in opposite directions, from the nucleus of this galaxy and there is some evidence of condensations in the radio arms. This is a very important observation for the theory of ejection formation of spiral structure.

## FIRST INDICATION OF ASSOCIATIONS BETWEEN LOW AND HIGH REDSHIFT OBJECTS. THE EJECTION HYPOTHESIS.

What I'm going to describe now is the way in which the original associations of high redshift with low redshift galaxies came to my attention.

I had assembled my Atlas of Peculiar Galaxies and checked for coincidences between peculiar galaxies and radio sources. A few coincidences had been noticed, but nothing of especial importance. Then Sersic mentioned to me that a southern peculiar galaxy had some radio sources <u>around</u> it that seemed to be associated. I then proceeded to systematically check the distribution of 3CR radio sources around my Atlas galaxies. I found a number of apparently significant associations.

Now the 3C catalogue lists sources down to nine flux units, so most of its sources are about nine to twelve flux units. 3C 65 and 3C 66 are two very bright radio sources, both about thirty flux units. Statistics tell you that you would only expect two such bright sources to occur randomly in 350 square degrees of sky, but they are seen only about $2\frac{1}{2}$ degrees apart in seven square degrees of sky. This strongly indicates that they form a physical pair. Atlas of Peculiar Galaxies number 145 is a very unusual object which falls right between this pair and which I believe is associated with these sources. This type of elliptical galaxy, with an explosive cloud of gas and possibly stars, will be seen repeatedly to be associated with double, ejected radio sources. 3C 66 is identified with a faint, sparse chain of galaxies having a redshift about 900 km/sec greater than that of Atlas number 145. 3C 66 is not optically identified. An objection that has been raised about 3C 66 is that it is a group of ellipticals which couldn't be ejected from an object like Atlas number 145. This is not an insurmountable objection because 3C 66 is a poor group of only about four ellipticals without fainter companions. I think that evidence indicates that this is the type of group that can be closer and intrinsically fainter than its redshift indicates.

In the vicinity of Atlas numbers 100 to 150 the very disturbed ellipticals and spirals tended to be bracketed by 3C radio sources. The violence and disequilibrium suggested by some of these galaxies is remarkable. Swirling dust, jets, shells and even "smoke rings" are seen associated with these galaxies. Forbidden oxygen $\lambda 3727$ is seen associated with the shred of one such galaxy (Atlas 142) reminiscent of the counterjet in M87.

The next example is 3C 216 and 3C 219, a double radio source bracketing a torn-up, peculiar galaxy. 3C 219 is a QSO with a high redshift, much larger than that of the central galaxy. There is also a large galaxy nearby and I'll describe its relationship with the radio sources and peculiar galaxy later.

When you look at fainter sources (in the Parkes Catalogue) you tend to find fainter sources bracketing fainter galaxies, and the more peculiar and disrupted the galaxy, the more likely that it is bracketed by a double radio source. I checked the Atlas of Peculiar Galaxies for associations and this latter fact emerged. My test involved checking for pairs of radio sources aligned to better than 30°. I was pleased when Lynden-Bell took the Atlas and checked on not the alignment aspect, but only the closeness or how often paired radio sources fell within a given (angular) distance of a peculiar galaxy. From this independent test he concluded that the significant correlations obtained between the 3C radio sources and the Atlas galaxies could be only reproduced by one random sky in a hundred. This independent confirmation at the 99% significance level was surprising and pleasing to me, although Lynden-Bell himself argued against the significance of the result, saying, if I remember correctly, "only" a 99% certainty.

The next thing I did was to go to a fainter flux level using the Parkes radio measurements. I searched the Parkes catalogue for conspicuous double radio sources and then checked between the sources for peculiar galaxies. Again quasar, N galaxies and some blank fields were found in the radio source positions bracketing peculiar galaxies. One of the most interesting central galaxies found in this way was NGC 7541, a bright spiral Shapley-Ames galaxy. It has a slightly curved feature leading out of its nucleus which probably represents the path of ejection of the radio sources bracketing NGC 7541. This feature must be relatively young because differential rotation in a spiral galaxy tends to wind linear features into curves in one or two rotation periods. The feature is just slightly curved by the same amount that would be expected in a fraction of a rotation of the line of the radio sources 3C 458 and 3C 459. We might expect emission lines in the spectrum of this object, but in fact I found nothing but absorption lines characteristic of an early type population of stars with ages of about $10^9$ years. This is probably older than the linear feature's age and leads me to speculate that perhaps star formation precedes gas and dust's accumulation into features.

I tested the reality of the associations selected by combining all the correlations of peculiar galaxies with double radio sources and double Parkes sources with peculiar galaxies. A strong pattern emerged. The strongest radio source of a pair tended to fall closer to the central galaxy, regardless of whether it was a QSR or a blank field. The more distant sources tended to have flatter spectra. Most important, E galaxies tended to be associated most often with radio galaxies and only occasionally with quasars, while spirals were most often associated with quasars. This confirmed the earlier empirical result that galaxy ejection takes place from ellipticals, and secondary quasar ejection is a secondary or tertiary ejection process from spirals and young galaxies. I'll continue with this subject in my next lecture.

# Lecture 3

Last time I talked about how the mechanism of ejection was suggested by the observations. You remember I showed a considerable sample of peculiar spiral and elliptical galaxies and pointed out the large number of alignments of galaxies in chains. I also noted the large numbers of radio sources about such galaxies and briefly mentioned the statistical significance of such associations.

Let me give you some numbers to begin with. I took the galaxies which were most correlated with radio sources, namely the E galaxies with associated nonequilibrium material, numbered 100 to 150 in the Atlas of Peculiar Galaxies, and analyzed some of their statistics. There were 35 3C radio sources between Atlas numbers 100 to 150 that fell in clear areas of the North Galactic Hemisphere. Choosing $\pm 30°$ as a successful alignment of two out the three nearest sources we would predict 4.6 accidental cases out of our sample of 35. In fact, 15 alignments were found. Restricting oneself to the two nearest sources, 2.7 cases are predicted and 11 were actually found. Thus these associations appear quite statistically significant.

Next, I studied the Parkes double radio sources which go to a fainter flux limit than the 3C sources. I identified the peculiar galaxies between the double sources and only did statistical tests after I had assembled all the data. Out of the twelve identifications, eleven central galaxies fell closer to the stronger radio source. Only in the least certain identification did this not happen. Combining this data with that of the previously investigated Atlas Peculiars and 3C sources, twenty out of twenty-eight central galaxies were found to fall closer to the stronger radio source. The second characteristic noted was that in the combined data, in seventeen out of twenty-six cases, the flatter radio spectrum was further away from the central galaxy. Third, the spectral indices of a pair of sources tended to be both above or below the average; sixteen out of twenty-four sources showed this behaviour.

## ASSOCIATIONS OF QSO'S AND BRIGHT GALAXIES

The 3C survey showed up a majority of peculiar ellipticals as centers of radio pairs, while the Parkes survey indicated a

majority of peculiar spirals as centers. These data were combined and then divided into two classes: those sources with central spirals and those with central ellipticals. The following radio source associations are made with each type of central galaxy.
a) With Central Spirals: 12 quasars, 6 galaxies, 8 blank fields
b) With Central Ellipticals: 4 quasars, 12 galaxies, 8 blank fields

It is clear that the spirals are accompanied by twice as many quasars as galaxies and the ellipticals by three times as many galaxies as quasars. This indicates that quasars are preferentially ejected from spirals. If indeed the quasars are ejected from spirals, then we should expect their masses to be less than that of either spirals or ellipticals. This conclusion fits with later evidence. In the Parkes higher frequency survey, the majority of objects between double radio sources were spirals. I'll argue in coming lectures that spiral arms are both a peculiarity and a sign of recent ejection in a galaxy.

The optically identified ejecta all seemed to have higher redshifts than the central galaxies. Therefore I decided to concentrate exclusively on the highest redshift objects, the quasars, in order to get the most critical test of anomalous redshift. This required a complete, homogeneous sample of quasars free from selection effects.

I assembled this catalogue by only using quasars above declination -20°, the area accessible to northern optical telescopes. Second, I excluded objects within 20° of the galactic plane to avoid galactic obscuration. Third, I used only the well-established 3CR and Parkes radio catalogues, and I could transform between their frequencies to be sure that my sample was complete down to my cutoff flux level. Finally, only quasars with measured redshifts were used. Both Parkes and 3CR had been published large enough to ensure fairly complete identification of the QSR's contained in them.

Optical identifications, using the Palomar sky survey, can be made down to about 21.5 apparent magnitude. It is found that very few optical identifications of quasars with radio sources are made beyond 19th magnitude at this radio flux level. These suggest that we are running out of a nearby concentration of these objects as we go to fainter magnitudes and hence to larger volumes of space.

The crux of this analysis is the assumption that quasars' redshifts z will not yield their distances. Some other distance criterion is needed. If the intrinsic absolute magnitude range of quasars is less than their apparent magnitude range, due to differing distance, then we can get a rough estimate of a quasar's distance from its apparent magnitude. One of the central goals of this study was to test this hypothesis.

The quasars have to be divided into magnitude ranges for this study, and since there is a dip in the plot of number of quasars N versus photographic magnitude $M_{pg}$ at $M_{pg} = 17$, I have arbitrarily chosen as my faintest group all the quasars with $M_{pg} \geq 17$. My brightest quasars are arbitrarily selected to be those for which $M_{pg} \leq 16.2$, and the intermediate brightness quasars have $16.2 \leq M_{pg} < 17$. There are ninety-three quasars in my sample.

The intermediate brightness quasars plotted on the sky are seen to be aligned through the region of the Virgo cluster of galaxies. A few QSR's are isolated near M81. There also seems to be some tendency for quasars to spatially pair when plotted on the sky. The south galactic hemisphere around 1 hour right ascension is dominated by M31 and M33. The quasars in this region are scattered about, but again have a tendency to spatially pair. The closest pairs of quasars tended to resemble each other in apparent magnitude, spectral index and radio brightness, but not in redshift.

The implication is that if intermediate brightness quasars are associated with the brightest galaxies like M81 or M87, then the fainter quasars (with $M_{pg} \geq 17$) should be associated with a somewhat fainter group of galaxies. Indeed, there is a striking similarity between the faint quasars' distribution on the sky and that of the bright Shapley-Ames galaxies. It's almost as if the bright Shapley-Ames galaxies were the core of the quasar distribution.

The next step is to numerically test these similarities. The faintest Shapley-Ames galaxies are, on the average, about 10 degrees away from the nearest quasar. Considering brighter and brighter classes of galaxies this distance steadily decreases, and the 18 brightest galaxies ($M_{pg} < 10.6$) are, on the average, only 5.7° away from the nearest faint quasar. The same number of quasars was randomly scattered on the same area of sky and the average distance to the bright galaxies was computed as

10.7°. In ten trials the random distribution gave average distances to the brightest Shapley-Ames galaxies as low as 9.3° and as high as 12.6°. This seems to indicate that the faint quasars really are associated with the brightest Shapley-Ames galaxies. This was the preliminary conclusion at least. However, I now believe that the quasars may be instead associated with peculiar, disrupted galaxies, which are in clusters dominated by the brightest Shapley-Ames galaxies. These peculiar galaxies are spatially close to the bright Shapley-Ames galaxies and hence the correlation between faint quasars and bright Shapley-Ames galaxies.

In general, QSO's are aligned across and closer to bright galaxies than would be expected for a random distribution. This is similar to what happened for pairs of blank field radio sources, which are conventionally interpreted as having been ejected from a central galaxy. In our case the optically identified objects fit the four empirical rules of ejection given at the beginning of the course, and therefore, in my opinion, the most natural explanation is the ejection hypothesis. If it were not so, explaining the effect would be a most difficult problem.

A test on the QSR's fainter than $V = 17.0$ magnitude in the North Galactic Hemisphere with the 31 brightest galaxies in the North Galactic Hemisphere reveals that the average distance from bright galaxies is 6° instead of the 11° which are expected for a random sample. We can perform a similar statistical analysis of the expected number of alignments of pairs of QSO's with bright galaxies. We find that many QSO's fall within 40° of the line determined by an ejecting galaxy and the other QSO in the pair. More precisely, the expected number of alignments for a random distribution is $6.2 \pm 2.5$, while the observed number is 13.

Therefore we conclude there is a significant tendency for the faint QSR's to be aligned across, as well as falling closer to, the bright galaxies in the sky. Burbidge, E.M., Burbidge, G.R., Solomon and Strittmatter found that there is an excess of QSO's closer than about 7 to 10 arc minutes to a bright galaxy. A plot of the observed number $N(r)$ versus the angular distance $r$ from any galaxy might be schematically represented as in figure 3.1. The rough behaviour of $N(r)$ is that for random associations we expect: smaller numbers at small angular distances (where statistics are poor), and larger for larger $r$ (where statistics are good). My results show there is an excess of the observed

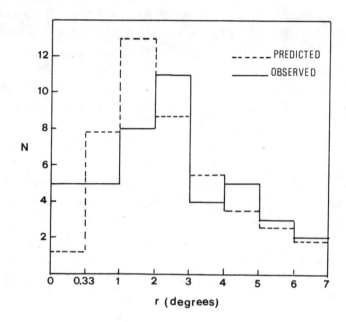

Figure 3.1 Distribution of angular separations between 3C QSO's and their nearest neighboring bright galaxy. The predicted distribution is also shown.

over the expected number all the way up to r≈6°. Large fluctuations probably appear for small r, and Burbidge et al. might have based their deduction on one of these fluctuations, thus coming to the conclusion that QSO's permanently sit at that distance from a bright galaxy. This is hard to reconcile with my ejection hypothesis where they escape from the parent galaxy. We can reconcile these results, however, if we say the ejection velocity is about 1000 km/sec, and so these objects get out to 10 kpc in $10^7$ years (a figure we will show to be a plausible evolution time scale). After this time, I will argue that QSO's start evolving into compact galaxies, or young spirals. In other words these objects stop being QSO's at an age of about $10^7$ years. There are also indications that the objects they are evolving into are also capable of emitting QSO's in secondary ejections. Quasars seem to be preferentially associated with young spiral galaxies.

We can perform a similar statistical test again on the south galactic hemisphere galaxies. The average galactic magnitude here is about one magnitude fainter than in the north galactic hemisphere. There are four fewer QSR's in this region of sky and our statistics are correspondingly poorer, but we see that the random QSR's give a higher mean separation from bright galaxies than the real QSR's. We also see that the quasars in the south galactic hemisphere with average magnitude about 18.2 are about half a magnitude fainter than those in the north. This is an important result in view of the average fainter magnitudes of the bright galaxies in the southern hemisphere.

One faint ($M_{pg}$ > 17) group of quasars in the south galactic hemisphere form a particularly tightly knit group. Under close scrutiny it is found that there are four quasars on a very well defined line in this region of the sky. In fact, three of the four objects fall off the line by only 10 minutes of arc over the 7° length of the line. At the origin of this line is Atlas number 157, a highly peculiar exploding galaxy. The parameters of the four QSR's with increasing distance from Atlas 157 are listed in table 3.1. For objects 1, 2, and 4, $z = 0.72 \pm 0.05$. Of twenty-eight faint QSR's in the south galactic hemisphere, three out of the four quasars in this redshift range fall on this straight line, a most unlikely happening. Two other characteristics are noticeable in table 3.1. First the radio flux decreases, and second the spectral index increases as we move away from Atlas 157. This behaviour is exactly as found earlier by Sharp for radio sources around Shapley-Ames galaxies and by myself for radio sources associated with peculiar galaxies and spirals (see beginning of lecture).

The object with redshift $z = 2.11$ is about a magnitude fainter than the average apparent magnitude of the rest of the quasars on the line. If we assume that all four quasars are at the same

Table 3.1

| QSR's | $M_{pg}$ | z | $S_{408}$ | α |
|---|---|---|---|---|
| 1 | 18 | .77 | 5.0 | −1.1 |
| 2 | 17.5 | .67 | 4.5 | −0.6 |
| 3 | 18.4 | 2.11 | 3.5 | −0.7 |
| 4 | 17.3 | .72 | 3.9 | −0.3 |

distance, then we must conclude that the object with $z = 2.11$ is about a magnitude fainter in intrinsic luminosity than the other three quasars. I want to maintain that this is the most natural explanation of the $z = 2$ quasar cut-off. Instead of invoking very different conditions in the universe at $z = 2$, as is conventionally done, we merely assume that at larger $z$ we are looking at fainter quasars and hence at a smaller volume of space. This neatly accounts for the observed cut-off: in a small volume we should see few quasars.

The brightest QSR's (now with $M_{pg} < 16.5$) give us the most striking and interesting results. While most of the faint quasars were in the north galactic hemisphere, most of the bright quasars were in the south. There are some extraordinarily good pairings in redshift: (see table 3.2). The two lowest redshift objects ($z = .202$ and $z = .200$) have recently been measured by Miley and he finds their radio diameters to be outstandingly large, larger than two minutes of arc. These two objects are also very bright, (15.54 and 15.76 apparent magnitude). Finally, these objects are an arc distance of 150° apart on the sky. These facts led me to conclude that these two objects are the most likely candidates for ejection from our own galaxy.

Looking at table 3.2, we also see pairs of quasars on opposite sides of M31 with remarkable correlations in redshifts, and I'd like to suggest that these are further examples of ejection. Next time I'll go into spatial distributions in the local supercluster and elaborate on the ejection hypothesis.

## Table 3.2
### DISTANCE ON SKY TO BRIGHTEST (V=16.5 MAG.) QSR'S FROM M31

| | OBJECT | V | Z | $S_{408}$ | $S_{750}$ | α | ARC DISTANCE FROM M31 | POSITION ANGLE FROM M31 |
|---|---|---|---|---|---|---|---|---|
| | 1. 3C 249.1 | 15.72 | 0.311 | --- | 3.7 | -0.8 | 61° | 6° }|
| | 2. 2251+11 | 15.80 | 0.323 | 3.7 | --- | -0.9 | 38° | 226° }|
| | 3. 3C 48 | 16.20 | 0.367 | --- | 25.5 | -0.79 | 14° | 122° }|
| | 4. 3C 351 | 15.28 | 0.371 | --- | 5.2 | -0.73 | 65° | 330° }|
| | 5. 3C 232 | 16.78 | 0.530 | --- | 2.1 | -0.57 | 97° | 34° }|
| | 6. 2128-12 | 15.98 | 0.501 | 1.5 | --- | +0.20 | 69° | 231° }|
| | 7. 3C 345 | 15.96 | 0.594 | --- | 7.9 | -0.34 | 82° | 318° }|
| | 8. 0405-12 | 14.79 | 0.574 | 9.3 | --- | C | 71° | 126° }|
| | 9. 3C 263 | 16.32 | 0.652 | --- | 8.1 | -0.87 | 72° | 6° }|
| | 10. 2344+09 | 15.92 | 0.677 | 2.7 | --- | -0.50 | 34° | 205° }|
| (1) | 11. 2135-14 | 15.54 | 0.202 | 10.0 | --- | -0.8 | 70° | 228° }|
| | 12. 0837-12 | 15.76 | 0.200 | 5.7 | --- | -0.9 | 120° | 79° }|
| | 13. 003+15 | 16.40 | 0.45 | 2.6 | --- | -0.6 | 26° | 201° |
| | 14. 3C 454.3 | 16.10 | 0.86 | --- | 13.4 | -0.16 | 34° | 231° |
| (2) | 15. 2145+06 | 16.47 | 0.367 | 3.0 | --- | C | 52° | 241° |
| | 16. 0232-04 | 16.46 | 1.43 | 3.2 | --- | -0.17 | 52° | 143° |
| | 17. 3C 95 | 16.24 | 0.61 | 11.6 | --- | -0.9 | 71° | 131° |
| (3) | 18. 3C 94 | 16.49 | 0.96 | 9.9 | --- | -0.8 | 72° | 130° |
| | 19. 3C 334 | 16.41 | 0.56 | 6.3 | --- | -0.77 | 103° | 307° |
| | 20. 1354+19 | 16.20 | 0.72 | 6.0 | --- | -0.7 | 117° | 340° |
| (4) | 21. 1510-08 | 16.52 | 0.361 | 3.0 | --- | 0.0 | 134° | 303° |
| | 22. 3C 273 | 12.80 | 0.16 | --- | 46.0 | -0.24 | 137° | 5° |
| | 23. 1217+02 | 16.51 | 0.24 | 1.7 | --- | -0.9 | (~133°) | ~8° |
| | 24. 1004+13 | 15.15 | 0.24 | 3.3 | --- | -0.8 | --- | --- |

(1) ARC DISTANCE ON SKY = 150°
(2) PROBABLY ASSOCIATED WITH 3
(3) PROBABLY ASSOCIATED WITH 7
(4) POSSIBLY ASSOCIATED WITH 3 AND 15

Lecture 4

OBSERVATIONAL DIFFERENCES BETWEEN SOUTH AND NORTH GALACTIC HEMISPHERES

We have previously seen that there is a general increase of the numbers of low z, bright apparent magnitude QSO's in a direction roughly coincident with M31 and its companions. Of course this direction is also toward the centroid of our Local Group of galaxies, of which M31 is the dominant galaxy. We may attempt to explain this fact as well as the concentration of the fainter QSR's in the North Galactic Hemisphere in terms of the galaxy Supercluster, a concept put forward by de Vaucouleurs. Fig. 4.1 shows a simplified view of our Supercluster, which is roughly 1000 times larger than the Galaxy, and is centered on the main body of the Virgo Cluster. The sizes are all very approximate: what is important to notice is that our Galaxy is perhaps 10 Mpc off the center and is oriented in such a way that the North Galactic Hemisphere looks toward the center of the Supercluster, where we see many faint galaxies and faint QSO's. In contrast, the South Galactic Hemisphere looks toward the center of the local group, where M31 sits, together with the brightest QSO's, and a few moderately bright nearby galaxies. Also, as we have seen, there is a tendency for lowest redshift (~0.2) and highest

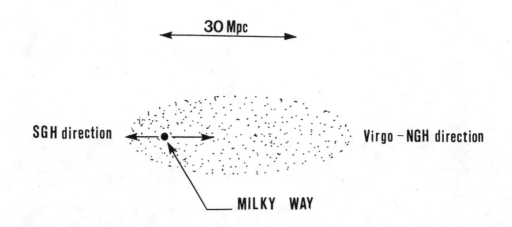

Figure 4.1 Schematic cross-section of the local supercluster showing location of Local Group and approximate dimensions.

# QSO's Observations and the Redshift Problem

(~2) objects to have the lowest absolute magnitude, which might suggest that they belong to the local group. This idea is reinforced by a concentration of z=1.8 to 2.0 redshifts in the SGH. (See Wills and Lynds, 1972)

One way to investigate the Supercluster concept is to make use of Faraday rotation discussed by Reinhardt and Thiel (1970) and by Kawabata et al. (1969). Their procedure is schematically represented in figs. 4.2 and 4.3.

The rotation measure (RM) essentially measures the integral $\int n_e B dl$. The result is that there is a peak centered at galactic latitude b~0°, as we can expect from the simplest model of the galactic magnetic field. There are, however, some Faraday rotations which depart from the expected curve, at high positive b. The authors thought that they had measured a metagalactic magnetic field, and tried to show that the RM increased with increasing z, as one should expect if redshift were cosmological. The observed behaviour of RM vs z was however disappointing (fig. 4.3).

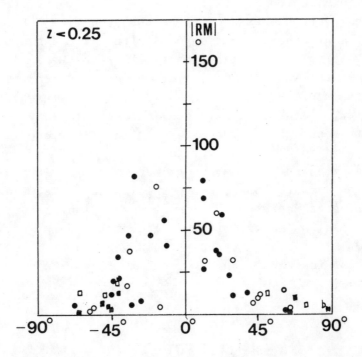

Figure 4.2

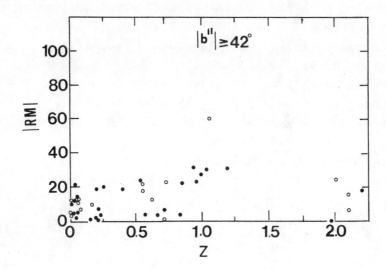

Figure 4.3

At low z one can indeed observe an increase, but the trend is not followed through, and there is a catastrophe at high z. This is, however, the kind of catastrophe we like beacuse it supports the hypothesis that objects with $z\sim2$ are nearby. Moreover, one can see a significant difference between the values of the RM in the NGH and SGH. A difference between the North and South Galactic Hemispheres is also observed in the average QSO redshift. Both trends are summarized in table 4.1.

The difference in behaviour between the two galactic hemispheres appears in almost all plots one can make, and can be easily accounted for by the Supercluster model. In the case of the rotation measure, one can hypothesize that the galaxies in the Supercluster represent condensations of ionized matter and magnetic field in analogy with the clouds in the galaxy. As the line of sight passes through the Supercluster it encounters more of these ionized patches and the average rotation measure

Table 4.1

|   | NGH | SGH |
|---|-----|-----|
| $\bar{z}$ | 0.66 | 0.56 |
| RM | 21±4 | 11.3±4 |

# QSO's Observations and the Redshift Problem

builds up out of positive and negative rotations like a random walk problem.

One might note in passing the standard interpretation of the results in table 4.1: The difference between the average redshifts is termed as not statistically significant, and the rotation measure difference is interpreted as due to the fact that the galactic magnetic field is oriented in opposite directions on the opposite sides of the galactic plane, and the sun is slightly above the plane. Therefore, while in the southern galactic hemisphere direction there is some cancellation between the $|R.M.|$, no cancellation is present in the opposite direction. But this will not explain the z dependence of $|R.M.|$, if it is present.

## SOME PROVOCATIVE EXAMPLES OF EJECTION

a) <u>By spiral galaxies alone</u> (Fig. 4.4)

This spiral galaxy is from the Atlas of Peculiar Galaxies and is peculiar for at least three reasons:

(I) It does not appear to lie on a plane, but rather seems to be deformed in a helical shape;

(II) There is a long narrow filament B with nothing at the end;

(III) There is a disturbance A, which is perhaps the direct cause of the filament. The main point of this example is to

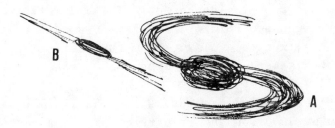

Figure 4.4

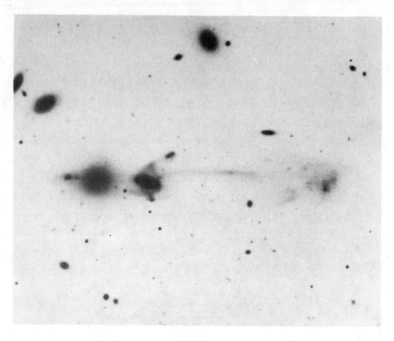

Figure 4.5

show that there can be a pure, straight ejection from a spiral and the source of activity is not necessarily placed in the nucleus.

b) <u>By spiral and elliptical galaxies</u>

(i) NGC 3561 (Fig. 4.5)

This example also has a historical importance, since it is one of the first pointed out by Ambartsumian. There are various peculiar features in the picture.

(A) First, a QSO with $z = 2.2$. There is no visible connection with the main galaxies, and this might be a projection effect.

(B) Next, the so-called "Ambartsumian Knot". Zwicky took its spectrum, and found evidence for (OII)$\lambda 3727$ emission. An absolutely straight and thin (<2") filament points toward the nucleus of the elliptical galaxy, from the knot.

This object has been recently discussed by Stockton (1972).

(ii) Hercules Cluster

This cluster, containing galaxies such as IC1182, has also been examined by Stockton and is particularly interesting: an enlarged portion is shown in Fig. 4.6. It contains the original Ambartsumian object (A) (the case which first suggested to him the ejection hypothesis). (A) has a continuum

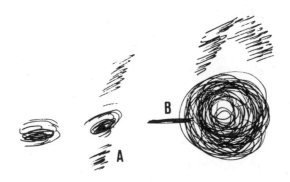

Figure 4.6

spectrum; (B) is a thin straight filament which also may have a continuum spectrum.

(iii)   I ZW 96 (E Galaxy, Fig. 4.7)

This is perhaps the best example in favour of the ejection hypothesis, since the alternative possibility is that galaxies collide, and filaments appear because of tidal forces.  In this case, however, we have two filaments, and it is extremely unlikely

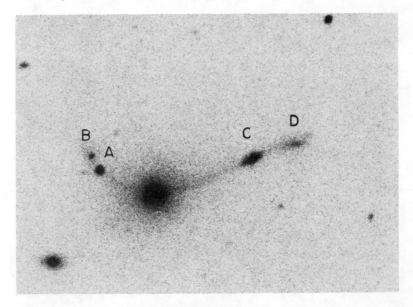

Figure 4.7

that two objects collide on opposite sides of the main galaxy
at exactly the same time. Object (A) is stellar in appearance,
but is a galaxy which shows absorption lines at one hundred
km/sec or so greater than the main galaxy. Objects (AB) and
(CD) show the tendency for doubleness of the ejected objects,
which, apparently, tend to undergo fission. The object is also
reminiscent of the chains of galaxies.

We now go into what could be called "Hard core ejection cases".
c) <u>3C 371</u> (Fig. 4.8)

3C 371 is an example of ejection from "N" or "compact" galaxies. In this sketch we have plotted 3C 371 and all the galaxies in its vicinity. This is a cluster of richness 0, according to the rules defined by Abell, and is like those included by Sandage in the Hubble diagram. It falls, however, well below the Hubble relationship. We can see a rough chain extending in the direction of the arrow. An isodensitometer plot shows that objects (A), (B) and (C) are connected. The last connection might well be caused by an accidental blending of the isophotes; if, however, we take the point of view that the three objects are connected, we may notice that this is another case of double ejection. Moreover, the redshift of object (B) has an excess of about 300 km/sec over the central galaxy, a definite excess, though not enough to consider the two objects as physically not associated on a conventional view.

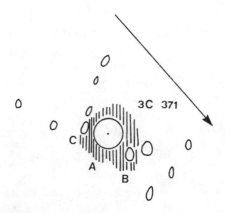

Figure 4.8

# QSO's Observations and the Redshift Problem 37

Figure 4.9

d) <u>Objects at the end of spiral arms</u>

In a preceding paper (Arp, 1969) six such objects were studied in detail. We shall see only a few here. These objects whose most well known example is M51, were arranged in a sequence which suggested evolution of the companion from small to large.

Fig. 4.9 shows the first case, and might be taken as an indication that the ejected companion (A) is very compact at the beginning. A wake leads back into the ejecting galaxy; its redshift is -100 km/sec with respect to the central galaxy. In the original paper it was suggested that the redshift is not due to rotation, but to a real velocity of separation along the minor axis. A motion of 100 km/sec would lead to a distance of 1 Kpc in $10^7$ years, which is just enough to get fractional rotation and curvature into the path. Object (A) shows a few emission lines and a featureless continuum. We do not know about the intrinsic redshift of the exciting body inside.

Fig. 4.10 shows three more cases.

Fig. 4.10(a) shows Atlas 58, a good example of an ScI galaxy, with rather large z. It has narrow, thin spiral arms, one of which is broken and contains a spheroidal E galaxy. If we take the redshift of the system to be cosmological, it turns out that the dimensions are really big: the diameter is 120 Kpc, the absolute photographic magnitude of the main galaxy is -22, while for the companion the corresponding values are 3 Kpc across, and $M_{pg}$ = -18.5.

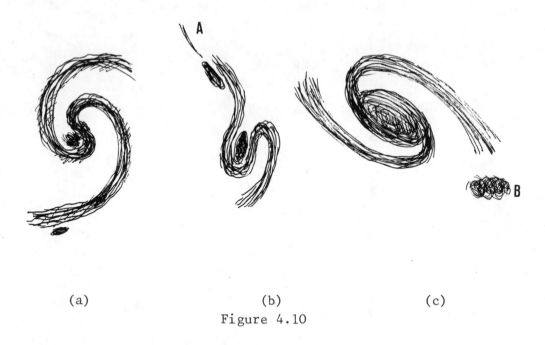

(a) (b) (c)

Figure 4.10

Conservative astronomers think that it is a normal galaxy, but the fact that it is so big as to contain a normal galaxy as a feature of one of the arms is something against which my conservative nature cries out.

Fig. 4.10(b) shows a large companion at the end of one arm, plus the feature (A) which we have called a "shred".

Fig. 4.10(c) shows NGC 7752 and 7753. The companion (B) shows a lumpy structure, and evidence for expansion. At the present rate of expansion the system would dissipate in a few times $10^7$ years. We have a sort of a rotation clock in order to check this result, because it has been estimated that it takes $10^8$ years for a spiral arm to wind up completely, and, therefore, the arm connecting the two galaxies has an age of about $10^7$ years. We shall now call attention to a very important result, obtained by Eric Holmberg. He made a statistical study of the distribution of the companions of the brightest apparent magnitude spiral galaxies from the Schmidt Survey Palomar Sky Atlas and checked it against selected control fields. He found quite unexpectedly that there is an excess and that the companions are generally distributed in a cone centered along the minor axis. In his opinion this was a demonstration of the process of ejection. Moreover, he argued that the reason they were distri-

buted along the minor axis was that the original ejection was
isotropic but that the disk of the galaxy stopped ejection in the
plane. I then argued that if ejecta were in fact stopped by the
disk, we should see, in some cases, companions near disks where
there was some disturbance or attachment of the companion by
luminous material. It then became clear that this was a natural
explanation for the whole class of Atlas spirals with companions
in the ends of arms. In turn this would naturally suggest an
ejection origin for all spiral structure.

The classical point of view on this problem centers on the
density wave theory. Much work has been done, but the theory
does not give any real reason for the origin of a symmetrical
two arm disturbance. The theory can only claim to preserve
it for a few rotations once we already have it.

The ejection hypothesis could overcome the classical objection
to evolution between galaxy forms. The fact is that the conservation of angular momentum forbids one flat system to be converted into a spherical one. One might then think that the ejection
of objects in different directions can fill up the volume.
Dramatic confirmations of the ejection hypothesis for spirals
come from the radio sources recently observed at Westerbork
in the interiors of spiral galaxies like NGC 4736 and NGC 4258.

Figure 4.11

Finally, Fig. 4.11 shows NGC 7309, a spiral galaxy which has the extremely unusual property of having three well developed, conspicuous spiral arms. A plot of all radio sources in the area shows the surprising result that there are just three radio sources spread around NGC 7309. One of the radio sources is a QSR with $z=0.6$ and the other two blank fields. The implication is clear that the three spiral arms have been formed by the ejection of the three radio sources.

# QSO's Observations and the Redshift Problem

Lecture 5

Let me begin this lecture by saying a few words about my purpose in compiling the Atlas of Peculiar Galaxies. If we really wanted to find out how normal galaxies behave, the ideal thing to do would be to perform experiments on them: perturb them, collide them with each other, and so forth, and then see how they react. Since we can't make such experiments, we have to take the ones in the sky that nature has given us. The arrangement of the galaxies in the Catalogue was an attempt to arrange these experiments in some kind of graded sequence, so we can see the progression of characteristics they show. One whole section of the Atlas consisted of objects apparently exploding or ejecting other objects; most of yesterday's lecture was devoted to establishing this idea using various examples.

Before we go on, I would like to illustrate two more examples of this ejection. Fig. 5.1 shows an Atlas peculiar, a barred spiral galaxy with an inner ring. Two very thin and well defined spiral arms are seen to emanate from a more or less arbitrary position on this ring. As we discussed yesterday, I would

Figure 5.1 Sketch of an Atlas peculiar, a barred spiral galaxy with arms beginning on an internal ring

Figure 5.2  Atlas Peculiars   Note the straight outer arms of the one spiral.

argue very strongly that the whole spiral phenomenon is intimately connected with the ejection processes.  Whether in fact you can have ejection from points in a galaxy other than the nucleus is an open question, but it is difficult to think of another explanation for this galaxy.

Fig. 5.2 shows a galaxy with two arms that end in very long straight filaments, apparently associated with some companion objects.  Again, a natural explanation would be that this was originally an ejection in equal and opposite directions and the inside parts of the ejecta have been wound up by differential rotation.  I would further argue that both of these galaxies are nonequilibrium systems, which will not remain unchanged for very long.

GALAXIES WITH HIGH-REDSHIFT COMPANIONS ON THE ENDS OF SPIRAL ARMS

I have tried to establish the concept of ejection of companions on the end of spiral arms by using cases with very obvious connections, and with small redshift differences.  With these objects as precedents, let us now consider that subclass of galaxies with connected companions in which the companion has a

QSO's Observations and the Redshift Problem    43

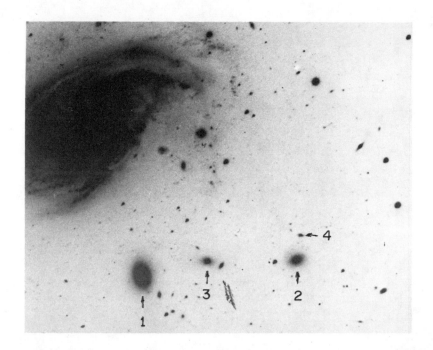

Figure 5.3 The disturbed spiral galaxy NGC 772, and its brightest nearby companions. The radial velocities of NGC 772 and companion 1 are 2400 km/sec, while those of companions 2 and 3 are 19,000 and 20,000 km/sec respectively.

much higher redshift.

Fig. 5.3 shows NGC 772, a very disturbed Sb galaxy. The original photograph was a very deep three hour exposure on a IIIa-J emulsion, so that the central parts of the galaxy are overexposed, and the disruption there is not visible. The redshifts of the central galaxy and the brightest companion 1 are both about 2400 km/sec, and the disruption of the spiral galaxy is probably due to the presence of this companion. However, there are also filaments between NGC 772 and the companions 2-4.

Now, the redshift of galaxies 2 and 3 are about 19,000 km/sec and 20,000 km/sec respectively, much higher than the redshift of NGC 772. Furthermore, as the original plate and isodensity tracing of the plate suggest, there appears to be a filament connecting galaxy 4 with 2. This is almost identical to the situation in NGC 4151, shown in the first lecture, where the primary ejecta has an even stronger secondary ejecta. Again, I do not yet have

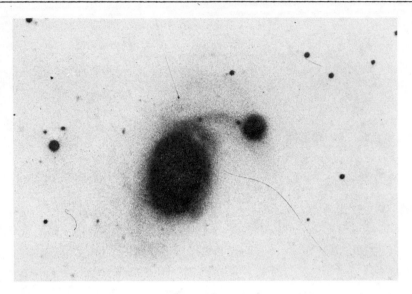

Figure 5.4  The Seyfert Galaxy NGC 7603, and its high-redshift companion.

the redshift of the secondary system. Finally, I would like to point out that this whole field seems full of peculiar galaxies and small galaxies which could be interpreted as debris. It is hard to judge the significance of this visual suggestion, since it is so hard to take control fields, but I think that the effect is important. This was one of the first systems found where there were connected companions with relatively high redshifts.

Another example of this type is shown in Fig. 5.4. This is the Seyfert Galaxy NGC 7603, which Dr. Hoyle has already mentioned. NGC 7603 has a redshift of 8000 km/sec, and the companion 16,000 km/sec. This is too great a velocity difference to pull out a tidal tail, since the galaxies pass each other too quickly. Further, NGC 7603 is a Seyfert Galaxy, very disturbed near its center. Now, one could argue that an internal explosion is responsible for the disruption, and the companion just happened by chance to lie in the background, along the line of sight to the optically connecting arm. But I think the crucial argument against this is the following. First, the companion is itself peculiar, it has a high surface brightness core, and a faint diffuse halo. So it is not typical of field galaxies, making the coincidence with the already rare Seyfert Galaxy even more improbable. Most important, the original plate has a dark arc near the arms connecting to the main galaxy, arguing even more strongly for a physical connection. Finally, I would

QSO's Observations and the Redshift Problem       45

like to emphasize the fact that the companion's spectrum seems to be due to normal starlight: H and K absorption and the G band are visible. Thus any explanation of the anomalous redshift should make each star radiate at the longer wavelength required.

Another interesting object of this kind is NGC 1199, an elliptical galaxy with an irregular, high-surface brightness object near it but not optically connected to it. This companion shows emission lines, and is bisected by an absorption lane. It is not at all a typical field galaxy, but is highly irregular, typical of bright galaxy companions. However, the redshift of NGC 1199 is only 3000 km/sec, whereas the redshift of the companion is 13,000 km/sec. Because of the unusual nature of the companion I feel that this is another example of a high redshift companion, even though the case for their being connected is not as strong as in previous examples.

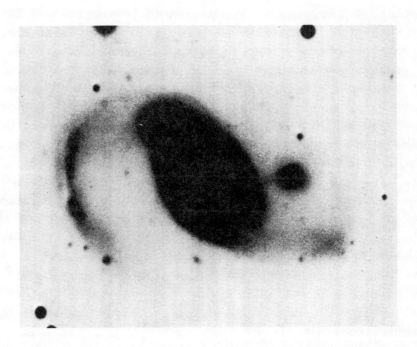

Figure 5.5  NGC 4319 and Markarian 205, from a four-hour exposure on a baked III a-J plate (200"telescope).

One of the best examples of a galaxy with a high redshift companion is shown in Fig. 5.5. The galaxy is NGC 4319, an obviously perturbed spiral. Forty arc seconds to the south is Markarian 205, which I would classify as a QSO, a radio-quiet quasar. The reason for this classification is that the object is completely stellar on a normal photograph (ie: unresolved on a scale of ~1"), and it has a redshift which indicates that it is extragalactic. Thus it fits the usual definition of a QSO. On a deep 200" IIIa-J plate, one can see a connection between NGC 4319 and Markarian 205. This shows up even more strongly on a narrow-bandpass H$\alpha$ picture. Now, the redshift of NGC 4319 is 1700 km/sec, and that of the QSO is 21,000 km/sec, so that there appears again an anomalously larger redshift. There is at present some doubt about whether the connection is real, whether it might perhaps be a photographic effect, or a background galaxy. I have tried to reproduce the observation by superposing previously unpublished ultraviolet plates; these also show the connection.

Further, I have superposition printed two additional H$\alpha$ pictures, both of which faintly show the filament. These observations make it unlikely that photographic effects are responsible, and I would argue strongly that the filament has too low a surface brightness and too narrow a form to be a superposed background galaxy. This, then, is my best case of a legitimate QSO connected to a low-redshift galaxy. Note that it fits in well with the four empirical rules of ejection which I discussed earlier.

Another association of a spiral galaxy with a quasar is illustrated in Fig. 5.6. The galaxy involved is NGC 5055, located between the radio sources 3C 285 and 3C 280.1. The presumption is already very strong that these objects are associated because the pair of radio sources brackets the galaxy. It is interesting, thus, that 3C 280.1 turns out to be a quasar of high redshift. A deep 200" photograph further shows straight luminous filaments on the west edge of NGC 5055 which point along the line of sight to the quasar. Of course, in judging the significance of this alignment, one could consider how much rotation the galaxy should show in the time taken for any ejecta to get as far away from it as is the QSO, if we know the distance of the galaxy, the speed of ejection and the time of ejection.

A final example is IC 1746, perhaps associated with a nearby PHL quasar, first noticed by the Burbidges. On the Sky Survey

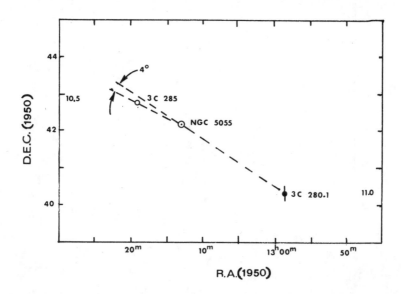

Figure 5.6 Diagram of sky around NGC 5055 showing radio sources aligned to within 4° across central galaxy.

prints, there seemed to be a luminous connection, but this was later resolved to be a faint compact galaxy between the quasar and the galaxy. The situation seems very peculiar and unusual but needs further investigation.

This completes the examples of companions with high redshifts connected to galaxies of relatively low redshift. Next, I would like to say a few words about clusters of galaxies.

CLUSTERS OF GALAXIES

Fig. 5.7 shows areas of high galaxy density in the Virgo Cluster, and in each region the mean redshift is given. There are two things significant about this plot. First of all, the brightest radio sources have been indicated by arrows; north to south, these are 3C 274, M84, NGC 4261 and 3C 273. It seems noteworthy that in this concentration of bright galaxies we also find a concentration of the brightest radio sources. Three of these are usually thought to be in the Virgo Cluster, and the fourth is 3C 273, the optically brightest quasar. It is, by this same reasoning, suggested to be a member of the Virgo Cluster also.

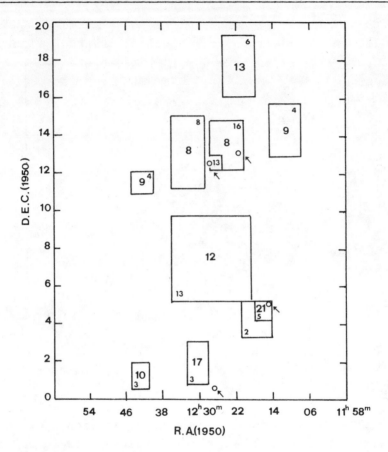

Figure 5.7 Radial velocities of Virgo Cluster galaxies within pictured rectangles have been averaged. Mean velocities within each rectangle are indicated in hundreds of km/sec. Number of galaxies involved in each average is written small in each rectangle. Arrows point to radio sources (open circles) in region which are greater than 9 (3C R) flux units.

The other thing I want to point out is this. The distribution of redshifts in the boxes in Fig. 5.7 could support a picture of either expansion of the galaxies from the center of the cluster or expansion with rotation. This would in turn support a picture of clusters originating from ejection or fragmentation of a central object.

Let us consider four nearby groups of galaxies: the Leo group, and those dominated by NGC 383, 7619, and 5846. I'll use data directly from Humason, Mayall and Sandage (1956), who listed these as groups because of their compactness, and their similar

# QSO's Observations and the Redshift Problem 49

redshifts. When we plot measured redshift against apparent
magnitude, we find (with a great deal of scatter) a tendency for
the fainter galaxies to have a higher redshift. Again, for the
NGC 253 group in the southern hemisphere, the faint galaxies
have 300-500 km/sec more redshift than the brighter galaxies.
Examination, then, of all these groups picked out by previous,
independent observers shows fainter members with systematically
high redshifts.

Next, I would like to review Tifft's (1972, 1973) works on the
Coma and Virgo Clusters. Tifft has measured nuclear magnitudes
in both these clusters, and plotted their relation with redshift.
He finds (Fig. 5.8) that generally the redshift increases with
the faintness of the nucleus. But he also identifies bands in
this diagram, and claims that these are statistically significant. The status of this work is still open to debate, but I
would like to make the following observations. Tifft claims
that it is the nucleus of the galaxy that evolves in the redshift-magnitude diagram, and that the surrounding material does
not; this is the reason why he measures nuclei only. I would
tend to agree with this view.

Quite apart from the possible quantization of the z-m relation into bands, the general trend is significant. In both Coma
and Virgo we find the same effect, namely that disk galaxies
have higher redshift than ellipticals. In Virgo, for instance,
the mean redshift of the ellipticals is 900 km/sec, and of the

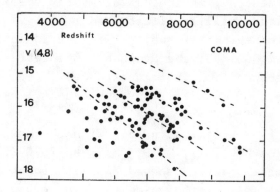

Figure 5.8  The redshift, nuclear-region magnitude diagram
for the Coma cluster. Lines indicate the primary bands in
the cluster.

spirals 1800 km/sec. This is too large a difference to be accidental; moreover, we feel that the spirals and ellipticals are at the same distance. So here we have two classes of objects which differ only in morphological type and in redshift, and we argue that the redshift is due to a non-distance effect. In Coma, the situation is slightly different in that there are only E and S0 galaxies, but again we find that the S0 galaxies have 700 km/sec higher mean redshift than the E galaxies. Since this involves only the core of the cluster, contamination of the field is negligible.

I would summarize by saying that in every case where we can test the association of galaxies, whether statistical association or association in cluster, or by connections of filaments, or identified as companions, we get the same redshift anomaly to a greater or lesser degree. Less luminous, younger galaxies have higher redshifts.

## NEIGHBOURHOODS OF NEARBY GALAXIES

Now I would like to start on a new subject, namely the neighbourhoods of nearby galaxies. If we accept the conclusion of the data we have presented, then we cannot use the redshifts of galaxies to determine their distances. To study the properties of clusters and groups and so forth, we need some different criterion for deciding that two objects are at the same distance. We will now consider the neighbourhoods of nearby spiral galaxies, where such techniques will be illustrated.

I would like to begin this subject with an in depth discussion of the area around NGC 7331 and Stephan's Quintet. Here we can see the physical connection in greatest detail, and so this object will be a sort of model for our subsequent looks at other nearby galaxies.

Stephan noted four galaxies in 1877 with the Marseille Telescope, these being NGC 7317, 7318, 7319 and 7320 (Fig. 5.9). Later photographic resolution showed NGC 7318 to be double, and showed further that all of the galaxies were highly disturbed, presumably due to their mutual interactions. In 1957, when Ambartsumian wrote his fundamental paper for the Solvay conference, four redshifts were available for NGC 7317-7319 (Fig. 5.9). He used these redshifts, ranging from 5700 to 6700 km/sec, to argue that the system was unstable on a short timescale, and

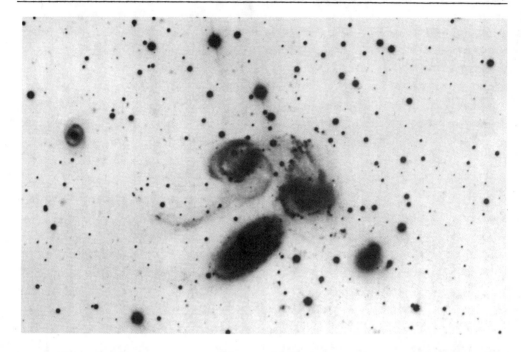

Figure 5.9

that, therefore, these galaxies were young. Then in 1962, E.M. Burbidge obtained the redshift of NGC 7320, which, very surprisingly, turned out to be 800 km/sec. The conclusions at the time were that either NGC 7320 was only accidentally projected on the other galaxies in the Quintet, or that the group was unstable, and flying apart with an unprecedentedly large velocity. The probability of the accidental projection was estimated to be one part in 1500. It was also noted at that time by van der Bergh that the redshift of NGC 7320 was almost exactly the same as that of the Sb galaxy NGC 7331, about 30' to the north. He suggested that NGC 7320 was in fact a companion to 7331. Now, I think we can accept this explanation; then the crux of our problem lies in the question of whether the other galaxies are at this same distance also, or whether they are background objects.

In order to investigate this question, I made a study of the HII regions in the system. A narrow-band Hα picture at the low redshift system picks out twenty-four measureable HII regions in NGC 7320. Then, a similar Hα picture at 6000 km/sec, which does not show the low-$z$ HII, picks out all the HII regions in NGC 7318a and b. Thus it is routine to differentiate the HII regions that belong to 7320 and those that belong to 7318a and b. Interestingly, no other galaxy in Stephan's Quintet shows any HII regions. Now, even by looking at the photographs, it is clear that the sizes of the HII regions in the two galaxies are similar. The diameters are actually measurable, being typically 2"-5" of arc; we found that both the sizes and the distribution of sizes of the HII regions were the same in the two galaxies. This would indicate, via the usual HII region technique for measuring distance, that the two systems are at the same distance. Alternatively, I can turn the argument around, to make the situation clearer. If we compute the absolute size of a 5" HII region in the large redshift system, using a Hubble constant of 50 km/sec Mpc to get the redshift distance, this diameter turns out to be 2800pc. Now, according to Sandage, the largest observed size of a HII region is about 600 pc., and my work on M31 and M53 confirms this. So, those HII regions would have to be five times the size of the largest HII regions as usually observed. In fact, they would be as big as a small galaxy. I therefore think that these two systems are both at the nearer distance.

Now, I would like to mention the most disturbing feature about this system, which I cannot fit well into my picture. During August of 1971, a supernova appeared in the southern arm of NGC 7319. Searle inspected my spectrogram of the Supernova and found that it showed the 6000 km/sec redshifted spectra of a type I supernova, about 20 days after maximum. However, the magnitude was clearly fainter than NGC 7319, whereas, if it were really at the distance of NGC 7331, it should have become about as bright as that galaxy when it was at maximum. There are three possible explanations for this problem. One is to take the conventional view that the high redshift galaxies are at their cosmological distances, in which case the magnitudes come out as they should. The second is to say that only NGC 7319 is in the distance, but NGC 7318a and b, which contain the HII region, are not. This cannot be ruled out, but I would like to argue that the obvious interaction between 7319 and 7318 means that they are near each other. If you accept the proposition that the galaxies are all at the same distance, then you would have to say that the SN is much less luminous than we believe these objects to be. This raises the question of whether or not anomalous redshifts can also affect the luminosities of stars.

Perhaps the most important argument for Stephan's Quintet to be at the distance corresponding to 800 km/sec are the optical and radio connections with NGC 7331. A IIIaJ plate of the group shows a trail emerging from NGC 7320, and curving up toward a small barred spiral of 6000 km/sec which would represent a sixth member of Stephan's Quintet. I would argue that this is an obvious interaction tail. Now the most important thing to notice about this tail is that it is resolved into bright knots (whereas, for instance, the filament to the north, that is emerging from 7319, is not resolved). This means that it must be as close as NGC 7331 and NGC 7320, confirming its status as a tidal filament. I would suggest that the galaxy responsible for this tail is 7318. (See Fig. 5.10)

I have obtained an isodensity trace of this region, and it clearly shows the tail curving upward toward a nearby faint galaxy which also has a redshift of ~6000 km/sec.

Now let us take a look at the wider surroundings of Stephan's Quintet, and examine its connection with NGC 7331. Fig. 5.10 shows the optical appearance of this region. Looking first at the galaxies, we see Stephan's Quintet as a compact group ~30' SW of 7331. Near NGC 7331 are a number of faint companion galaxies, which all have redshifts between 6000 and 8000 km/sec. Now the conventional interpretation would have both NGC 7331 and 7320 superimposed on two separate clumps of background galaxies at somewhat different redshift. This double coincidence seems very implausible to me.

My suggestion, especially in view of the precedent of the previous observations I've mentioned, is that Stephan's Quintet has been ejected from NGC 7331. In order to check on this, I went to the early radio map of the area of NGC 7331 by de Jong (1966) (Fig. 5.11). Being unaware of the existence of Stephan's Quintet, his map extends only about half way to the group, but it clearly shows the beginning of a radio connection starting from 7331. I was very interested in this as evidence for the ejection hypothesis (though it would also support a gravitational interaction between 7331 and 7320 alone). I therefore observed this area at 13 cm with the 210-foot telescope at Goldstone, and found that the area was so rich in radio sources that I had to cut off the survey at the relatively high flux level of .07 f.u. There then remained half a dozen sources near NGC 7331, which is a considerable excess over the number found in the

QSO's Observations and the Redshift Problem 54

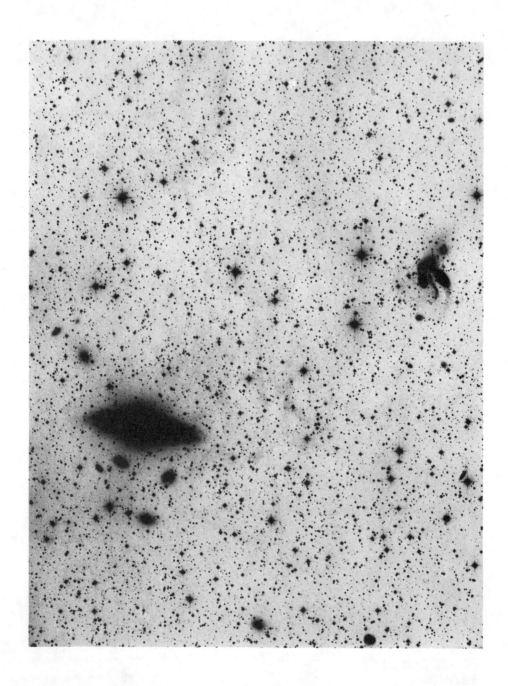

Figure 5.10 Photograph showing the region between NGC 7331 (lower left) and Stephan's Quintet, showing the new optical features believed to connect the two systems.

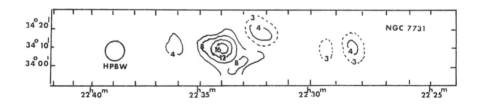

Figure 5.11

general field. My resolution was only about 10', but some sources were clearly complex. The association of radio sources with 7331 makes more plausible the connection of 7331 and Stephan's Quintet, and I would suggest that it supports the ejection hypothesis.

Fig. 5.10 also shows the results of a very deep optical study of the region around 7331, done with J. Kormendy (1972). We wanted to get as deep as possible, and since it is then particularly important to have independent confirmation, we both obtained 3 deep IIIaJ plates of the area. These were then added together photographically to give two independent and very deep photographs. Many of the features common to both photographs have been shown in the picture. Finally, a stack of all 6 plates was made; this goes deeper than any single deepest plate that I have obtained, even with the 200". Stacking the plates diminishes very strongly any plate noise (scratches, pinholes, airplane tracks and the like), and emphasizes very faint real filamentary features which we think represent optical connections between NGC 7331 and the area of Stephan's Quintet. However, we differ in our interpretation of this connection, in that I think it is an ejection remnant, while Kormendy believes it to be the remnant of a tidal filament, which is now old enough to be dispersing, and therefore is very faint.

This completes my discussion of NGC 7331 and Stephan's Quintet, as the situation stands at the moment. It then seems profitable to ask whether or not there are any other such systems in the sky (i.e. multiple interacting non-equilibrium systems),

Figure 5.12  Multiple system VV 150

and, if there are, to ask further whether they are associated with nearby spiral galaxies.

I would nominate for this category the system VV 150 (Fig. 5.12), previously shown by Sargent. Twenty arc minutes away is NGC 3718, a very bright spiral showing conspicuous distortion. It is remarkably close to VV 150. Then there is the chain of peculiar galaxies, studied by the Burbidges, of which two have redshifts of about 5000 km/sec. Amazingly enough in the same 200-inch field is the edge of the very large, low redshift galaxy NGC 247.

Finally, Table 5.1 then summarizes the available data on the outstanding examples of small interacting groups of galaxies. All of these are seen to show similarities to NGC 7331 and Stephan's Quintet, and are near bright galaxies as I have discussed.

## Table 5.1
### MULTIPLE, INTERACTING GALAXIES

| Group Redshifts (km/sec) | Associated Galaxy Redshift |
|---|---|
| 1. Stephan's Quintet<br>800, 5400, 6700, 6700, 6700 | NGC 7331 (11.2mag Sb), 30' NE<br>800 km/sec |
| 2. VV 150<br>7900, 8000, 8100, 8300 | NGC 7318 (12.4mag SBb), 7' N<br>1100 km/sec |
| 3. Burbidge Chain<br>6200, 6300 | NGC 247 (10.7mag S), 20' SW<br>−28 km/sec |
| 4. NGC 907, 899, IC223<br>1600, 1800 | NGC 908 (11.1mag Sc), 33' S<br>1700 km/sec |
| 5. VV 116<br>6500, 6600, 6800, 6900 | NGC 2974 (12.7mag E), 87' NE<br>2000 km/sec |
| 6. Seyfert's Sextet<br>4100, 4400, 4500, 4600, 19,900 | NGC 6052 (13.0pec), 72' ENE<br>− km/sec |

Lecture 6

THE RELATIONSHIP BETWEEN QUASARS AND NORMAL GALAXIES

A definition for quasars may be stated as follows:
  (i) Not resolvable on a scale of approximately 1" of arc, i.e. under optimum conditions it appears completely stellar down to the present maximum resolution of ~1".
  (ii) It has an extra-galactic redshift, i.e. z > 300-400 km/sec.

At the time when quasars were first discovered a very obvious point was considered, namely that if one has stellar-like objects with extra-galactic redshifts, the existence of similar but somewhat more diffuse objects should be considered. These would

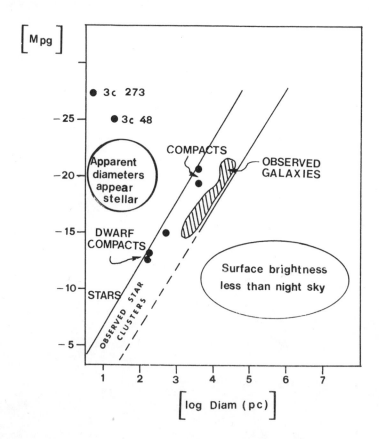

Figure 6.1

# QSO's Observations and the Redshift Problem

constitute something of a transition configuration with regard to quasar and normal galaxies. The observation that galaxies come in varying degrees of condensation lent strength to this idea. At this time Zwicky detected his compact galaxies: these objects appear stellar on short exposure plates but long exposures reveal a faint nebulous envelope. These objects seem to stand between quasars on the one hand and normal galaxies on the other (Arp, 1970). The spectra display emission lines and blue continua. One may postulate a sequence of objects with regard to continuity of properties on the basis of the two plots in Fig. 6.1 and 6.2. In Fig. 6.1 3C 273 and 3C 48 have been plotted assuming a cosmological redshift. If they are assumed to be local, however, and are moved to smaller $M_{pg}$ a more plausible sequence is evident.

In the next graph (Fig. 6.3) B-V (corrected for extinction) is plotted against the measured redshift. In the case of ellipticals, K corrections are applied on going to a higher redshift (Fig. 6.4), thus moving them to the right and making them appear redder. K corrections have not been applied to the quasars as

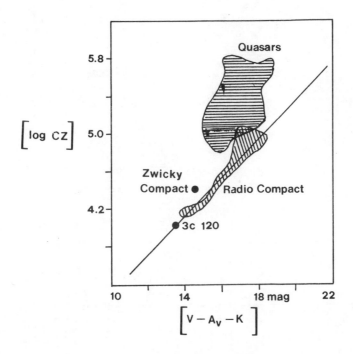

Figure 6.2

# QSO's Observations and the Redshift Problem

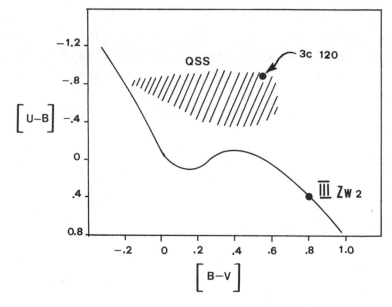

Figure 6.3

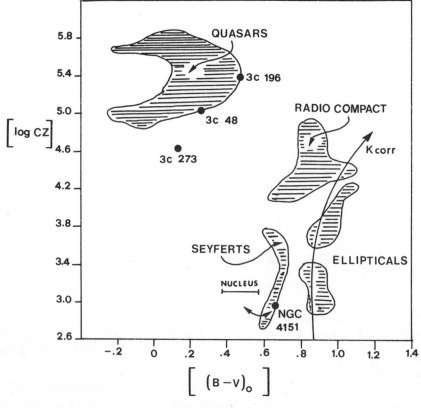

Figure 6.4

their energy distribution appears roughly flat and the necessary K correction is uncertain.

On the basis of this data one can say that there definitely does appear to be a rough continuity between quasars and normal galaxies by way of the compact objects. The essential point of disagreement is over the trend in luminosity, namely, whether one goes to higher or lower luminosity in progressing towards the quasars.

A case which illustrates this well is the central region of the Coma cluster. On a long exposure plate the faint outer envelope of the central galaxy is visible, along with the companion galaxies. On a medium exposure the central galaxy is somewhat less conspicuous relative to the companions. On a very short exposure plate the companions are still visible while the central object is almost invisible. One concludes that the companion galaxies in this group (the lower luminosity galaxies) are more compact and that they have a higher surface brightness. This is found to be generally true. If it is contended that quasars are the most luminous objects known (which they would be if their redshifts are taken to be cosmological) then an incongruity arises: in going from lesser to greater luminosities the objects become more and more spread out, but then the quasars are encountered with enormous luminosities and highly compact configurations. Some quasars identified as being in clusters turned out to be slightly fuzzy on deep 200" plates: they are thus compact galaxies. In the case of low redshift bonafide quasars claimed to be in clusters one should see these clusters; closer examination reveals that, at most, groups are seen.

It is clear in the following plot (Fig. 6.5) that a slope steeper than the Hubble slope is obtained if one wishes to establish a continuity between quasars and normal galaxies and to place the quasars at cosmological distances.

It has been argued that QSO's have intrinsically higher luminosity and that in progressing to lower redshifts the QSO luminosity becomes the same as that of an E galaxy at the same redshift. This argument points to the strange situation that two very different objects, viz. E galaxies and QSO's, have the same intrinsic luminosities. To avoid this it has been argued by Sandage and others (from the standpoint of cosmological redshifts) that at the point of comparable luminosities we are seeing QSO's in the interior of E galaxies. On going further out

# QSO's Observations and the Redshift Problem

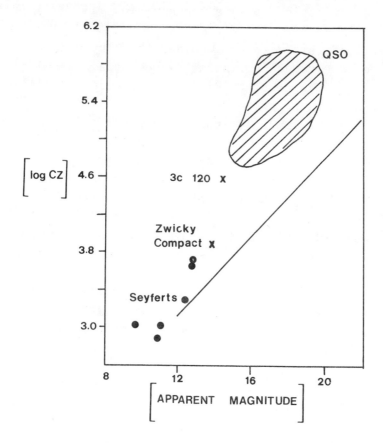

Figure 6.5

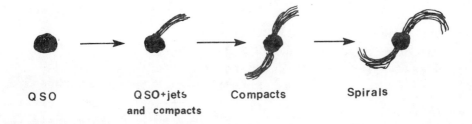

Figure 6.6

# QSO's Observations and the Redshift Problem 63

the QSO's dominate the galaxy in which they are located. This would constitute a transition from normal galaxies to QSO's. In the opinion of the lecturer a different situation seems to be actually observed: a sequence of forms as in Fig. 6.6 is observed. This scheme can be fairly well supported observationally. If the Sandage argument is correct one would expect to see transition cases in connection with large E galaxies. That is, E galaxies with quasar cores which are not observed.

## THE NEIGHBOURHOOD OF GALAXIES

Stephan's Quintet is found to lie in the neighbourhood of NGC 7331. Turning to the neighbourhoods of other multiple interacting systems one always finds large galaxies nearby and in addition the interacting systems always have higher redshifts than these large galaxies. A similar situation exists in the case of double interacting systems as well and, being more numerous, they provide opportunity for systematic study of this phenomenon.

Refering now to table 6.1, it is seen that 22 out of the 35 systems listed have double interacting systems in the field. The galaxies listed were taken from the Shapley-Ames catalogue of bright spirals; the range 10 magn. to 11.2 magn. was chosen since at fainter magnitudes the companions would be fainter and closer to the parent galaxy and trouble would arise in seeing them; at brighter magnitudes one would encounter closer systems with a large dispersion over the sky. For the purposes of this discussion, groups falling within ~45' of arc of the main galaxy are regarded as lying in the field. Control fields 6° away from the main galaxy were studied and it was found that these interacting systems are about ten times more numerous in the original than in the control field. In general, the systems have considerably higher redshifts than the central galaxies.

Some systems of interest are sketched below.
1) South of NGC 2841 an interacting double is found. The redshift for the system is around 8000 km/sec, while that of NGC 2841 (some 30' to the north) is about 1000 km/sec. (Fig. 6.7)
2) The neighbourhood of NGC 2903. (Fig. 6.8)
The central galaxy displays a disturbance in the rotation curve of the interior. Radio maps of the region do not at present show evidence for radio interaction, but more resolution is

TABLE 6.1

## SURVEY OF NON-EQUILIBRIUM FORM GALAXIES
## IN AREAS NEAR BRIGHT SPIRAL GALAXIES

In this table are considered bright spirals with 10.0 mag.<
< $m_H \leq$ 11.2 mag. at declinations $\delta$ > -21° and $|b^{II}|$ > 20°, with
omission of Virgo cluster galaxies.

NGC 3623 has been omitted because together with NGC 3627 and
NGC 3628 they form a disturbed triplet, but there is a nonequilibrium 45' SE of NGC 3628.

| NGC | $M_H$ | TYPE | NON-EQUILIBRIUM TYPES | REMARKS |
|---|---|---|---|---|
| 157 | 11.1 | Sc | 30' N peculiar double | Atlas 127 further out along minor axis |
| 247 | 10.7 | S | 20' NE peculiar chain | chain + double 40+58' SE |
| 908 | 11.1 | Sc | 33' N peculiar triplet | |
| 1232 | 11.1 | Sc | 45' ESE peculiar double | Atlas 332 chain 2½° S |
| 2403 | 10.2 | Sc | 23' ESE peculiar double | |
| 2683 | 10.8 | Sc | | small, sharp spiral 50' SE |
| 2841 | 10.8 | Sb | 35' S peculiar double | |
| 3310 | 10.9 | Sb | | small face-on galaxy, bright arc |
| 3521 | 10.3 | Sc | | |
| 3556 | 11.0 | Sb | | many companions |
| 3893 | 11.0 | Sc | | M51 type companion + high surface brightness companion |
| 4038 | 11.0 | S | 42' NE peculiar double | Antennae galaxy-part of chain |
| 4244 | 11.0 | Sb | | probably irregular galaxy, not Sb |
| 4258 | 10.2 | Sb | 23' WNW peculiar double | many companions |
| 4414 | 11.1 | Sc | 10' SE peculiar group | chain of 4 compacts, 50' NW |
| 4559 | 10.7 | Sc | 45' ESE connected group | |
| 4565 | 10.7 | Sb | 70' E Atlas 34 | edge-on should be brighter |

QSO's Observations and the Redshift Problem 65

| NGC | $M_H$ | TYPE | NON-EQUILIBRIUM TYPES | REMARKS |
|---|---|---|---|---|
| 4605 | 10.9 | Sc | 1' SE peculiar double<br>13' NE galaxy with tail | second companion along minor axis |
| 4699 | 10.5 | SBb | 52' SSE chain of 3 | rich area |
| 4725 | 10.8 | SBb | 23' NE Atlas 159 | counter-companion west |
| 5055 | 10.5 | Sb | | many peculiar galaxies within 1° |
| 5194 | 10.1 | Sc | | many peculiar galaxies within 2° |
| 5248 | 11.1 | Sc | 43' W peculiar chain | |
| 2903 | 10.3 | Sc | 33' WSW peculiar group | peculiar companion 42' ENE opposite |
| 628 | 11.2 | Sc | | |
| 1084 | 11.2 | Sc | 36' SSW peculiar double | |
| 1087 | 11.2 | Sc | | comparable to NGC 1068; many peculiar companions |
| 2976 | 11.2 | Sc | 40' NW peculiar triplet<br>45' SE peculiar double | |
| 4030 | 11.2 | Sb | 19' SE irregular galaxy | low surface brightness, probably low z |
| 4088 | 11.2 | Sc | 53' SE 4 peuliars | disrupted S; wide pair of galaxies 63' + 52' |
| 4151 | 11.2 | Sb | 28' SE connected pair | 4145 in 30' N making pair with peculiar many faint peculiars in the region |
| 7331 | 11.2 | Sb | 30' SW Stephan's Quintet<br>59' NE interacting group | Nearly straight line across NGC 7331 |

needed before any conclusive statement can be made.
3) <u>The neighbourhood of NGC 4038.</u> (Fig. 6.9)
 NGC 4038, the "Antenna System", is one of a chain of galaxies which form a line through this system. All bright galaxies in the area are found in this chain. Both of the objects labelled "pec" at the upper end of the chain have stellar-like cores. The lower of these objects is extremely peculiar, with a redshift

Figure 6.7

of 10,000 km/sec and a spectrum showing strong emission lines.

NGC 4027 is a Coma-shaped galaxy. This non-equilibrium configuration will not be maintained for long so that some estimate of the time scale for the phenomenon can be obtained. A similar object, NGC 4618, is found in another chain which features NGC 4736 (one of the brightest galaxies in the sky), NGC 4626, NGC 4485 and NGC 4490. An interesting feature of this chain is the

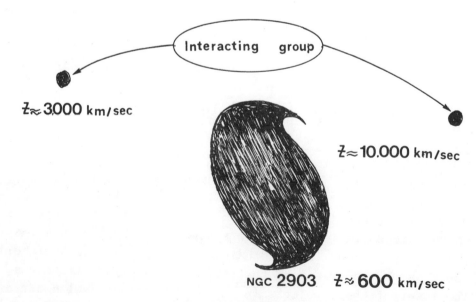

Figure 6.8

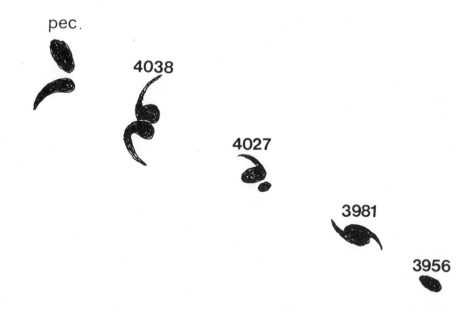

Figure 6.9

tendency for objects to pair. (Fig. 6.10)

Independent evidence for the validity of these chains exists:
1) Almost all of the galaxies are peculiar and were catalogued separately by Vorontsov.
2) The tendency for doubling (cf. tendency for compact galaxies

Figure 6.10

TABLE 6.2

REDSHIFTS IN CHAINS

| GALAXY | MAG. | TYPE | V (km/sec) | REMARKS |
|---|---|---|---|---|
| NGC 4736 | 9.0 | Sb | 300 | peculiar rotation curve |
| NGC 4618 | 11.5 | S pec. | 484 | VV 73 Atlas 23 |
| NGC 4625 | 13.5 | S pec. | 550 | |
| NGC 4490 | 10.5 | I | 570 | VV 30 Atlas 269 |
| NGC 4485 | 12.9 | S pec. | 795 | |
| NGC 4038 | 11.0 | S | 1650 | VV 245 Atlas 244 |
| NGC 4027 | 11.6 | S pec. | 1909 | VV 66 Atlas 22 |
| NGC 3981 | 12.7 | S pec. | 1890 | VV 8 Atlas 289 |
| NGC 3956 | 12.6 | S: | | |
| VV 269 | 15 | pec. | 14,500 | VV 269 |

to double).
3) The redshifts of the galaxies are similar enough to force the conclusion that they are located at the same distance.(Table 6.2) Over and above this the companions show systematically higher redshifts relative to the central galaxy.

A further interesting class of object is illustrated by the two examples below. The first, an extremely peculiar object

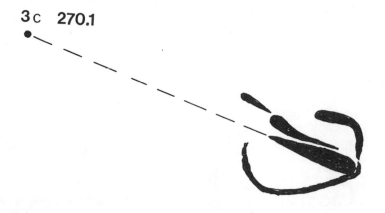

Figure 6.11

# QSO's Observations and the Redshift Problem 69

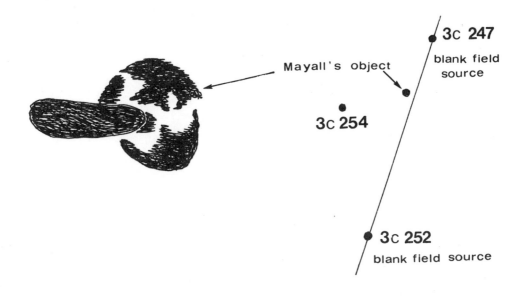

Figure 6.12

found in the neighbourhood of a large bright galaxy, shows a marked "smoke ring" structure (Fig. 6.11). The QSO 3C 270.1 is located 20' of arc away and could be a case of secondary ejection.

A similar example is that of Mayall's object, which is associated with radio sources. (Fig. 6.12).

## THE EVIDENCE FOR ANOMALOUS REDSHIFTS

1) Lines and pairs of optically identified radio sources.
2) Statistical association of quasars and bright galaxies.
3) High redshift objects connected to low by filaments.
4) Faster than light radio expansions.
5) Systematically higher redshifts for companion galaxies and young spirals.
6) Association of peculiar doubles and non-equilibrium systems.

With regard to point(4) it is interesting to note that if objects like 3C 273, 3C 279 and 3C 120 are moved into distances indicated by their association with other objects, the expansion velocities fall to between 10 and 20 thousand km/sec. This is comparable to the expansion velocities observed in various active galaxies.

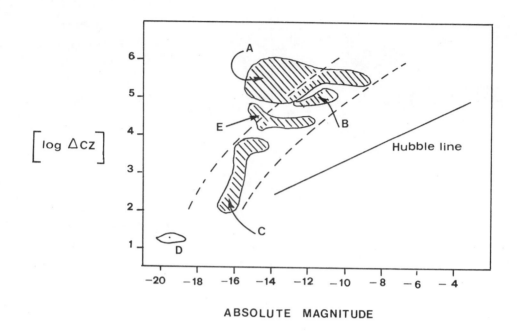

Figure 6.13

Notes to Fig. 6.13.
1) Absolute magnitudes are taken from the object with which the secondary is associated.
2) D is the representative galaxy relative to which the secondary redshift is measured.
3) Region A: QSR's at highest excess redshift. Some have been identified with individual galaxies whose distances must be guessed. It is possible that the left hand sector of this region could be moved a little to the right for this reason.
4) Region B: QSR's tentatively identified with the Local Group. They have been given a mean distance of $2\frac{1}{2}$ Mpc around the Local Group. Absolute magnitude computed with this distance lies in range -11 to -13.
5) Region C, E: Region E contains the N galaxies, the compact galaxies and Stephan's Quintet. The lower part of Region C is occupied by dwarf companions such as M 32, NGC 205: objects with a slight excess redshift.

QSR's which have been identified with different objects and

for which two different ways of assessing distances have been used, still fall in the same region of the plot. It is interesting to discuss the Hubble diagram from Humason, Mayall and Sandage's paper (1956) at this point. In drawing up their diagram, the authors included all nebular types, presumably also low luminosity systems. The photographic apparent magnitudes were K corrected and corrected for galactic obscuration.

There is a clear excess of points above their line for magnitudes greater than about 12.5. The classical explanation is that on going out to increasing z one sees increasing volumes of space. Thus at low z small volumes are seen and only the most common low luminosity systems will be recorded; these fall below the Hubble line. At higher z larger volumes are being considered and intrinsically more luminous systems are seen: therefore one finds systems to the left of the Hubble line. This argument appears to have two shortcomings: firstly, arguments of this nature could be invoked to explain points anywhere in the redshift-apparent magnitude plane; secondly, the reasoning should apply to all regions spanned, especially to those further out. The fact of the matter is that objects are found to fall below the line at the upper end of the relationship.

I have no observations which challenge the Hubble relation as defined by the largest galaxies in clusters of galaxies. All the associations discussed above have been concerned with individual galaxies or groups of galaxies (as distinct from rich clusters). All the evidence presented has dealt with systems dominated by one large galaxy and compact companions.

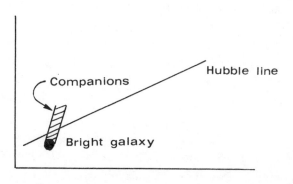

Figure 6.14

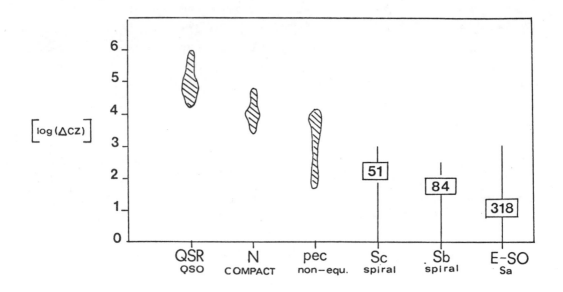

Figure 6.15

If we have a bright galaxy the companions seem to lie in a region as shown in Fig. 6.14. Thus the companions with excess intrinsic redshifts could be used to explain the systematic deviation above 3000 km/sec in the 1956 Hubble diagram referred to earlier.

Finally, there appears to be a correlation between excess redshift and morphological type which hints at an evolutionary sequence in which morphological change towards E-SO type galaxies is accompanied by decay of the intrinsic redshift.

Notes to Fig. 6.15.
1) QSR, QSO: Ages around $10^6$ years. Strong tendency to be double.
2) N, Compact: Stage lasts $\sim 10^7$ years. Doubling tendency.
3) Peculiar: e.g. Stephan's Quintet. $10^7$-$10^8$ years from dynamical arguments. Doubling tendency still evident.
4) Sc spirals: In the picture of the ejection hypothesis, ages $\sim 10^9$ years.
5) E-SO: Ages $\sim 10^{10}$ years.

6) The numbers for Sc, Sb spirals and E-S0 galaxies are taken from Jakkola's paper in Nature in which it was shown that the Sc spirals had systematically higher z's than Sb and the Sb systematically higher z's than the E.

TABLE 6.3

EXCESS REDSHIFTS OF SPECIFIC OBJECTS

| OBJECT | CLASS | $\Delta Z$ | MAG. | $M_v$ | COMMENTS |
|---|---|---|---|---|---|
| PKS 0237-23 | QSR | 2.22 | 16.6 | -14.2 | near NGC 908, m-M < 30.8 |
| 3C 280.1 | QSR | 1.66 | 19.4 | -9.9 | NGC 5055 |
| 3C 270.1 | QSR | 1.52 | 18.6 | -8.5 | NGC 4395 |
| 3C 268.4 | QSR | 1.40 | 18.4 | -11.8 | NGC 4138 |
| 3C 275.1 | QSR | .56 | 18.7 | -11.8 | NGC 4651 |
| 3C 279 | QSR | .54 | 17.8 | -13.0 | south extension Virgo cluster |
| 3C 232 | QSR | .53 | 15.8 | -15.1 | NGC 3067, m-M < <30.9 (see also below) |
| PKS 2344+09 | QSR | .67 | 15.9 | -11.1 | |
| 3C 263 | QSR | .65 | 16.3 | -10.7 | |
| PKS 0405-12 | QSR | .57 | 14.8 | -12.2 | |
| 3C 345 | QSR | .59 | 16.0 | -11.0 | in Local Group, mean distance assumed 2.5 Mpc m-M = 27.0 |
| PKS 2128-12 | QSR | .50 | 16.0 | -11.0 | |
| 3C 232 | QSR | .53 | 15.8 | -11.2 | |
| PKS 2251+11 | QSR | .32 | 15.8 | -11.2 | |
| 3C 249 | QSR | .31 | 15.7 | -11.3 | |
| 3C 351 | QSR | .37 | 15.3 | -11.7 | |
| 3C 48 | QSR-N | .37 | 16.2 | -10.8 | |
| 3C 459 | N-Comp. | .22 | 17.6 | -15.3 | near NGC 7541 m-M < < 32.9 |
| Markarian 205 | QSO | .070 | 15.0 | -16.0 | NGC 4319 m-M < 31.2 |
| 3C 390.3 | N-Comp. | .057 | 15.4 | -11.6 | Local Group m-M > 27.0 (2.5 Mpc) |
| 3C 371 | N | .051 | 14.8 | -12.2 | |
| 3C 120 | Comp-Sey | .032 | 14.1 | -12.9 | |
| Comp. NGC 772 | Comp | 17,300 | 18.0 | -14.0 | NGC 772 m-M = = 32.0 |
| Comp. NGC 772 | Comp E | 17,800 | 17.0 | -15.0 | NGC 772 m-M = = 32.0 |
| VV 172 E | Comp | 37,000 | 18.0 | <-18 | VV 172 m-M << 36 |
| Comp. NGC 7603 | pec N | 8,100 | 18.0 | -16.6 | NGC 7603 m-M < < 34.6 |
| Comp. NGC 1199 | compact | 10,300 | 18.0 | -14.1 | NGC 1199 m-M = 32.1 |

TABLE 6.3 cont'd

| OBJECT | CLASS | ΔZ | MAG. | $M_V$ | COMMENTS |
|---|---|---|---|---|---|
| NGC 7317 | pec | 6,700 | 15.2 | -15.0 | assoc. with NGC |
| NGC 7318A | | 6,700 | 14.9 | -15.3 | 7331 |
| NGC 7318B | | 5,400 | 14.4 | -15.6 | m-M = 30.2 |
| NGC 7319 | | 6,700 | 13.6 | -16.6 | |
| NGC 4151 | Seyfert | 721 | 11.2 | -15.9 | NGC 4244 m-M = 27.1 |
| M 82 | | 234 | 9.4 | -17.9 | M 81 m-M = 27.3 |
| NGC 5195 | | 109 | 11.1 | -18.7 | M 51 m-M <29.8 |
| NGC 2976 | SA59 | +81 | 11.1 | -16.2 | |
| IC 2574 | SX59 | +91 | 11.2 | -16.1 | |
| M 32 | Comp E | +85 | 9.5 | -14.7 | |
| NGC 205 | | +62 | 10.8 | -13.4 | M 31 m-M = 24.2 |
| NGC 185 | | +58 | 10.9 | -13.3 | |
| M 33 | Sc | +57 | 7.8 | -16.4 | |
| VV 150 | | | | | |
| NGC 247 comp. | | | | | |
| NGC 4138 | | | | | |
| NGC 2903 | | | | | |
| Arp 324 | | 2,700 | 18 | -17 | m-M = 35 |
| NGC 4625 | | 250 | 14 | -15.4 | |
| NGC 4618 | | 184 | 11.5 | -17.9 | NGC 4736 |
| NGC 4490 | | 270 | 10.5 | -18.9 | m-M = 29.4 |
| NGC 4485 | | 495 | 12.9 | -16.5 | |
| pec. gr. W of | | | | | |
| NGC 2913 | | 9,560 | 17 | -12.5 | m-M = 29.5 |
| NGC 2916 | | 3,020 | 15 | -14.5 | |
| NGC 3981 | | 240 | 12.7 | -17.3 | |
| NGC 4027 | | 260 | 11.9 | -19.1 | NGC 4038 = N.C |
| pec NE | | 13,900 | 16 | -15 | 16.5 Mpc  m-M< 31 |

## REFERENCES

Arp, H. C., 1966, "Atlas of Peculiar Galaxies", California Institute of Technology, Pasadena.

Arp, H. C., 1969, Astro. Ap., $\underline{3}$, 418.

Arp, H. C., 1970, Ap. J., $\underline{162}$, 811.

Arp, H. C., 1971, Science, $\underline{174}$, 1189.

Arp, H. C., 1971, Ap. L., $\underline{9}$, 1.

Arp, H. C., Kormendy, J., 1972, Ap. J. Letters, $\underline{178}$, 101.

De Jong, M. L., 1966, Ap. J., $\underline{144}$, 553.

Humason, M. L., Mayall, N. U., Sandage, A. R., 1956, A. J., $\underline{61}$, 97.

Kawabata, K., Fujimoto, M., Sofue, Y., Fukui, M., 1969, Publ. Astron. Soc. Japan, $\underline{21}$, 293.

Miley, G. K., 1971, M.N.R.A.S., $\underline{152}$, 477.

Reinhardt, M., Thiel, M. A. F., 1970, Ap. L., $\underline{7}$, 101.

Stockton, A., 1972, Ap. J., $\underline{173}$, 247.

Tifft, W., 1972, I.A.U. Symposium No. 44, 367.

Tifft, W., 1973, Ap. J., $\underline{179}$, 29.

Wills, R., Lynds, D., 1972, Ap. J., $\underline{172}$, 531.

Big Bang Cosmology

Lectures by

**K. Brecher**

**Massachusetts Institute of Technology**

Notes by the lecturer

# Big Bang Cosmology

Lecture 1

It was thought to be appropriate, for this summer school on non-standard cosmology, for someone to review the current view of orthodox Big Bang Cosmology and the arguments for and against it. I will consider both its geometrical and physical aspects. What are the simplest facts which we think relate to the universe as a whole?

1. The distribution of galaxy clusters out to a redshift $z \simeq 0.2$ is both spatially isotropic and homogeneous. The isotropy is seen in Fig. 1.1, where the clusters in Abell's catalogue are plotted on the sky. Further, when the number of clusters is plotted as a function of z, no particular inhomogeneity in $N(z)$ is seen (though one must emphasize that with galaxies only about 1% of the "volume of the universe" has so far been examined).

As we shall see, "isotropy" of the universe in the large can be established to a higher degree of accuracy by considering the quasars and the so-called microwave background radiation.

2. The next most important point to be incorporated into cosmology is the observed general recession of galaxies with respect to us (Hubble's Law). In the 1930's Hubble found that the dimmer the galaxy, in general, the greater was the shift

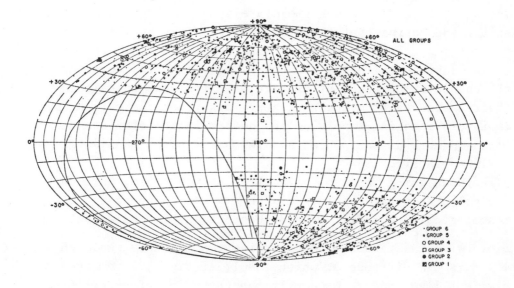

Figure 1.1

Big Bang Cosmology                                                      80

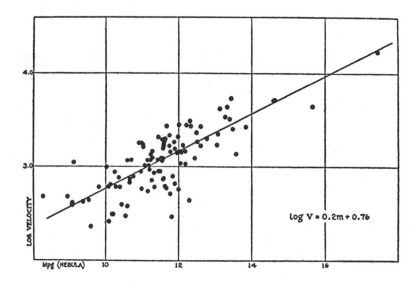

Figure 1.2

in its spectral line toward longer wavelengths, that is, that redshift $z \equiv \frac{\lambda - \lambda_o}{\lambda_o}$ is unversely proportional to magnitude m. Noting that m is the log of the apparent brightness, we see in Fig. 1.2 that Hubble's original data fit a straight line with slope 1 (Fig. 1.2). z was immediately interpreted as a Doppler shift, indicating that the dimmer (larger m) the object, the greater its recession velocity with respect to us.

In order to calibrate the Hubble law, that is, to determine the absolute magnitude of distant galaxies, and thus learn their distances from their relative magnitude there is a long chain of reasoning involving the determination of distances to more and more distant objects. The basis of the subject is outlined in Fig. 1.3

Without going through the entire logical chain, note that one calibrates the absolute distance scale independent of any cosmological arguments. The distance to nearby galaxies, combined with their redshifts determine the distance scale $cH_o^{-1}$, which is the distance at which z = 1, if Fig. 1.2 remained a straight line, and is presently thought to be of order $10^{10}$ light years. The present Hubble plot includes clusters of

# Big Bang Cosmology

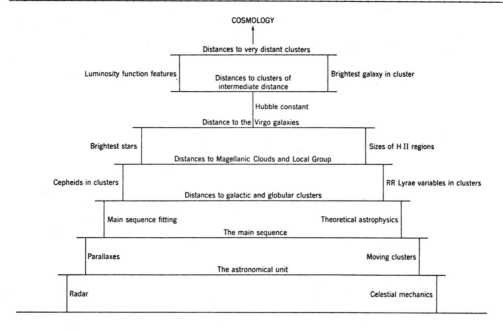

Figure 1.3

galaxies out to a redshift of $z \simeq 0.46$, at which point there may be indications of deviation from a linear relation (Fig. 1.4).

3. Another traditionally mentioned point at the beginning of any cosmological discussion is the darkness of the night sky. If the universe were filled with a density $\rho$ of point sources of absolute luminosity L <u>and</u>: (a) $\rho$ and L are constant in space and time; (b) there are no systematic motions of the sources; (c) space is Euclidean; (d) the laws of physics are true, so that, for example, the relative luminosity I of an object at distance R is given by $I \propto L/R^2$, then the night sky should have brightness F given by

$$F \propto \int_0^R (L/r^2) r^2 dr = LR.$$

As $R \to \infty$, $I \to \infty$. The night sky should be infinitely bright, which it isn't. This is Olbers paradox. (Note that even for finite size sources, one expects the night sky to be bright.) Therefore one or more of the assumptions are false. From what has already been said, at least (b) is false, and

# Big Bang Cosmology 82

## REDSHIFT-DISTANCE RELATION

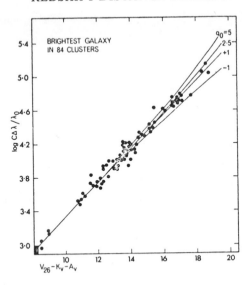

Figure 1.4

since a systematic motion producing a redshift appears to exist, with redshift proportional to distance, $I \propto R^{-(2+a)}$ (where a>1) and F converges to a finite value. As we will now see, assumption (c) is also false.

Let me now outline the standard approach to understanding the expanding universe, in the first instance, from a Newtonian point of view. Consider an expanding fluid medium of density $\rho$, pressure p. If we choose an origin of co-ordinates O, then a point at the dimensionless co-ordinate position r at time $t_0$ can be taken to be a distance $l_0 = R(t_0)r$, from O. At a later time t, the point will still be at the position r, but with $l(t)/l(t_0) = = R(t)/R(t_0)$. $R(t)$ is a function of time which sets the scale of distance between particles in the medium, and is determined by the dynamical equations for a self gravitating fluid. Defining $R(t) \equiv R$, $R(t_0) \equiv R_0$, one can write the force, continuity, and field equations as:

Force Equation: $\quad \dfrac{R}{R_0} \vec{a} = - \vec{\nabla}\phi - \dfrac{1}{\rho} \vec{\nabla} p \qquad (1.1)$

Continuity Equation: $\quad \dfrac{R}{R_0} \dfrac{\partial \rho}{\partial t} = - \vec{\nabla} \cdot (\rho \vec{u}) \qquad (1.2)$

# Big Bang Cosmology

Field Equation: $\left[\dfrac{R_o}{R}\right]^2 \nabla^2 \phi = 4\pi G\rho - \Lambda c^2$  (1.3)

where $\vec{a}$ and $\vec{u}$ are the local acceleration and velocity of the fluid, $\phi$ is the gravitational potential, and $\Lambda$ is a constant to be considered shortly. If one assumes that one is sufficiently far from an "edge" of the fluid, so that it appears isotropic, then $\vec{\nabla}p = 0$. Further, if it is homogeneous, one can assume that $\rho$ is independent of r, so $\rho = \rho(t)$. Under these circumstances, taking the gradient of eq.(1.1), combining with (1.3) one finds:

$$\ddot{R} + \dfrac{1}{3}(4\pi G\rho - \Lambda c^2)R = 0 \qquad (1.4)$$

where $\ddot{R} = d^2R/dt^2$. (1.4) can immediately be integrated to yield

$$\dot{R}^2 = \dfrac{1}{3}(8\pi G\rho + \Lambda c^2)R^2 - kc^2, \qquad (1.5)$$

where k is a constant of integration.

Equation (1.5) is a simple, first order differential equation governing R(t), containing three arbitrary parameters $\rho$, $\Lambda$, and k as written above. This leads to the first problem of classical cosmology: to in some way determine these three parameters (or any other linearly independent set of parameters). Given $\rho$, $\Lambda$, and k one can then describe the past and future properties of this simple expanding fluid, the universe. Such a determination is traditionally made by observations, though some people rely solely on deep theoretical insight or divine inspiration for guidance in this problem. In our wisdom, let us for the moment set $\Lambda = 0$. Then equation (1.4) simply says that $\ddot{R}/R \propto \rho$, and (1.5) says that given R and $\dot{R}$ one can determine k. Thus $(R, \ddot{R}, \Lambda)$ are an alternative set of parameters to use. Notice that since $\rho = \rho(t)$ one must specify $\rho(t_o)$ as a parameter (or, equivalently, $(4/3)\pi R(t)^3 \rho(t)$ which by assumption is the constant mass of the fluid contained within a volume $(4/3)\pi R^3(t)$). There are then two complementary approaches to specifying R(t). One is to try to add up all of the mass density locally to determine $\rho(t_o)$, the other is to determine $\ddot{R}$ from geometrical measurements. In either case (for $\Lambda = 0$) one must also specify the Hubble expansion parameter $H_o = \dot{R}_o/R_o$.

# Big Bang Cosmology

It is convenient to choose $q_o \equiv - H_o^{-2} R_o^{-1} (d^2 R_o/dt^2)$ as the second parameter ("deceleration parameter"). Alternatively one can choose the density parameter $\sigma_o = (4/3)\pi G \rho_o H_o^{-2}$. Note that $q_o$ and $\sigma_o$ are related by the relation $q_o = \sigma_o - (1/3)\Lambda H_o^2$, so that for $\Lambda = 0$, $q_o = \sigma_o$. (Note that this discussion is exact only for $p \equiv 0$. In reality, as long as $p \ll \rho c^{-2}$ nothing above need be modified, and is false only in the earliest phases of expansion of the medium.)

It is not possible here to do justice to the relativistic version of the above picture. However, I wish to outline relativistic cosmology just to indicate the roles of $\Lambda$ and $p$. As was shown by Einstein, the equivalence of gravitational and inertial masses of bodies leads inexorably to a description of space and time which is non-Newtonian and non-Euclidean. It begins by describing the connection between space time events $ds^2$ by a metric $g_{ik}$

$$ds^2 = g_{ik} dx^i dx^k, \qquad (1.6)$$

where $g_{ik}$ is in general a function of the $x^i$. One constructs the Ricci tensor $R_{ik}$, and calling the stress-energy-momentum tensor of a system $T_{ik}$, Einstein was lead to the field equation of general relativity

$$R_{ik} - \tfrac{1}{2} g_{ik}(g^{ik} R_{ik}) = \frac{8\pi G}{c^2} T_{ik} - \Lambda g_{ik}. \qquad (1.7)$$

One also finds the equivalent of the force equation (1.1) as

$$\frac{d^2 x^i}{ds^2} = \Gamma^i_{kl} \frac{dx^l}{ds} \frac{dx^k}{ds} \qquad (1.8)$$

where the Riemann-Christofel symbol $\Gamma^i_{kl}$ is a function of $g_{ik}$ and its first derivative. We are now in a position to see what $\Lambda$ means. If one considers a system with slow motion, so that $\partial/\partial t \ll \partial/\partial x$ ($v \ll c$), and consider weak fields, so that

$g_{ik} \simeq \eta_{ik} + \varepsilon\gamma_{ik}$, $\varepsilon \ll 1$ ($\eta_{ik}$ = Minkowski metric), then equation (1.8) becomes

$$\frac{d^2\vec{x}}{dt^2} = -\tfrac{1}{2} c^2 \varepsilon \vec{\nabla} \gamma_{oo} . \qquad (1.9)$$

If we identify $-\tfrac{1}{2} c^2 \varepsilon \vec{\nabla} \gamma_{oo}$ with the Newtonian term $-\nabla\phi$ one has $g_{oo} = 1 + (2\phi/c^2)$. Now the solution to equation (1.7) for the $g_{ik}$ outside a point mass M is given by the Schwarzschild metric

$$ds^2 = (1 - \frac{2GM}{c^2 r} - \frac{\Lambda r^2}{3}) c^2 dt^2 - (1 - \frac{2GM}{c^2 r} - \frac{\Lambda r^2}{3})^{-1} dr^2 $$
$$- r^2(d\theta^2 + \sin^2\theta d\phi^2) \qquad (1.10)$$

Identifying $(1 + \frac{2\phi}{c^2}) \simeq (1 - \frac{2GM}{c^2 r} - \frac{\Lambda r^2}{3})$, then for $M \to 0$ one has $\nabla^2\phi = -\Lambda c^2$. But this is simply Equation (1.3), showing the Newtonian limit of (1.7), and also that $\Lambda c^2$ could be identified as a negative mass density, i.e., $\Lambda c^2 \equiv -4\pi G\rho_c$. This uniform background mass density, while allowed by the theory, might just as well be taken to be 0. (It was originally introduced by Einstein to cancel the real mass density to produce a static universe. Despite the subsequently observed expansion of the universe, $\Lambda$ has persisted to the present day.)

The role of pressure in relativistic cosmology is as follows. Suppose one considers a medium which is again homogeneous, isotropic, and can be described by a metric $g_{ik}$. One is lead, independent of the field equations (1.7) to the Robertson-Walker line element

$$ds^2 = c^2 dt^2 - R^2(t) \left[ \frac{dr^2 + r^2 d\theta^2 + r^2 \sin^2\theta d\phi^2}{(1 + kr^2/4)^2} \right], \qquad (1.11)$$

where R(t) is simply a function of t, and k is a constant. If

# Big Bang Cosmology

the medium is a perfect fluid (e.g., no shear or vorticity,) with pressure p (due to random motion with respect to the locally co-moving co-ordinates) then

$$T_{ik} = (\rho + p/c^2) u_i u_k + g_{ik} p/c^2. \tag{1.12}$$

Substituting (1.11) and (1.12) into (1.7) one finds

$$8\pi G\rho = (3/R^2)(kc^2 + \dot{R}^2) - \Lambda c^2 \tag{1.13}$$

$$\frac{8\pi Gp}{c^2} = -\frac{2\ddot{R}}{R} - \frac{\dot{R}^2}{R^2} - \frac{kc^2}{R^2} + \Lambda c^2 \tag{1.14}$$

A unique solution for R(t) will, therefore, require specification of p (say $p = p(\rho)$) even if $\vec{\nabla}p = 0$.

Well, what do the solutions R(t) look like? Examples are shown in Fig. 1.5. Particularly notable are the Einstein model already mentioned (R(t) = constant), the expanding Friedmann models, and the oscillating models. The Lemaitre universe, containing a quasi-static epoch ($\dot{R} \ll c$ but not 0) has also received interest of late because of the possible excess of QSO redshifts with $z \simeq 1.95$. The properties of some of the models are summarized in Table 1.1. The steady state universe has

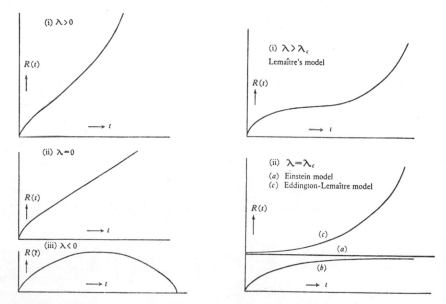

Figure 1.5

Table 1.1

| Model | M | Λ | k | $q_o$ | $\sigma_o$ | R(t) |
|---|---|---|---|---|---|---|
| Einstein | >0 | >0 | +1 | >0 | >0 | R=constant |
| Milne | 0 | 0 | -1 | 0 | 0 | R=ct |
| Open Friedmann | >0 | 0 | -1 | >0 | >0 | $R=(\frac{9}{2}GM)^{1/3} t^{2/3} \longrightarrow R=ct$ |
| Closed " | >0 | 0 | +1 | >0 | >0 | oscillates |
| " with $p=(1/3)\rho c^2$ | >0 | | | | | $R \propto t^{1/2} \longrightarrow R \propto t^{2/3}$ |
| Einstein-deSitter | >0 | 0 | 0 | 1/2 | 1/2 | $R=(\frac{9}{2}GM)^{1/3} t^{2/3}$ |
| deSitter | 0 | >0 | 0 | -1 | 0 | $R=R_o \exp(t-t_o)/T; T=(3/\Lambda)^{1/3}$ |
| Steady State | | | - | | | as above |

been included, but note that it does not satisfy the conservation of matter that the other models do. I have included pressure in one of the above, just to underscore the fact that pressure, even if $\vec{\nabla}p = 0$, acts as another source of gravitation to slow down the expansion rate. We can now understand also the physical significance of the constant k. It specified whether the kinetic energy of the expansion $\frac{1}{2}M\dot{R}^2$ is greater or less than the gravitational potential energy $GM^2/R$, where M is the "mass" of the universe, and is equal to $\rho_o R_o^3$ (to within a numerical coefficient). Note that for k = 0, the system can just expand to infinity ($\dot{R} \to 0$ as $R \to \infty$). Note also that the deSitter model expands forever, having no beginning and no end, being driven by the cosmological constant $\Lambda$. Further, if one wishes to have such a universe with matter of finite density, one must continuously create it, and this leads to the notion of matter creation in the steady state universe (which postulates that the universe appears the same not only in space, but in time as well.) Before proceeding to the observational tests of these different models, it is easy to show (McVittie, 1965) simply from the invariance of $ds^2$ in (1.11) that photons will be redshifted by an amount

$$z \equiv \frac{\lambda_o - \lambda_e}{\lambda_o} = \frac{R(t_o)}{R(t_e)} - 1$$

where $\lambda_e$ is the wavelength of a photon emitted at cosmic time $t_e$ and $\lambda_o$ is the observed wavelength at cosmic time $t_o$. Note also that one can define an "age of the universe" T, for models with a beginning, as

$$\tau = \int_o^T dt = \int_o^{R_o} \dot{R}^{-1} dR$$

For $q_o = \frac{1}{2}$, $\tau = (2/3)H_o^{-1}$ and for $q_o = 0$, $\tau = H_o^{-1}$. Thus the inverse of the observable Hubble parameter is a measure of the time since R = 0.

Big Bang Cosmology                                                         89

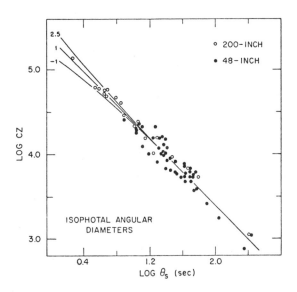

Figure 1.6

What are the observational tests from which one can learn something about the actual universe?

1. The first test we have already mentioned, namely, that the universe is expanding. From the magnitude-redshift relation for brightest cluster galaxies, one could in principle determine the value of $q_0$. In particular, one finds that the bolometric magnitude is given by

$$m_{bol} = 5\log(1/q_0^2) \left\{ q_0 z + (q_0 - 1)\left[(1 + 2q_0 z)^{1/2} - 1\right]\right\} + c \quad (1.15)$$

or, neglecting second order terms in z (z << 1)

$$m_{bol} \simeq 5\log z + 1.086(1 - q_0)z + \ldots \quad (1.16)$$

From figures (1.2) and (1.4) we can see that, for $z < 0.1$, $m_{bol} \simeq 5\log z$. At larger redshifts, out to $z = 0.46$, there is too much scatter in the data to draw a meaningful conclusion about $q_0$. It is, however, gratifying that the slope of the curve is to within statistical uncertainty equal to 5.

2. The only other (thus far) systematically studied property of distant objects is their angular size. In principle, if

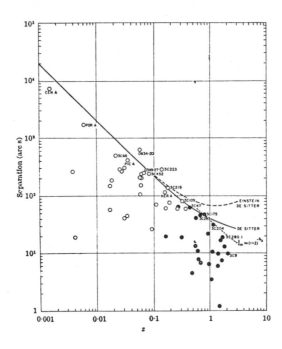

Figure 1.7

galaxies (or clusters of galaxies, or HII regions) were all of equal linear size, one could expect their apparent angular size $\Theta$ to decrease with redshift. In relativistic cosmology, the $\Theta$-z test should contain information about $q_o$. In fact, Sandage finds that the isophotal size $\Theta_s$ for a standard (illuminated) meter stick (length d) should vary with z as:

$$\Theta(\text{isophotal}) = \frac{dH_o q_o^2}{c\{q_o z + (q_o - 1)[(1 + 2q_o z)^{1/2} - 1]\}} \quad (1.17)$$

For example, for $q_o = 1$, $\Theta = \text{const} \times z^{-1}$, and for $q_o = -1$, $\Theta = \text{const} \times z^{-1}(1 + z)^{-1}$. Observationally, studying the same cluster galaxies used to determine the m-z plot, Sandage finds

$$\log \Theta_s = -0.986(\pm .027) \log cz + 5.331(\pm .010)$$

consistent with $q_o = +1$ (see Fig. 1.6). However, the uncertainties in defining the angular size of galaxies, no less in measuring them, leaves one with little confidence in the value of $q_o$ so determined. Before going on, note that a similar test

# Big Bang Cosmology

has been applied to the "double bubble" radio sources. Such a test can be applied to QSO's with very large redshifts, but suffers from the possibility that the QSO redshifts may not indicate their distance and that the linear separation of the bubbles may vary with z in some odd way. Still, one sees in Fig. 1.7 a general decrease in angular separation $\Theta$ with z. The scatter in the data, however, is too great to make a realistic test of the value of $q_o$.

3. Finally, in discussing geometrical tests of cosmology, we should consider the infamous log N - log S curve. It was pointed out by Hubble that if one counted all objects brighter than a given limiting brightness, in the optical say, then, in a static Euclidean universe, since the number of sources out to which one is looking N is porportional to $R^3$, $N(S \geq S_o) \propto R^3$, and since the luminosity of sources $S \propto 1/R^2$, then $N \propto S^{-3/2}$. Thus the slope of the curve d log N/d log S = -3/2. In relativistic cosmologies, one expects a slope $\beta$ = d log N/d log S = = -3/2 + $\mu$ (S,$q_o$,z), where $\mu$ varies with flux density and depends on $q_o$ and the limiting redshift z to which one can see, and $\mu > 0$. This assumes that there have been no changes of either the luminosity or number density of sources with time. This, then, seems as another test of cosmology and could in principle be used to determine $q_o$. Such a test has not been applied to optical galaxies, at least in **part** because one cannot observe faint enough (or in large enough numbers) for the test to be significant.

However, with the discovery of radio sources, this test was again taken up. In 1962, with the publication of the revised 3C catalogue of radio sources, containing about 300 galaxies, such a test was done. The, by now, well known result was, for sources with radio flux > 9 flux units (1f.u. = $10^{-26} W/m^2 Hz$), that $\beta$ was greater than 1.5, and in fact $\beta$ = 1.8. (see Fig. 1.8)

How is one to understand this result? Clearly no information about $q_o$ can be derived from this observation, since <u>all</u> relativistic cosmological models must have $\beta \leq 1.5$. However, taken at face value, the curve does have great astrophysical significance which, in a back-handed way, is supposed to have cosmological implications. What a steep slope implies is either that there is an excess of weak sources, or a deficit of strong

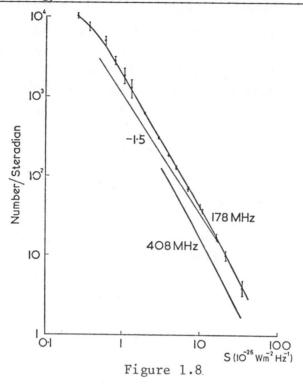
Figure 1.8.

ones. In the former case, this could result either from an evolutionary effect in luminosity or density of sources with cosmic time: sources were either more abundant or brighter in the past. If this were the case, one could immediately eliminate the Steady State Theory of cosmology, which in the simplest form does not allow for variations in properties on a time scale $H_o^{-1}$.

Well, the hue and cry over this simplest result has been heard over the past ten years and cannot be fully repeated here. For recent reviews of the interpretations of this curve see the papers by Brecher, Burbidge, Strittmatter (1971) and by Kellerman (1972), and references contained therein.

Briefly, however, it's worth asking just what are the radio sources. If they include a population of galactic objects, clearly they are useless as a cosmological indicator. Of the 300 or so 3C radio sources, about 39 are known to be QSO's. For a variety of reasons it is thought that there are no more than 5-10 more QSO's in the 3C catalogue. Thus the known QSO's form a more or less complete sample of objects: all radio QSO's with $F \geq 9$ f.u. at 178 MHz. The value of $\beta$ for

# Big Bang Cosmology 93

them is 1.6 ± 0.3, not inconsistent with any model including the non-evolving steady state theory. The roughly 60 known radio galaxies in the 3C catalogue have a slope $\beta \simeq 1.26 \pm 0.3$, again consistent with almost any model. However, if the remaining objects are all galaxies, and have redshifts $z \leq 0.5$ then the steepness of the log N - log S slope must be due to some intrinsic property of the radio source luminosity distribution rather than to a cosmological effect.

As has been emphasized by both Jauncey and Kellerman, other difficulties creep in simply in calculating $\beta$. For example, at low flux levels, say 1 flux unit, there have been systematic differences between the Cambridge and Molongolo source fluxes, with Cambridge tending to overestimate them. Such an effect tends to steepen the curve. Further, in computing $\beta$ from a given set of data one should calculate the slope differentially, and then subtract 1, in order not to compound the confusion due to weak sources.

However, the most severe blow to the credibility of the cosmological significance of the source counts comes from the variation of $\beta$ over the sky. If one divides the radio sources into Northern and Southern galactic hemispheres, as was done recently by Yahil (1972) one finds that the slope of the counts is different in the two parts of the sky (cf. Fig. 1.9). Unless

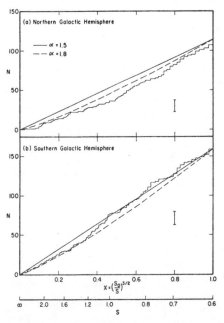

Figure 1.9

the universe is anisotropic with respect to our Galaxy, this is very difficult to understand, and adds further doubt to the cosmological significance of the log N - log S curve.

This completes my discussion of the geometrical aspects of cosmology. In the next lecture, I will try to put some meat on the bones here presented.

# Big Bang Cosmology

## Lecture 2

So far I have tried to derive the necessary parameters to specify the cosmological model in a purely geometrical way. As has been already stressed, instead of $H_0$, $q_0$ and $\sigma_0$, one could as well specify $\rho$, $\Lambda$, $k$. If $\Lambda = 0$, then $q_0 = \sigma_0$, and, given the observed value of $H_0$, the correct model is specified if one can determine either $q_0$ (geometrically) or $\sigma_0 = (4\pi/3) G \rho_0 H_0^{-2}$ by estimating the current local value of $\rho_0$ which, by assumption, is the mean density of the entire universe. One must of course determine $\rho$ by averaging over a suitably large volume. Assuming that the redshift implies distance, one need only count all galaxies in the local neighborhood and multiply by their masses to determine $\rho_0$. However, the estimate of the average galaxy masses also depends on the distance estimate. For example, if one assumes a value for the mass to light ratio for galaxies, M/L, and determines M by measuring L for an individual galaxy, then $M \propto L$. But from Hubble's law, since observed brightness $I \propto L/R^2$, where R is the distance to the object ($R \ll cH_0^{-1}$), then $L \propto R^2 \propto H_0^{-2}$ or $M \propto H_0^{-2}$, so that for N objects of mass M to a distance R, $\rho_0 \equiv NM/(4\pi/3)R^3 \propto H_0$.

Alternatively, one can determine masses by measuring the dispersion velocity of galaxies in a cluster. Assuming that the cluster is bound, the Virial theorem determines the mass of the cluster M from the relation $GM/r \simeq <v_{rms}^2>$, where r is the cluster radius, and $v_{rms}$ is the directly measureable dispersion velocity of cluster members. But $r \propto H_0^{-1}$, so that $M \propto H_0^{-1}$. Therefore $\rho_0 \propto H_0^2$. It then follows that the dimensionless parameter $\sigma_0 \propto \rho_0 H_0^{-2}$ can in this second way be determined independent of the exact value of $H_0$. Both methods give essentially the same value. Oort (1958) gives

$$\rho_0 = 3.1 \times 10^{-31} (H_0/75)^2 \text{ gm/cm}^3;$$

with the current estimate of $H_0 \simeq 50$ km/sec Mpc (Sandage,

1972) one finds $\rho_o \simeq 1.4 \times 10^{-31}$ gm/cm$^3$ in visible galaxies. Further this implies that $\sigma_o \simeq .015$, or a factor of 30 less than the value of $\rho$ needed to "close the universe", or make $k = +1$. If one believes the universe to be closed, this leads directly to the problem of the missing mass needed to close it. (Note that $\rho_{closed} \simeq 2 \times 10^{-29} (H_o/100)^2$ gm/cm$^3$). I wish to emphasize that other components of the universe are not tested in this way, for example the universe could be filled with non-luminous matter, which in practice means ionized or unionized hydrogen gas, neutrinos, or spirits.

What can one now say about the physics of the universe? Equations (1.3) and (1.4) can be combined to give the energy conservation equation

$$\frac{d\rho}{dt} = -3(\rho + \frac{p}{c^2}) \frac{1}{R} \frac{dR}{dt} \qquad (2.1)$$

If $p = 0$, one has the solution to (2.1) $\rho \propto R^{-3}$. But for $p \neq 0$, the kinetic plus rest energy within a volume is not constant, and matter within "does work on the universe" in expanding. For radiation with $p \simeq (1/3)\rho c^2$, $\rho \propto R^{-4}$. Notice that for a black body radiation field with energy density $\rho \propto T^4$, where T is the temperature of the radiation, we have $T \propto R^{-1}$; this also says that the number of photons is conserved. Notice that for $\Lambda = 0$, from eq. (1.13), as $R \to 0$, $t = (6\pi G\rho_m)^{-1/2}$ for a matter filled universe of density $\rho_m$, where t is the "free fall" time. For a radiation filled universe, $t = (32\pi G\rho_r/3)^{-1/2}$. With these simple points in mind, Gamow asked (1948) what would be the conditions for elements to form in the early stages of a Big Bang universe. One is immediately lead to consider a hot beginning, in order that nucleosynthesis could occur. In the very crudest approximation, if the universe were filled with neutrons and protons, nucleosynthesis would be possible only when there was a significant probability of neutron capture by a proton, that is when

$$\sigma n v t \simeq 1, \qquad (2.2)$$

# Big Bang Cosmology

where $\sigma$ = radiative capture cross section for neutrons by protons, n = particle density, v = thermal velocity of particles, and t = expansion time scale of the universe. Now if $T \to \infty$ as $t \to 0$, then when the universe cooled to a temperature of about $10^9$ °K nuclear reactions could take place. As we shall see later, at early times $\rho$ is dominated by $\rho_r = aT^4/c^2$, and then $\rho_r$ is connected with t. Noting that $v \simeq (2kt/m)^{1/2}$, the condition (2.2) implies that at the time of nuclear reactions, $n \simeq 10^{18}$ cm$^{-3}$. If the present value is $n_o \simeq 10^{-6}$ cm$^{-3}$ (roughly the Oort value), then the ratio of the radius of the universe at the time of element formation, $R_o$, to the present value $R_p$, is simply $R_o/R_p = (n_o/n_p)^{-1/3} \simeq 10^{-8}$. Therefore

$$T_p = T_o R_o/R_p \simeq 10^9 \times 10^{-8} \simeq 10 \text{ °K}$$

Thus Gamow <u>predicted</u> in 1948 that there should be a relic black body radiation field whose current temperature is about 10 °K pervading the universe. Doing the above more rigorously, his students Alpher and Herman predicted $T \simeq 5$ °K. Notice that this would introduce a new dimensionless ratio into cosmology, $aT_o^3/kn_o \simeq 10^8$, which is the ratio of heat capacities of radiation and matter or the photon/baryon ratio. Before proceeding to a discussion of the observational evidence concerning such radiation, let me simply outline Gamow's history of the universe, or more particularly, the thermal history of the hot big bang. In Table 2.1 we have outlined the events occurring at various ages (and temperatures) of the universe. Notice that at $t \to 0$, the event horizon, or the region over which various elements of the universe are causally connected shrinks. That is the radius of the universe $R > ct_{expansion}$, so that for all elements of the universe to have known precisely when to start expanding, without having been able to communicate would seem to be a violation of physics at the outset. At $10^{-2}$ seconds, the universe, in thermodynamic equilibrium, was filled with baryons, anti baryons, etc. At $t = 1$ sec, all anti-matter

Table 2.1

HISTORY OF THE UNIVERSE

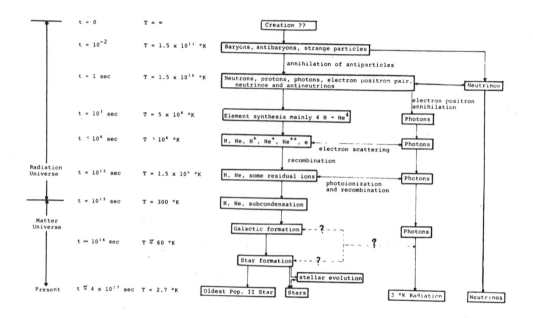

recombines to leave a net excess baryon number, with an equal number of electrons, plus $e^+ - e^-$ pairs, photons, and neutrinos. At $t \sim 10^2 - 10^3$ seconds nucleosynthesis occurs. It must stop at $10^3$ seconds, of course, because at that point neutrons have decayed. Up to $t \sim 10^{15}$ seconds the universe evolves as in Table 2.1. The question of galaxy formation, as we shall see, is very mysterious. Notice that nowhere is the neutrino density specified, which introduces yet another free parameter into physical cosmology.

In Fig. 2.1 is plotted the above history of the universe, assuming that this universe is closed. Notice that prior to $z \simeq 10^4$ the average density of the universe is radiation dominated. Note also that when the universe cools to $10^4$ °K, hydrogen recombines, and the matter, which had previously maintained the same temperature as the radiation because of Thompson scattering

# Big Bang Cosmology

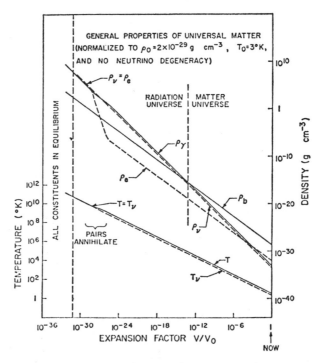

Figure 2.1

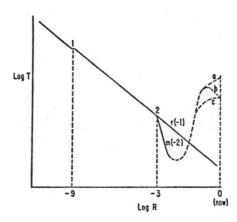

Figure 2.2

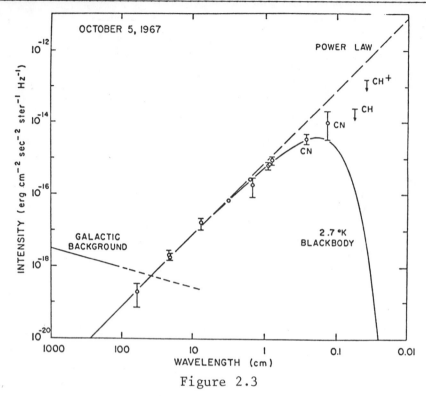

Figure 2.3

with the photons, begins to cool faster than the radiation. This is simply because for any gas it is the momentum which scales as R. Thus for non-relativistic matter $T \propto p^2 \propto R^{-2}$. Fig. 2.2 has a hypothetical curve of T(t) for matter which is subsequently re-heated after the formation of galaxies. Notice also that eventually the energy density of the universe is dominated by the rest energy of matter ($z \simeq 10^3$).

What is the observational evidence for such a picture? As everyone probably knows, in 1965 Penzias and Wilson discovered such a radiation field. The observational data as of 1967 is shown in Fig. 2.3. In the Rayleigh Jeans regime, the spectrum fits a $\nu^2$ power law. However, at the peak of the spectrum ($\sim 1$mm), there existed only upper limits, derived indirectly from the existence of emission lines seen in interstellar molecules. The first direct measurements in the short wavelength region also gave an excess radiation above the black body spectrum. The observations of Weiss and Meuhlner, however, are consistent with a 3 °K spectrum. They find the total integrated flux at present under the band covering the entire blackbody

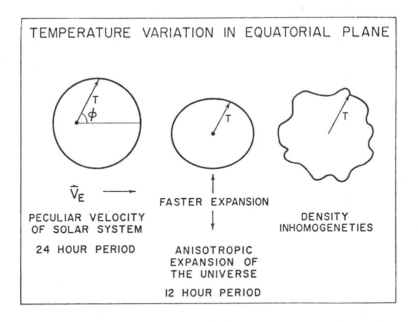

Figure 2.4

curve, yielding a temperature $T \lesssim 3.2$ °K, not inconsistent with the spectral points giving $T = 2.7$ °K throughout the rest of the curve. It appears, therefore, that the microwave radiation has a truly blackbody character, and is, therefore, probably the original fireball radiation, diluted to a present temperature of 2.7 °K.

If the microwave radiation is primordial, it can be used to test the isotropy of the universe at the time of last scattering, $z \simeq 10^3$ (for a closed universe). It also can be used to measure, in addition, our motion with respect to the local co-moving co-ordinates. Finally, it would test how many sources would be required to produce it, if it were not truly primordial. The types of anisotropy expected are shown in Fig. 2.4. For example, if the expansion of the universe were initially faster in one direction than in another, this should result in a 12 hour variation of the temperature over the sky. Observationally the upper limit in $\delta I/I$ is

$$\delta I/I \sim \delta T/T \sim 10^{-3}$$

Therefore, we can conclude to great precision, < .1%, the

universe is isotropic. Furthermore, the upper limit on $\delta I/I$ implies that our random motion with respect to the local zero momentum frame of the microwaves is less than 300 km/sec. Notice that this is close to, but greater than, our local rotation velocity about the center of the galaxy of 250 km/sec. Any further reduction in $\delta I/I$ (observationally) will be very embarrassing implying, perhaps, that we are at the center of the universe after all!

Upper limits on beam to beam fluctuations are also interesting. They give $\Delta T/T \leq 10^{-3}$ as well (for a beam 15° in diameter). If one wishes to attribute the microwave radiation to a distribution of point sources, then, for example, for N sources uniformly distributed about us, one expects RMS fluctuations in $\delta T/T$ of order

$$\delta T/T \simeq (4\pi/N\Omega)^{1/3}$$

For $\delta T/T \simeq 10^{-3}$, $\Omega \simeq 10^{-2}$ steradians one needs more than $10^{12}$ sources, which is greater than number of visible galaxies in the universe. More recent limits imply $N > 10^{15}$, requiring source models to postulate an entirely new class of object to produce the background. This is a further argument in favor of the cosmological origin of the 3 °K background.

Assuming now that the 3 °K radiation is indeed of cosmological significance, we can go back and ask to what extent Gamow's original hope of element formation in the Big Bang is realized. At temperatures greater than $10^9$ °K one has an equilibrium between neutrons, protons, electrons, and neutrinos of the form

$$p + e^- \leftrightarrow n + \nu$$
$$p + \bar{\nu} \leftrightarrow n + e^+$$

As the universe expands and T falls below $10^9$ °K, the neutron formation will cease and the neutrons to proton ratio will be frozen out. At that point, nuclear reactions can begin. From then on, until neutrons begin to decay, various reactions will occur to produce heavier elements. For example one has

$$n + p \rightarrow D^2 + \gamma$$

Big Bang Cosmology                                                            103

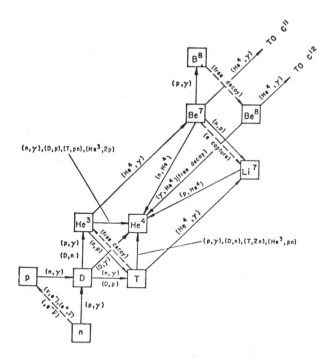

Figure 2.5

$$D^2 + D^2 \rightarrow He^3 + n$$
$$D^2 + D^2 \rightarrow T^3 + p$$
$$He^3 + n \rightarrow T^3 + p$$
$$T^3 + D^2 \rightarrow He^4 + n$$

In principle, one could produce more massive elements, B, Be, Li, C, N, O, but detailed calculations show that very little survives these extreme temperatures except H, $He^4$ and perhaps a little deuterium. Part of the reason for this is the non-existence of a stable nucleus of mass 5 upon which further reactions could proceed. The reaction network is shown in Fig. 2.5. The amount of helium produced, observed to be about 25% by mass (10% by number) of the mass of the observable stars, galaxies, etc., is uniquely determined if the universe has always been isotropic, homogeneous, and if the initial neutrino flux was small. A calculation of $n_{He}/n_H$ requires, therefore, knowledge of:

(a)  S = photon/baryon ratio (now);
(b)  initial neutrino flux $n_\nu$;

# Big Bang Cosmology

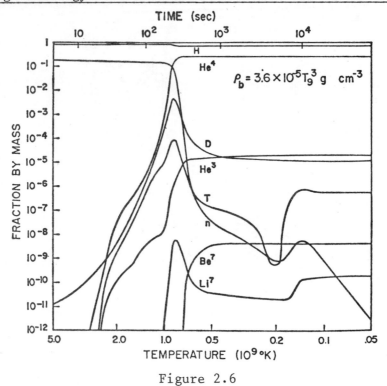

Figure 2.6

(c) initial anisotropy at t = 0.

For $n_\nu = 0$, the results appear in Fig. 2.6, where the evolution of the various species with time (temperature) and with assumed S are given. Thus Gamow's original dream of creating the elements in the Big Bang is only partially realized.

Finally, let me briefly discuss the origin of structure in the universe. The usual approach to studying the origin of galaxies within the framework of big bang cosmologies is to ask whether or not an otherwise homogeneous universe is unstable to density fluctuations away from uniformity. In particular, a hot gas containing fluctuations will be unstable to the gravitational collapse of condensations. In a static (Newtonian) model, a region of size $r_j$ will collapse if its pressure cannot counter balance its self gravitational attraction, or if

$$\frac{3}{2} kT < \frac{GMm}{r_j} = \frac{4\pi}{3} \frac{G\rho}{c^2} r_j^2 \, mc^2 \qquad (2.3)$$

or

$$r_j^2 > (9/8\pi)(kT/mc^2)(c^2/G\rho)$$

where M and $\rho$ are the mass and density of the region at temperature T contained in the region of scale size $r_j$, the Jeans' length. This instability, known as Jeans' instability, will lead to collapse with an exponentially growing density $\rho \sim \exp(t/t_o)$, where $t_o \sim (G\rho)^{-1/2}$.

Returning to Big Bang models, at decoupling of the matter from radiation, $M_j = (4\pi/3) r_j^3 \rho \simeq 10^5 - 10^6 M_\odot$ and for all later times, $M_j < 10^6 M_\odot$. Thus it would appear, even if there is sufficient time for fluctuations to increase their relative density contrast $\delta\rho/\rho$ with respect to the mean density, that galaxy size masses will **not** form. However, it is interesting that $M_j$ at decoupling is roughly a globular cluster mass, and these might themselves later coalesce to form galaxies.

In order to answer the question will such masses have time to form at all, we must answer two questions:
1) What is $\delta\rho/\rho(t)$?
2) What is $\delta\rho/\rho$ initially (at t = 0 or some subsequent time)?

As has already been mentioned, in a static medium $\delta\rho/\rho \sim \exp(t/t_o)$. This rapid density evolution occurs because nothing inhibits collapse. In Big Bang models, however, until decoupling of matter from the radiation, the radiation drag exerted by the radiation on the matter is sufficient to keep collapse from occurring at all, so that $\delta\rho/\rho \simeq$ constant until decoupling. (That is, charged particles will not be able to have a large proper motion with respect to the zero momentum frame of the radiation.)

However, even after decoupling, because the medium is expanding, one will not find that $\delta\rho/\rho$ grows exponentially. To see this, consider a region of density $\rho'(t)$ in which there are perturbations away from uniformity so that

$$\rho(\vec{x},t) = \rho'(t)\left[1 + \delta(x,t)\right] \qquad (2.4)$$

Let the velocity of a fluid element be given by

$$\vec{v} = R^{-1}\dot{R}\vec{r} + \vec{u}(\vec{r},t), \qquad (2.5)$$

where $\vec{u}$ is a perturbation away from the general expansion velocity.

The matter conservation, Euler, and gravitational potential equations become respectively

$$\frac{\partial \delta}{\partial t} + \frac{r}{R} \dot{R} \frac{\partial \delta}{\partial r} = - \vec{\nabla} \cdot \vec{u} \tag{2.6}$$

$$\frac{\partial \vec{v}}{\partial t} + (\vec{v} \cdot \nabla) \vec{v} = - \rho^{-1} \vec{\nabla} p - \vec{\nabla} \phi \tag{2.7}$$

$$\frac{\partial \vec{u}}{\partial t} + \frac{\dot{R}}{R} (\vec{u} + r \frac{\partial}{\partial r} \vec{u}) = - \vec{\nabla} \psi - \frac{kT}{m} \vec{\nabla} \delta \tag{2.8}$$

where $p = \rho kT/m$, and $\psi$ is the perturbation away from the uniform gravitational potential $\phi$. Taking the divergence of (2.8) and combining with (2.6) one finds

$$(\frac{\partial}{\partial t} + \frac{r}{R} \dot{R} \frac{\partial}{\partial r} + \frac{2}{R} \dot{R})(\frac{\partial}{\partial t} + \frac{r}{R} \dot{R} \frac{\partial}{\partial r}) \delta = 4\pi G \rho' \delta + \frac{kT}{m} \nabla^2 \delta \tag{2.9}$$

Transforming from locally Minkowski co-ordinates $\vec{r}$ to co-moving co-ordinates $\vec{x}$, where $\vec{x} = \vec{r}/R$, (2.6) and (2.9) become

$$(\partial \delta/\partial t)_x = - \vec{\nabla}_x \cdot \vec{u}/R(t) \tag{2.10}$$

and

$$\left[\frac{\partial^2 \delta}{\partial t^2}\right]_x + \left[\frac{2}{R} \dot{R} \frac{\partial \delta}{\partial t}\right]_x = 4\pi G \rho' \delta + \frac{kT}{mR^2} \vec{\nabla}_x^2 \delta \tag{2.11}$$

One can now Fourier decompose $\delta$ so that one has $\delta = \delta(t) \exp(i\vec{k} \cdot \vec{x})$. Assuming now an Einstein-deSitter universe, so that $R^{-1}\dot{R} = (2/3)t^{-1}$, the above equations become

$$\frac{\partial^2 \delta}{\partial t^2} + \frac{4}{3} \frac{1}{t} \frac{\partial \delta}{\partial t} = (2/3)\delta/t^2 \tag{2.12}$$

with solution

$$\delta = At^{2/3} + Bt^{-1} \tag{2.13}$$

Therefore, density perturbations grow only algebraically in time, not exponentially. Changing the cosmological model only changes the growth rate to $\delta \propto t^a$, $1/2 \leq a \leq 1$. Clearly even starting fluctuations at decoupling where $R/R_o \simeq 10^{-3}$, one can have a relative growth of density perturbation by a factor of $10^3$. But for a universe of $10^{80}$ baryons in which one wishes galaxies with $10^{70}$ baryons to form, one expects statistical fluctuation $\delta\rho/\rho \sim N^{-1/2}$ initially. Thus the relative growth of fluctuations is not nearly enough to allow $\delta\rho/\rho \sim 1$ to appear. Thus in order for galaxies to grow from density fluctuations, one must postulate them initially.

To conclude the review of cosmology, let me summarize the successes and failures of the canonical Big Bang Cosmologies:

I. Successes of Big Bang Cosmologies
 a) Prediction of expansion of universe
 b) Prediction of 2.7 °K relict radiation
 c) $He^4$/H ratio resulting from hot models

II. Inconclusive tests of specific models
 a) ln N - ln S curve
 b) $\Theta(z)$, all other "geometrical" tests

III. Failures of the Big Bang
 a) origin of galaxies
 b) origin of magnetic fields, angular momentum, elements beyond He.

## REFERENCES

Bondi, H., 1961, "Cosmology" (Cambridge U. Press, Cambridge).

Brecher, K., Burbidge, G. R., Strittmatter, P. A., 1971, Comments Ap. Space Phys., $\underline{3}$, 99.

Burbidge, G. R., 1971, Nature, $\underline{233}$, 36.

Gamow, G., 1948, Phys. Rev., $\underline{74}$, 505.

Kellerman, K. I., 1972, A. J., $\underline{77}$, 531.

McVittie, G. C., 1965, "General Relativity and Cosmology" (U. of Illinois Press, Urbana).

Oort, J. H., 1958, "La Structure et l'Evolution de l'Universe", Bruxelles, Solvay Conference.

Peebles, D. J. E., 1971, "Physical Cosmology" (Princeton U. Press, Princeton).

Sandage, A., 1972, Ap. J., $\underline{178}$, 1.

Sciama, D. W., 1971, "Modern Cosmology" (Cambridge U. Press, Cambridge).

Weinberg, S., 1972, "Gravitation and Cosmology" (John Wiley and Sons, Inc., New York).

Yahil, A., 1972, Ap. J., $\underline{178}$, 45.

Observational Problems of High Energy Astrophysics

Lectures by

E. M. Burbidge

University of California, San Diego

Notes by

R. Chevalier, M. Rosenberg, T. Thuan

# Observational Problems of High Energy Astrophysics

Lecture 1

This will be in the form of an introductory talk, unlike the specialized lectures already presented and those which will follow. This lecture will be concerned primarily with normal galaxies. To have a perspective, I list a division of things one may consider in the general cosmological problem:

I. Normal Galaxies - Later defining a normal galaxy, and deciding just how "normal" is normal.
II. Abnormal Galaxies.
III. QSO's - to be defined.
IV. Intergalactic medium, about which, unlike the first three categories, there exists more theory than definitive observation.

Under I, normal galaxies, one may group the observational information into:
1. content
2. structure
3. organization
4. redshifts

## 1. Content

Normal galaxies like our own, are comprised of stars, which make up the greater part of the galaxy's mass, provide most of its light, and give the main body of information about the galaxy. The gas, hot or cold, comprises a few percent to zero percent the mass of a galaxy. The dust makes up less by mass than the gas. The structure of a normal galaxy is delineated by its stars and, also, its hot gas. The shape of a galaxy in the Hubble classification is spiral, elliptical, or irregular (Sargent will give a detailed description later).

## 2. Structure

It is difficult to define what a really normal galaxy is; however, one of the best known spiral galaxies is M51. Seen nearly face on, one sees the not completely regular spiral arms, comprised of highly luminous stars and hot gas excited by the stars, the dust lanes within the spiral arms, and one gets the impression of rotation. Indeed all spiral galaxies are rotating, as are many ellipticals to a lesser extent and irregulars. Rotation is a very general property of galaxies. For example, NGC 891, seen edge on, looks pretty much like a wide angle pic-

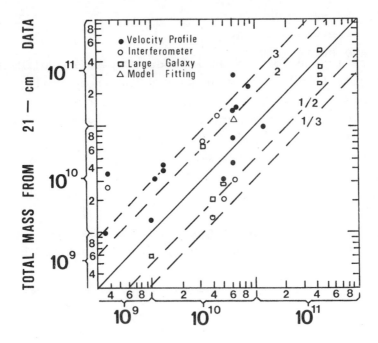

TOTAL MASS FROM OPTICALLY DERIVED DATA

Figure 1.1 Here are shown some determinations which have been obtained by two different methods; firstly, by studying rotations by means of the optical light from the galaxy, and secondly, by studying radiation coming from the cold gas, the neutral hydrogen gas in the galaxy.

ture of the Milky Way, and is flat in form, suggesting rotation, and has a not completely regular dust lane in the center, which, however, gives the impression of an overall regularity. An orientation in between face on and edge on is shown by NGC 3627, whose spiral arms are seen less clearly because of its inclination and which is not as regular as the latter two galaxies. The arms appear to be non co-planar, due perhaps to gravitational distortion by its two companions.

In elliptical galaxies practically all the matter is in the form of stars, with little or no gas, and apparently no dust. The classical example of an elliptical galaxy is M87, which by the way is not a normal galaxy; it's a strong radio emitter. An elliptical galaxy which is a member of the local group is NGC 185. One can see the resolution into individual stars, and its small dust patch.

To determine what is a really normal galaxy, define normal galaxy as one satisfying the two criteria:
i) All or most of the radiation is emitted via thermal processes (radiation from stars or the hot gas surrounding stars).
ii) The galaxy is in equilibrium under its own gravitational field and its own rotation. This implies that, observationally a) any motion is circular, ie., the motion is rotational in nature, and b) generally, the galaxy has a symmetrical structure. According to this definition, an example of "supernormal" galaxy is NGC 7814, an edge-on galaxy, with a thin and very even dust lane, and no impression of breaking up in the dust lane. This galaxy demonstrates very well an equilibrium situation.

There is a variety of forms of galaxies. Before turning to organization, consider the range of masses one is dealing with. To get some idea of the masses, consider first how one determines the mass in equilibrium situations. The rotation curve of the galaxy gives the run of density with distance from the center and thus the mass. The mass of the galaxy is also determined via 21-cm data. The orbital motion of double galaxies is used in a statistical sense to determine the masses of galaxies, but not individual masses. The virial theorem can be used to determine masses of elliptical galaxies in which it is difficult to measure rotation, and spherical ellipticals. It's difficult to get a lower limit for the mass range of galaxies, because the smaller-mass galaxies are more difficult to measure. The masses range, in units of solar masses, from $\sim 10^4$ to a few times $10^{12}$. The range of masses given in Fig.1.1 is less, because for the low mass galaxies the rotation is too small to be measurable, (one is limited by the amount of light coming and the amount of resolution), but one can extend the curve down, because one sees objects faint enough whose component of velocities can be seen with sufficient accuracy to indicate there must be lower masses; one can similarly extend the curve up into the range of massive ellipticals, where data come from the virial theorem, not rotation.

3. Organization

There is a hierarchy in the organization of galaxies just as there is a hierarchy in the organization of stars in our own galaxy. The stars in our own galaxy are grouped individually, as double stars, as members of small clusters and groups,

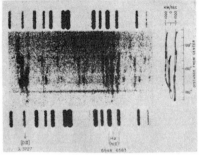

Figure 1.2

and into large groups.

Galaxies are grouped also into groups and clusters of galaxies (poor to very rich). For example, the central part of the Hercules Cluster is a fairly rich cluster with a fair number of spirals throughout.

4. Redshifts

This is the most important observational information for this summer school. Define

$$z = \frac{\lambda_m - \lambda_o}{\lambda_o}$$

where $\lambda_m$ is the measured wavelength of the radiation and $\lambda_o$ is the tabulated value (rest wavelength).

If the redshift is Doppler, for z small we get:

$$z \sim \frac{v}{c}$$

where v is the line of sight velocity. In measuring redshifts one is concerned with displacements, and as an example I show a spectrogram of the Crab Nebula, with a comparison spectrum of an He-Ar glow tube (Fig. 1.2). The nebula, the remains of the supernova of 1054 A.D., is hot ionized gas; it is in our own galaxy, so it doesn't have a bodily redshift. The Crab spectrum is produced with the long slit of the spectrograph across

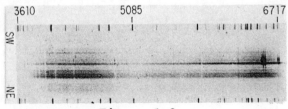

Figure 1.3

Observational Problems of High Energy Astrophysics    115

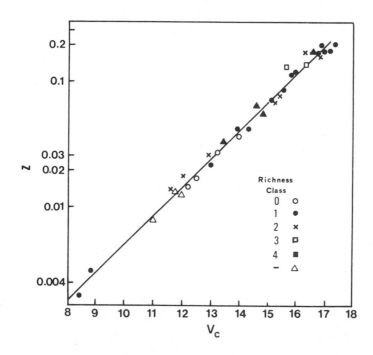

Figure 1.4 Red-shift-magnitude diagram for thirty-eight brightest members of clusters of galaxies. Abell richness classes are indicated; those clusters whose richness class is unknown are entered as triangles. The least squares best fitting line giving $q_0 = 0.64$ and $M_v = -22.50$ (for $H_0 = 75$ km $S^{-1}$ Mpc$^{-1}$) has been drawn in.

the nebula; one gets extended image of the spectrum lines, each point along the slit giving an individual point image along the spectrum. The bright strip (very dark on the negative) is continuous radiation upon which is superimposed absorption from a foreground star. The sky background produces straight undisplaced lines all across the slit. The dashed lines across the slit are the filaments or wisps of the nebula, the lines being split into two, even three, and bowed. The redshifted line, corresponding to recession, is from the back part of the nebula, which is expanding away from us, while the blue-shifted line is from the front part, expanding toward us, of the nebula. The distance between the splittings corresponds to an expansion velocity of ~ 1000 km/sec.

With galaxies, as well as finding shifts like these within the galaxy, one expects to find a total bodily shift of the whole galaxy. Hubble and Humason discovered this systematic effect for galaxies, that z is positive and increases with the faintness of the galaxy. To see how this comes about, assume z small, so that $z \sim v/c$ is applicable. Consider the spectrum of a galaxy with the slit placed across the whole galaxy (as with the Crab) (Fig. 1.3). The bright strip (dark on the negative) is radiation from the center of the galaxy; there are dark regions where the slit crossed dark regions of the galaxy, and light regions where the slit crossed light regions. The double lines here are not due to splitting, they are H$\alpha$ and a line of (N II). The lines are not straight, but are bent at the top and at the bottom, an evidence for motion within the galaxy: for example, if this is regular and symmetrical along the major axis, there is rotation, one side moving toward, the other side moving away. The absorption lines produced by stars in the galaxy are more difficult to measure: one sees the Balmer series of hydrogen, most of the light of this galaxy coming from medium hot stars in the galaxy, as well as emission lines of hydrogen, (O II), (N II), and (S II). If one measures the wavelength of the mean line of sulfur, and compares it with the rest line, one finds a **redshift** indicating that the whole galaxy is moving away from us.

This is what the observations look like that one measures to get the evidence that we'll be discussing here. Assuming the total bodily shift is Doppler, $z \sim v/c$ is positive and the redshift is due to recession. According to the inverse square law for intensity of radiation, $L \propto D^{-2}$ (D is the distance from the galaxy to us). Define

$$\frac{\ell_1}{\ell_2} = 10^{0.4(m_2 - m_1)}$$

Hubble and Humason found a correlation between magnitude and redshift such that, if one plotted log z vs. m for galaxies, through the points one could fit a straight line whose slope was 0.2 (Fig. 1.4). So one gets the relation

$\log z = \text{constant} + 0.2\, m$

From the definition of $\ell_1/\ell_2$, from $L \propto D^{-2}$ and the above relation for $\log z$, and as a consequence of the assumption $z \sim v/c$ one obtains:

$\log z = \text{constant} + \log D$

i.e., the redshift is proportional to distance.

The graph of $\log z$ vs. m has large scatter when measuring all galaxies. The scatter is due to intrinsic dispersion in magnitude of galaxies (horizontal scatter), but a small part could also be due to a vertical scatter. Such a dispersion in redshift may be due to some other sort of velocity (e.g. shear), other than expansion of the universe. There do exist other sorts of velocity within the universe than the expansion of the universe; because of the great distance of galactic clusters it is sufficient to consider that all galaxies in the cluster are at the same distance from us, so that the spread of velocities of the galaxies gives the individual kinetic energy of the galaxies. Incidentally, a measurement of the kinetic energy is a way to derive the individual masses of the galaxies; assuming equilibrium and using the virial theorem, one has the kinetic energy, the potential energy in terms of the unknown mass, and an equation to determine the average masses. Recently, an idea originally put forward by de Vaucouleurs, is being checked for nearby clusters, that is that an anisotropy between the North and South galactic hemispheres in velocity may be due to shear.

So the dispersion in the Hubble diagram of $\log z$ vs. m for individual galaxies indicates:
1. dispersion in m;
2. dispersion in velocity within clusters;
3. dispersion within cluster velocities.

We must check also for the possibility of non-velocity redshifts, which will be discussed in this school.

This covers the divisions under the heading normal galaxies. Going briefly into the topics II and III (the information about these subjects will be covered in later lectures) the definition of normal enables one to define an abnormal galaxy:
1. One whose radiation is derived in a fairly large amount

from non-thermal sources. The simplest way to detect non-thermal radiation is to see in what range of frequencies the radiation from normal galaxies is emitted, and to look for radiation in other frequency ranges. Because of the temperatures of stars, from about 3,000 °K to about 100,000 °K, the bulk of the radiation from normal galaxies is emitted at optical frequencies, with ultraviolet from the hot end of the distribution and near infrared from the cool end. Abnormal galaxies emit in the radio, long infrared, and x-ray regions, frequency ranges at which normal stars emitting thermally in normal galaxies won't be doing anything.

2. Another criterion is that there are shifts in the wavelengths of lines, which, if interpreted as velocities, indicate non-circular velocities in the gas. (Using circular in the wide sense to include elliptical orbits of stars within the galaxy). In galaxies which emit large amounts of non-thermal radiation, there are frequently non-circular velocities, indicating motions produced e.g., by explosive events in nuclei.

3. A less important factor is structural peculiarities, for some peculiarities may be produced by the gravitational interaction between close galaxies. However, if one observes the first two characteristics, the third lends weight to the classification as an abnormal galaxy.

All three characteristics point to the fact that in abnormal galaxies there was a large production of energy, to produce large amounts of non-thermal radiation. If the radio emission is due to synchrotron emission of high energy particles spiralling around magnetic field lines, the energies necessary for such radio emission are large, indicating a large energy release. Where one may tie down the location, it appears that energy release occurs in the central regions of galaxies (where the greatest mass concentration occurs).

In Table 1.1 I list a number of abnormal galaxies, in which we see that the energy released in the radio frequency range goes up to $1.5 \times 10^{43}$ erg/sec from Hydra A and $5.7 \times 10^{44}$ erg/sec from Cygnus A (computed using distances from the Hubble relation with H = 100 km/sec per Mpc).

The radio emission from the nucleus of our own galaxy is $10^{38}$ erg/sec, and from Andromeda M31 $1.9 \times 10^{38}$ erg/sec. Perhaps most galaxies have this amount of abnormality. Compare

Table 1.1

MINIMUM ENERGIES REQUIRED FOR SYNCHROTRON EMISSION IN GALAXIES

| Galaxy | Rate of emission (ergs/sec) | Total energy (electrons + mag. energy) (ergs) | Mean Value of H (gauss) | Total energy (protons + mag. energy) (ergs) | Mean Value of H (gauss) |
|---|---|---|---|---|---|
| Galaxy | $\sim 10^{38}$ | $\sim 3 \times 10^{54}$ (electrons) $\sim 10^{56}$ (mag. field) | — | $\sim 3 \times 10^{56}$ | $7 \times 10^{-6}$ (disk) $2 \times 10^{-6}$ (halo) |
| M31 | $1.9 \times 10^{38}$ | $2.1 \times 10^{55}$ | $8 \times 10^{-7}$ | $3.0 \times 10^{56}$ | $3 \times 10^{-6}$ |
| Magellanic Clouds | $1.3 \times 10^{37}$ | $2.5 \times 10^{54}$ | $1 \times 10^{-6}$ | $3.4 \times 10^{55}$ | $4 \times 10^{-6}$ |
| NGC 4038-39 | $2.1 \times 10^{39}$ | $1.7 \times 10^{56}$ | $2 \times 10^{-6}$ | $2.3 \times 10^{57}$ | $7 \times 10^{-6}$ |
| NGC 1068 | $7.5 \times 10^{39}$ | $3.2 \times 10^{55}$ | $2 \times 10^{-5}$ | $3.6 \times 10^{56}$ | $6 \times 10^{-5}$ |
| NGC 5128 (Central region) | $2.4 \times 10^{41}$ | $3.2 \times 10^{56}$ | $2 \times 10^{-5}$ | $4.4 \times 10^{57}$ | $9 \times 10^{-5}$ |
| NGC 5128 (halo) | $2.2 \times 10^{41}$ | $5.0 \times 10^{58}$ | $1 \times 10^{-6}$ | $7.0 \times 10^{59}$ | $5 \times 10^{-6}$ |
| NGC 1316 (Central region) | $8.0 \times 10^{40}$ | $2.1 \times 10^{56}$ | $2 \times 10^{-5}$ | $3.0 \times 10^{57}$ | $6 \times 10^{-5}$ |
| NGC 1316 (halo) | $1.6 \times 10^{41}$ | $1.8 \times 10^{58}$ | $1 \times 10^{-6}$ | $3.2 \times 10^{59}$ | $5 \times 10^{-6}$ |
| NGC 4486 (jet) | $2.3 \times 10^{42}$ | $1.7 \times 10^{54}$ | $2 \times 10^{-4}$ | $2.4 \times 10^{55}$ | $7 \times 10^{-4}$ |

Table 1.1 cont'd

| | Rate of emission (ergs/sec) | Total energy (electrons + mag. energy) (ergs) | Mean Value of H (gauss) | Total energy (protons + mag. energy) (ergs) | Mean value of H (gauss) |
|---|---|---|---|---|---|
| NGC 4486 (Central radio source) | $3.5 \times 10^{41}$ | $1.7 \times 10^{57}$ | $1 \times 10^{-5}$ | $2.4 \times 10^{58}$ | $4 \times 10^{-5}$ |
| NGC 1275 | $6.4 \times 10^{41}$ | $9.4 \times 10^{56}$ | $2 \times 10^{-5}$ | $1.3 \times 10^{58}$ | $8 \times 10^{-5}$ |
| NGC 6166 | $7.8 \times 10^{42}$ | $1.4 \times 10^{57}$ | $3 \times 10^{-5}$ | $1.9 \times 10^{58}$ | $1 \times 10^{-4}$ |
| Hydra A | $1.5 \times 10^{43}$ | $1.0 \times 10^{58}$ | $8 \times 10^{-5}$ | $1.5 \times 10^{59}$ | $3 \times 10^{-4}$ |
| Cygnus A | $5.7 \times 10^{44}$ | $2.8 \times 10^{59}$ | $4 \times 10^{-5}$ | $3.9 \times 10^{60}$ | $2 \times 10^{-4}$ |
| Coma Cluster | $1.0 \times 10^{41}$ | $2.9 \times 10^{59}$ | $2 \times 10^{-7}$ | $4.0 \times 10^{60}$ | $7 \times 10^{-5}$ |

# Observational Problems of High Energy Astrophysics

these values with the average emission of $10^{43}$-$10^{44}$ ergs/sec in the optical frequencies from thermal processes.

Computing the minimum total energy required to produce the observed radio emission due to the synchrotron process, one needs an energy source of $4 \times 10^{60}$ ergs in Cygnus-A, and $3 \times 10^{56}$ ergs in our own galaxy.

In later lectures there will be details about abnormal galaxies and QSO's, about which I have said nothing yet.

Lecture 2

In the last lecture I gave an outline of the normal galaxies and I will now turn to the abnormal galaxies.

The first and foremost characteristic in my definition of abnormal galaxies is that they are strong emitters of nonthermal radiation. Secondly, they can be recognized by the presence of large-scale, noncircular velocities, and thirdly, less importantly, by the presence of structural peculiarities. All these point to the fact that there must have been a violent energy release and the location appears to be the nucleus of the galaxy. Radio emission was the first recognized form of nonthermal emission and is still the most characteristic feature. The radio fluxes can be large and the energies required to generate these fluxes by synchrotron radiation in the form of high energy particles and magnetic fields are likewise very large.

Since many cosmological arguments are based on things that depend on the properties of the radio galaxies in particular (counts of these and so on), it is very important to try to understand what it is we are counting, what the physics of these objects is, what causes a galaxy to be of this nature, how long it will exist in this state, and so on. Therefore I will give a description from an observational point of view. You will see that there **are** a great variety of radio galaxies, but hopefully some pattern will emerge. The nearest we have gotten to understanding this is to say that something happens in the nucleus of a galaxy and rapid energy release is triggered.

I will talk about radio galaxies first and foremost. I will also talk about Seyfert galaxies: these may be one and the same thing. A Seyfert galaxy can be a radio galaxy and the two phenomena are probably manifestations of the same sort of physical process.

Perseus A ≡ NGC 1275 is a strong radio galaxy which is also a Seyfert galaxy. It is not symmetrical, as there is a bright area at the center and filamentary structure, which appears only on one side of the galaxy (Fig. 2.1), so it definitely fulfills the third criterion for an abnormal galaxy. The fact that it was first recognized as being a strong radio galaxy shows it fulfills the first criterion. The second

# Observational Problems of High Energy Astrophysics

Figure 2.1  Optical appearance of NGC 1275

criterion was shown some years ago, when it was found by Minkowski (1957) that the filaments have a very different velocity from that of the main body of the galaxy. The main body has a redshift which implies a recession velocity of about 5300 km/sec, which is consistent with the velocities of the other galaxies in the cluster in which NGC 1275 is the brightest member. The filaments show emission lines produced by hot gas which has a velocity of about 8300 km/sec, that is, 3000 km/sec greater than the central region. This is very large compared to the rotational velocities observed in galaxies (up to 300 km/sec) and the velocity dispersion in clusters of galaxies (usually up to 1000 km/sec). This galaxy is a representative example of a radio galaxy and since it is such a good case I will come back to it later. Since NGC 1275 is a Seyfert galaxy, I will now define this class. Seyfert galaxies have small, starlike, very bright nuclei which show up in short exposure photographs. The spectrum of the nucleus is the most characteristic feature of Seyfert galaxies: it has very strong, broad emission lines, indicating large scale velocities in the nucleus of the order of several thousand km/sec. The lines may characteristically be of high ionization.

This class of galaxies was first discovered in 1943 by Seyfert. Table 2.1 gives a list of Seyferts and some of their properties

Table 2.1

PROPERTIES OF SEYFERT GALAXIES

| NGG | Type | $\Delta V_r$ core | $\Delta V_r$ wing | log P (1400) |
|---|---|---|---|---|
| 1068 | Sb | 3000 | a) | 22.86 |
| 1275 | p | (4500) | a) | 24.69 |
| 3227 | Sb | | | |
| 3516 | E/Sa | 1400 | 8500 | <22.26 |
| 4051 | Sbc | 1200 | 3600 | <21.06 |
| 4151 | Sab | 1500 | 7500 | 21.68 |
| 5548 | S | | | |
| 7469 | Sa | 2000 | 8500 | <22.78 |
| VV144 | p | (1000) | 3000 | |
| 3C120 | | 3000 | a) | 24.83 |

a) Not present

(Burbidge, G. R., 1970). Most of the Seyferts are spirals. The velocity spread indicated by the emission lines is of order of a few $10^3$ km/sec. Many (but not all ) have been found to be radio emitters, NGC 1275 being especially strong. In some cases only an upper limit to the radio emission has been set.

A second example of a Seyfert galaxy is NGC 4151. A spectrum of the nucleus of NGC 4151 shows the characteristic broad emission lines. The spectrum from about 3700 to 4100 A° does not show any lines of high states of ionization, although such states are present; the highest in this spectral region is (Ne III) (the brackets indicate the line is forbidden). The gas is of low enough density so that many forbidden lines are seen. The spectrum shows an interesting absorption feature (Fig. 2.2).

In a normal galaxy the absorption lines are due to stars, but in this case the absorption is due to He I, which would not be produced by stars, but by gas that is cooler than the gas producing the emission lines and presumably closer to us, since it

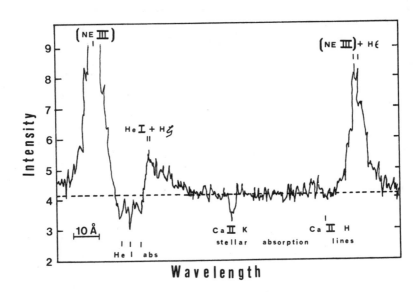

Figure 2.2 Tracing of NGC 4151 spectrum showing the triple He I absorption feature.

is absorbing radiation. The velocity of recession indicated by the emission lines is about 900 km/sec, but the absorption lines give three velocities (the line is triple), all of which are less than 900 km/sec. This suggests there are clouds of gas moving towards us in the line of sight which have been ejected by the galaxy. The absorption lines, which also occur in hydrogen, have changed with time on the order of a year, indicating changes in the amount and distribution of the gas coming towards us.

Strong infrared emission has been detected from a number of Seyfert galaxies and has been shown to also be nonthermal in character. Fig. 2.3 shows a plot of the total spectrum of two Seyfert galaxies, going from radio to ultraviolet optical frequencies. The flux starts out low in the ultraviolet and rises steeply towards the infrared. Between $10^{11}$ and a few x $10^{13}$ Hz the observations are very difficult to make, due to atmospheric extinction, so it is not clear what happens, but there does seem to be a peak in the infrared. The form of the spectrum in this region should give us a lot of information on the physical processes giving rise to the nonthermal emission.

## Table 2.2
### 10-MICRON FLUX DENSITIES

| Galaxy | Distance (Mpc) | 10μ Flux (W/m$^2$ Hz) |
|---|---|---|
| Galactic Center | 0.010 | $5.5 \times 10^{-24}$ |
| NGC 1068 | 13 | $2.6 \times 10^{-25}$ |
| NGC 1275 | 70 | $9.7 \times 10^{-27}$ |
| 3C 120 | 120 | $3.5 \times 10^{-27}$ |
| NGC 2782 | 33 | $1.6 \times 10^{-26}$ |
| M82 | 4.3 | $2.1 \times 10^{-25}$ |
| NGC 3077 | 4.3 | $3.3 \times 10^{-26}$ |
| NGC 4151 | 13 | $8.7 \times 10^{-27}$ |
| 3C 273 | 630 | $2.7 \times 10^{-26}$ |
| NGC 5236 | 4.3 | $6.0 \times 10^{-26}$ |
| NGC 7469 | 0.68 | $8.7 \times 10^{-27}$ |
| NGC 7714 | 39 | $3.0 \times 10^{-27}$ |

If the radiation were thermal, there would be a little peak in the optical, a Planck type curve. Table 2.2 shows the size of the infrared fluxes of Seyfert galaxies. The distances needed for a luminosity estimate are obtained using the Hubble relation with $H_o$ = 75 km/sec Mpc. The total infrared luminosities deduced are large compared to radio luminosities since the energy per photon is higher. It should be noted that flux measurements at a wavelength of about 100μ are presently very uncertain, but are crucial in obtaining total infrared luminosities.

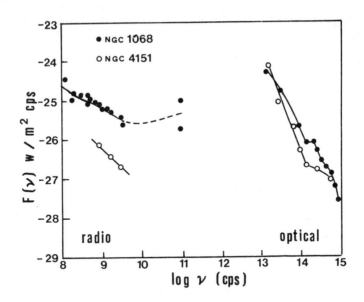

Figure 2.3 Total spectra of Seyfert galaxies

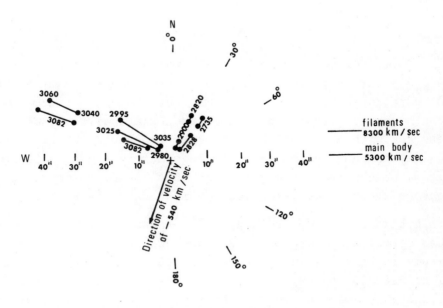

Figure 2.4 Velocities of the filaments around NGC 1275 relative to the central part

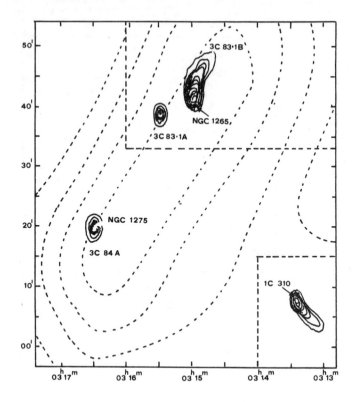

Figure 2.5   Radio contours of NGC 1275 and vicinity

I will now go back to NGC 1275 which has the advantages of being a strong radio source and of being near, so a fair amount of detail can be seen. We can map out where the filaments having the high velocity lie by a spectroscopic analysis (see Fig. 2.4). The gas with the displaced velocity runs into the center of the galaxy on one side only, which suggests that gas is being ejected from the center of the galaxy at this speed. Since some of the gas is farther out, the ejection must have started some time ago; that time can be determined and it turns out to be $\sim 10^6$ years. Gas ejection appears to have been continuous since then, and seems to be still in progress. If the direction of ejection is intrinsic to the galaxy and the galaxy is rotating, the ejected gas should sweep around in the sky; the observations are slightly suggestive of this. A typical rotation period for a galaxy is $10^8$ years, 100 times the time over which ejection has been occurring. As to what physical mechanism could give directed ejection we have no idea at present.

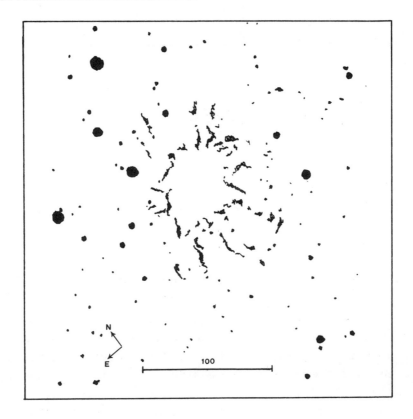

Figure 2.6  Photograph of NGC 1275

The ejection idea is supported by looking at the structure of the radio contours around Per A and the galaxies in its vicinity. There are two radio sources, less intense than Per A, but near to it, which are also members of the Perseus cluster. Their radio contours are not symmetrical and seem to point away from Per A (Fig. 2.5). It appears that high energy particles from Per A impinged on the two other objects and caused them to be radio sources. At a lower flux level, there is another contour which envelopes all these sources and suggests their association.

Lynds (1970) has taken a very interesting picture of NGC 1275 at Kitt Peak using an interference filter, which passes a relatively narrow wavelength band centered on H$\alpha$, set not at the rest wavelength, but at the wavelength corrected for a velocity of 5300 km/sec (Fig. 2.6). Thus only the gas with the same velocity as the galaxy will appear and not the higher redshift filaments. Yet a complex filamentary structure shows up which

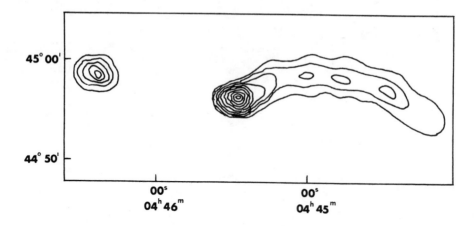

Figure 2.7  Radio contours of the sources 3C 129 and 3C 129.1

is very reminiscent of the Crab Nebula.  This observation disposes, once and for all, of the old idea that we are seeing the collision of two galaxies, since these filaments are all around the nucleus and there is no suggestion of an interaction starting on only one side.

The radio astronomers are also finding other evidence for ejection by galaxies; for example, the radio contours around 3C 129 and 3C 129.1 (Fig. 2.7).  They suggest that a series of explosions has occurred during the rotation of the parent galaxy.  Unfortunately the source is at fairly low galactic latitude, where optical obscuration is a problem and no optical identification has yet been made.

I will turn now to a variety of radio galaxies to indicate the number of different structures one can get.

The outer structure of M87 appears normal on a long exposure plate, but a shorter exposure will show the central activity associated with its radio emission.  There is a line of blobs of light forming the well-known jet.  It is nonthermal optical radiation, as can be deduced from its high polarization and the lack of lines in its spectrum; it is a source of continuum synchrotron radiation which requires high energy electrons in a stronger magnetic field than that needed for radio emission. Synchrotron radiation is normally only observed at radio wavelengths, where the energy per photon is less.  There is a diffuse radio source all around M87 as well as small diameter radio sources along the jet.  The ejection idea is again sup-

# Observational Problems of High Energy Astrophysics 131

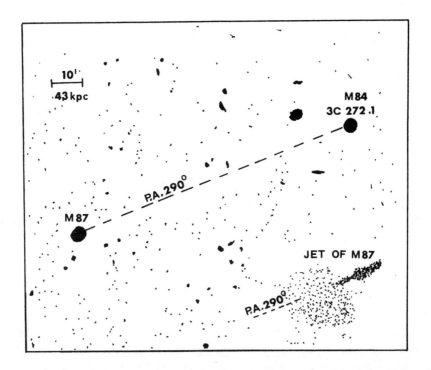

Figure 2.8  Orientation of M84 and jet relative to M87

ported when one looks at the objects in the vicinity of M87 on the Sky Survey plate. There is a galaxy, M84, at a position angle of 290° relative to M87 and which is also a radio source, 3C 272.1 (Fig. 2.8). The position angle of the jet relative to the nucleus of M87 is also 290°, so perhaps the situation is similar to what is occurring in the Perseus cluster. The galaxies around M87 are all in the Virgo cluster. M87 is also a nonthermal X-ray source.

NGC 5128 ≡ Cen A is in the southern sky and can only be optically investigated properly from the Southern hemisphere. It looks like an elliptical galaxy with a dust lane across it (Fig. 2.9). The dust lane is much thicker than those which characterize normal galaxies (and especially my "supernormal" galaxy!). Cen A was one of the first identified and strongest radio sources. One has the impression that there is a big energy input which is stirring up the dust, since it appears in large blobs. Hot gas is associated with the galaxy, particularly near the dust lane, and a velocity study of this gas shows that the galaxy is rotating about the long axis of the

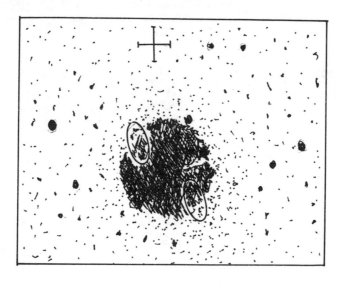

Figure 2.9 NGC 5128

elliptical body at right angles to the dust lane. There are also some non-circular motions present of order of several hundred km/sec, an order of magnitude smaller than those associated with NGC 1275. Besides being a radio source, the galaxy is an X-ray and an infrared source. It should be very profitable to study the galaxy at long infrared wavelengths since this radiation will reach us through the dust of the object.

A similar galaxy is NGC 1316, which is also a strong radio source, Fornax A. The galaxy is elongated, but in this case the axis of rotation is along the minor axis as one would normally expect from non-radio emitting galaxies. The radio distribution shows two lobes of radio emission centered on the optical object (Fig. 2.10), a very frequent structure for radio sources. Deep photographs of the optical object show luminous extensions which appear to be leading into the regions of radio emission.

M82 is not such a strong radio source, but is optically very interesting. It is obviously not symmetrical and was first classified as an irregular galaxy. It has very broken up dust, looking as if it has been stirred up by some source of energy within the galaxy. It is rotating about the minor axis, but Lynds and Sandage (1963) found a velocity gradient along the minor axis which they interpret as the ejection of

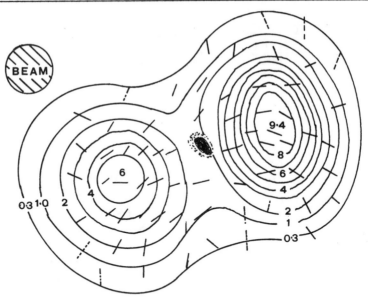

Figure 2.10   NGC 1316

gas. The radio emission does not extend around the whole object, but is fairly restricted in size. There appears to be polarization of the optical light indicating optical synchrotron radiation. An interference filter photograph centered on Hα at the redshift of this galaxy was taken by Lynds and Sandage, and the central region of the galaxy became much enhanced, with loops and structures appearing along the direction of the velocity gradient. Lynds and Sandage estimate that the galaxy is seen nearly edge on, so the measured ejection velocities of 100-200 km/sec imply a space velocity of about 1000 km/sec if a correction is made for projection. Another puzzling feature is that the emission lines seem to be polarized; a possible explanation is that it is electron scattering from a bright source at the center of the galaxy, but it is not understood at present. M82 might have a Seyfert type nucleus which we cannot see optically because it is behind the dust of the galaxy.

Continuing with our discussion of the variety of radio sources, we come to NGC 4782-3, which is a double galaxy with 40 arc second separation between the two components, both of which are ellipticals. There is no optical abnormality although there is fairly strong radio emission around the whole pair. It is not known which galaxy is responsible for the radio emission. There are no strange velocities in the system; in fact, there is no

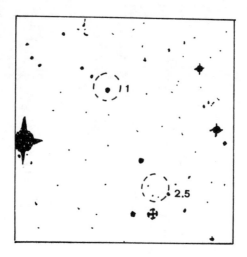

Figure 2.11   3C 33 radio source

hot gas observed at all. There is a velocity difference between the two galaxies which can be interpreted as the orbital motion of one galaxy about the other, suggesting a large mass for the system. The minimum mass obtained by assuming that the plane of the orbit is perpendicular to the line of sight and that the galaxies are at the ends of the orbital axis is a few times $10^{12}$ $M_\odot$. Short exposure photographs show that the luminosity distributions of the two galaxies are structurally different, one being more concentrated than the other. The fact that no hot gas is seen may mean that whatever event gave rise to the radio emission happened sufficiently long ago in the past that any hot gas around has dissipated or has cooled so as to become invisible. Stellar collisions have been suggested as a possible energy source in dense nuclei, but the time scale needed is too long to be applicable here in explaining the nuclear structural difference.

One of the strongest radio sources in the sky is Cygnus A which has the common appearance of two radio lobes centered on the optical galaxy. It is difficult to study optically, since it is at such low galactic latitude that it is considerably obscured by the dust of our own galaxy. There are small diameter structures in the radio lobes suggesting blobs of emission regions.

Another double radio source is shown in Fig. 2.11. It is particularly interesting because of the small angular size of the radio emission lobes. The galaxy has hot gas with a velocity gradient, but it is too far away to be observed in detail. The radio sources are strongest at the outer edges and are elongated in the direction of the parent object. What keeps the plasma giving rise to the synchrotron radiation so confined is a great puzzle.

Lecture 3

The remaining lectures will be mainly devoted to the quasi-stellar objects, and especially their spectra from which most of the information is derived.

The QSO's were first found as star-like objects, at the positions of radio sources where there were no peculiar galaxies. In the case of the first QSO's to be identified, the radio positions were accurate to about 10 seconds of arc. They were strong radio sources, since they belonged to the 3C radio catalogue which is a list of radio sources at 178 Mhz with radio fluxes greater than 10 flux-units. The very first quasar to be discovered was 3C48, a blue star-like object of 16th magnitude, with a spectrum showing broad emission lines, very reminiscent of lines produced by old novae. Attempts at the identification of these lines were unsuccessful until 1963 when M. Schmidt showed that the emission lines were in fact very familiar lines, but with a considerable redshift. The first observed spectral properties of the QSO's were:
1) The spectrum showed broad emission lines with a very large redshift.
2) UBV photometry showed a strong ultraviolet excess.
3) A strong infrared excess was later found.
The UV excess is shown in the 2-color diagram(Fig. 3.1).

Stars are not truly blackbody emitters, because of their spec-

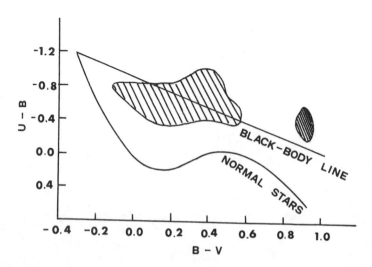

Figure 3.1

tral lines which change the character of the total radiation measured. The dip in the curve for the stars is due to the Balmer limit at λ3648A°. The quasars lie closer to the blackbody line than do normal stars, which, at first look, may be surprising. This fact arises because the graph covers a very small portion of the whole spectrum, from the visual to the ultraviolet. In the same general area as the QSO's are the White Dwarfs, their lines being less strong than those of normal stars. This led to the original suggestion that QSO's could be hot, highly evolved stars. The diagram suggests immediately a very useful way to look for an optical counterpart to the accurate position of a radio source: take two exposures of the same area, one through an ultraviolet filter and the other through a blue filter; objects which stood out in the ultraviolet photograph would be high temperature objects, white dwarfs or QSO's; if a radio source is present at the location of such an object, the presumption would be strong that the object might be a QSO. However, in the course of carrying out this procedure, it was found that there were many strong UV emitters at other than radio positions. Some of them, when looked at spectroscopically were found to have the same properties as the radio emitting QSO's. They were called radio quiet QSO's and found to be much more numerous, per unit area in the sky, than the radio emitting QSO's. Naturally, there is a whole range of radio intensities in QSO's, from radio quiet to very strong radio emitters. In fact, some of the QSO's that were originally identified as radio quiet, were later investigated at higher frequencies and found to be low radio emitters. Conversely, surveys, like the Palomar-Haro-Luyten survey, that were made years ago, at high galactic latitudes, in caps around the galactic poles in both hemispheres, in order to pick out white dwarfs and similar objects, listed objects that were later identified as QSO's. For instance, 3C9, one of the first quasars to be identified, is listed as a blue star in the PHL catalogue.

We turn now to the identification story. Where do we see emission lines? In gaseous nebulae, in H II regions. The physical conditions in these objects are well understood: they have an electron density $N_e \sim 10^4$ cm$^{-3}$ and an electron temperature $T_e \sim 10^4$ °K. The emission spectrum has been investigated in detail; lengthy tables of lines and relative strengths, abundance calculations are all standard and have been so since the 1930's. The spectrum is characteristically composed of H, He and forbidden lines. They are produced by a strong source of UV radiation, which in the case of a gaseous nebula in our galaxy, is a hot, young star. The quanta of energy higher than the Lyman limit ionize the hydrogen (most abundant element), giving a big supply of electrons. These electrons rapidly achieve a Maxwellian distribution and collide with ions of various elements such as O, N, C. In O and N in particular, there are low lying metastable ground levels, the transition frequency to the ground level of which is in the optical region of the spectrum, but which is forbidden. However, electrons which are collisionally excited can get up to these levels, and if the gas is of low enough density and is not in thermal equilibrium with the radiation field, the electrons could remain in this level long enough so that the forbidden transition can occur. In contrast, in the atmosphere of a hot star, the electrons would be excited to a higher level, giving rise to a permitted transition to the ground state.

In the early 1960's space telescopes were planned, to look at the UV spectra from gaseous nebulae in our galaxy. They were needed because of the atmospheric cut-off at wavelengths less than 3000 A°. For this purpose, Osterbrock and his students at the University of Wisconsin prepared a list of the strongest emission lines in the UV. The table was just finished in time for Schmidt and others, who were starting to realize that, since the first QSO identified had a big redshift, they were going to be faced with the task of identifying lines in the UV but shifted into the visual. Table 3.1 presents a list of the principal emission lines in the spectra of QSO's. Four of the strongest transition lines in the table are: Mg II $\lambda 2798$A°, C III) $\lambda 1909$A°, C IV $\lambda 1549$A°, Ly$\alpha$ $\lambda 1216$A°. For example, Ly $\alpha$ for a redshift z = 2 is shifted to 3648A°, using the formula: $\lambda_{obs} = \lambda_{emitted}(1 + z)$, which is in a nicely observable region. 3C 273 is among the first QSO's to be identified by Schmidt. Using the z determined from the Balmer lines, Schmidt identified a Mg II line, which had already been seen in rocket spectra of the sun. Going to slightly greater redshift, Schmidt found other lines, and so on, until a redshift of about 2. Two or more lines are needed to produce a good z determination, and most of the published redshifts are based on enough lines to be confident about a redshift determination. A very useful table,

## Table 3.1

### PRINCIPAL EMISSION LINES SEEN IN SPECTRA OF QSO'S
### (1200-5000 A°)

| Ident. | λ(A°) | Ident. | λ(A°) |
|---|---|---|---|
| Lyα | 1216 [a] | Mg II | 2798 [a][b] |
| NV | 1240 [b] | (Ar IV) | 2854, 2869 |
| S IV | 1397 [b] | (Mg V) | 2931 |
| O IV | 1406 | He II | 3203 |
| C IV | 1549 [a][b] | (Ne V) | 3346, 3426 |
| He II | 1640 | (O II) | 3727 [b] |
| O III) | 1664 | (Ne III) | 3869, 3968 |
| C III) | 1909 [a] | (O III) | 4363, 4959, 5007 [a] |
| C II) | 2326 | Balmer series | 4102, 4340, 4861 [a] |

(a) Strongest transitions
(b) Blends of doublets

---

listing all the transitions that have been seen, has been published by Lynds (1968) in the "Proceedings on Beam-Foil Spectroscopy" (Bashkin, Ed.). Table 3.2 shows a list of observed emission line strengths in various QSO's. The numbers fall diagonally across the table because the QSO's are listed in order of increasing redshift. For low z, the table shows the long end of the spectrum and for high z the UV end.

How do we identify a QSO spectrum? Lynds' table can be used to compute ratios of wavelengths of different lines. The ratios are then arranged in increasing order. If $\lambda_1$ and $\lambda_2$ are measured wavelengths of 2 lines in the spectrum:

Table 3.2

OBSERVED EMISSION-LINE STRENGTHS

| Object | Hα, $\lambda 6562$ | | Hβ+(OIII), $\lambda 4922$ | | MgII, $\lambda 2798$ | |
|---|---|---|---|---|---|---|
| | W (A°) | Jx10$^{-42}$ (ergs/sec) | W (A°) | Jx10$^{-42}$ (ergs/sec) | W (A°) | Jx10$^{-42}$ (ergs/sec) |
| Ton 256 | 572 | 15 | | | | |
| PKS 2135-14 | 748 | 51 | 133 | 12 | | |
| 3C 323.1 | 710 | 54 | 266 | 15 | 84 | 13 |
| 3C 249.1 | 744 | 67 | 190 | 24 | 75 | 29 |
| 3C 277.1 | 857 | 15 | 375 | 9.3 | 109 | 7.4 |
| PKS 2251+11 | 928 | 102 | 198 | 30 | | |
| 3C 48 | | | 410(Hβ) 264(OIII) | 17 40 | 15 | 3.3 |
| 3C 351 | 375 | 76 | 88 | 27 | | |
| 3C 334 | 803 | 108 | 155 | 33 | 71 | 37 |
| PKS 0405-12 | | | 133 | 110 | 57 | 119 |
| 3C 345 | | | 97 | 16 | 89 | 27 |
| 3C 380 | | | 248 | 40 | | |
| PKS 1354+19 | | | 432 | 98 | | |
| 3C 286 | | | 86(Hβ) 250(OIII) | 13 36 | 15 | 6 |
| PKS 2145+06 | | | 565 | 237 | | |
| CTA 102 | | | 201 | 50 | 112 | 50 |

$\lambda_1 = (1+z)(\lambda_1)_o$

$\lambda_2 = (1+z)(\lambda_2)_o$

Taking the ratio: $\lambda_1/\lambda_2 = (\lambda_1/\lambda_2)_o$. One can then use the table to find the appropriate corresponding ratio, and thus get a

Table 3.2 cont'd.

| Object | CIII, $\lambda 1909$ | | CIV, $\lambda 1550$ | | Ly$\alpha$, $\lambda 1216$ | |
|---|---|---|---|---|---|---|
| | W (A°) | Jx$10^{-42}$ (ergs/sec) | W (A°) | Jx$10^{-42}$ (ergs/sec) | W (A°) | Jx$10^{-42}$ (ergs/sec) |
| 3C 286 | 10 | 8 | | | | |
| CTA 102 | 32 | 48 | | | | |
| 3C 208 | 58 | 65 | 710 | 930 | | |
| 3C 446 | 20 | 9 | 230 | 130 | | |
| PHL 938 | 32 | 72 | 127 | 320 | 428 | 1270 |
| 3C 9 | | | 98 | 100 | 460 | 500 |
| PKS 0237-23 | | | 116 | 320 | 284 | 1670 |

tentative identification of the 2 lines. Supporting evidence has then to be looked for by seeing if other lines fit in the tentative scheme. Lynds has adopted another method for line identification. He slides a logarithmic plot of the wavelengths of the tabulated spectral features along a logarithmic plot of the measured wavelengths until he gets a fit. Because of lack of telescope time, not all the QSO's have been observed by independent observers, but when duplication has been made, there are very few instances where redshifts measured by independent observers have not been agreed upon.

The range of QSO redshifts now known extend from z less then 0.1 up to the largest redshift measured by Lynds and Wills (1970) z = 2.88. To convert the redshifts into velocities, the interpretation of the redshift comes in. If the velocities are considered cosmological, a particular cosmological model has to be considered. If the velocities are local, special relativity gives:

$$\frac{v}{c} = \frac{(1 + z)^2 - 1}{(1 + z)^2 + 1}$$

At z = 2, the formula gives: v/c = 0.8 and at z = 3, v/c $\simeq$ 0.9. Fig. 3.2 shows the magnitude-redshift plot for QSO's. The points

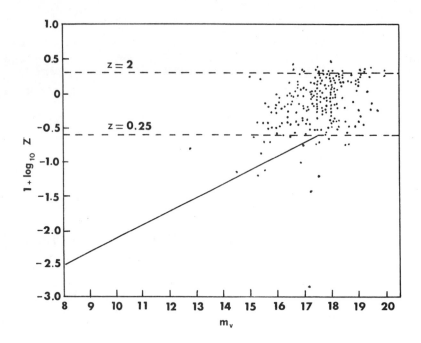

Figure 3.2 Redshift apparent magnitude relation for 235 quasi-stellar objects (very schematic) (Full line is the Hubble relation for normal galaxies)

show a considerable scatter, departing from a prolongation of the galaxy line established by Humason, Mayall and Sandage (1956). The vertical concentration of points at $m_v$ = 18 mag reflects the fact that there is still little accurate photometry on QSO's. (Actually, QSO's tend to be variable in the optical region.) The magnitude estimate is done by the observer who has made the identification. Another concentration is visible at $m_v$ = 17.5. The faintest QSO's have a magnitude of about 20. They are quite interesting objects to study photometrically. Some faint objects that were not seen on the Palomar Sky survey a few years ago, have turned up in more recent plates taken in the area of a radio source. To study faint objects, we need new methods of detection which would give higher quantum efficiency and more linear response than the standard photographic plate. Also, with the photographic plate, saturation problems prevent accurate spectrophotometry.

Lecture 4

The emission line spectra of QSO's were considered in the last lecture, which was concluded with examples of different redshifts. The redshifts are reasonably well-determined; at least, there hasn't been misidentification of the lines. One can't use only one line, as can be done for radio galaxies, e.g., assuming an emission line is due to (O II), because with QSO's one has no prior notion from the objects magnitude of whether it will have a z of .1 or 2. One has no notion of what region of the spectrum one is investigating, so that a search for more than one emission line is essential.

The new detectors spoken about, which provide new ways of getting the spectra of QSO's, have application in many problems in extragalactic astronomy, apart from the cosmological problems. The reason for going to newer detectors is the need for an instrument with 1). higher quantum efficiency and 2). a linear response to the signal, for which therefore one can do accurate sky subtraction. Two such new systems are the Image Dissector Super Scanner and the Digicon.

The Image Dissector Super Scanner has been developed by Wampler and Robinson at Lick Observatory, has been used by Wampler, Hazard, who made new identifications of QSO's, and Burbidge. The scheme consists of having one or more image tubes, with the final signal in the form of either light coming off the last phosphor screen, or converting this back to electrons. One scans with a small aperture over the face of the region giving the emission and therefore gets a record of intensity with position. If the final screen is a phosphor screen, as it is in the Wampler and Robinson instrument, the scan time must be short compared to the decay time of the phosphor and this was looked into carefully before the instrument was built.

There are three image tubes in series to give an amplification of $\sim 10^5$. The instrument has two apertures, each with 2048 channels, so one sets the object and sky in one aperture with sky alone in the other, scans, and then sets the object in the other aperture, so that any unevenness in the photocathode is removed. Since each aperture has 2048 channels there is reasonably good resolution; the dispersion you get depends on what grating is put in the spectrograph. In its final form, in which it is not in yet, there will be a range of possibilities. The data is taken digitally using a PDP8/I computer, and a cathode ray tube displays the counts during an integration. One sees the counts accumulating, so if you think the object is a possible QSO and it turns out to be a galactic star, you can abandon the observation as soon as you see absorption lines at rest wavelengths appearing. One may also continue on to get as good a signal to noise ratio as is needed. While you're doing the next integration, the computer subtracts the sky, and displays and adds successive runs. The estimated effective limit of the Super Scanner is m = 22, and it has already been used for m = 21 by Wampler, and Robinson. A calibration lamp is used during the course of a night's observation; the signal is divided by the calibration lamp to insure the correct relative intensities. There is also a program to correct for the extinction of our atmosphere.

In a spectrum of an object with low redshift one sees the pattern characteristic of the spectra of radio galaxies: two narrow emission lines of forbidden (O III), strong emission lines of H$\beta$, and the H$\delta$ line, another O III line, and H$\gamma$.

In the spectrum of a high z object (z$\sim$2) one sees the familiar C IV 1549A° line, and semi-forbidden C III 1909A°. Some of the most interesting, from a physical point of view, QSO's have absorption lines. If one looks at the spectrum of this object out to longer wavelengths, one sees a whole lot of absorption lines, e.g., out to 7500 A°.

The Digicon (digital image tube) was designed by McIlwain and Beaver at the University of California, San Diego. A linear array of silicon diodes of small dimension, about 1/10 mm. square, is put inside an image tube; the silicon diodes, used for years by nuclear physicists to detect electrons, are excellent for detecting the accelerated electrons produced in the image tube. Each of these diodes has its own small pre-amplification microcircuit inside the image tube. The beauty of the silicon diode is that you get accelerated electrons producing large numbers of electron-hole pairs in reverse-biased silicon diodes, generating a charged pulse of much higher energy than the thermal noise of the system. There is an external amplifier and discriminator for each diode; one thus removes the noise with the discriminator and one need not cool the whole system: it's used at room temperature. The prototype instrument has 38 channels; a 200 channel instrument is under construction.

Once one has gotten the redshift, made the identification, what can one deduce about the physical conditions in the region in a QSO producing the emission lines? Recall that these objects are strong emitters of non-thermal radiation, have very large redshifts, and if, as was originally done, one assumes z is cosmological in nature, they are very distant, and so a large release of energy is required to accelerate high-energy particles to spiral around magnetic field lines. The energies discussed up to now have all been computed using a Hubble constant H = = 100 km/sec Mpc. The problems involved with getting this amount of energy out have been magnified by the recent drop in the Hubble constant to H = 50 km/sec Mpc.

The conditions in the hot gas giving rise to the emission lines are $N_e \sim 10^6$ cm$^{-3}$, $T_e \sim 2.3 \times 10^4$ °K. Both the electron temperatures and the electron density are somewhat higher than values typical for gaseous nebulae in our own galaxy, and for gas in normal H II regions in other galaxies. The point about the electron density is important because, for example, in radio galaxies, where one sees the strong forbidden line of O II 3727A°, whose upper level is most easily collisionally de-excited, this is the first of all the forbidden lines you see to go, as in 3C 273, as the electron density increases. Whether the line is seen or not is indicative of $N_e$ variations from one object to another. As was pointed out also for radio galaxies and Seyfert galaxies, there is a characteristic inhomogeneity present in the gas producing emission lines, in that lines from neutral atoms appear next to lines from ionized atoms, needing upwards of 75 eV for ionization. Thus highly non-uniform conditions must prevail in the emitting gas and this must also be the case in QSO's. The

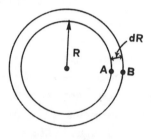

Figure 4.1

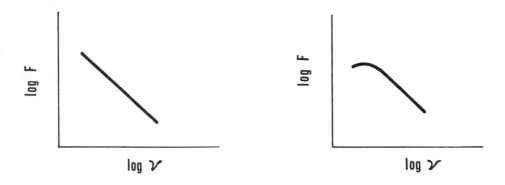

Figure 4.2a                Figure 4.2b

lines that one sees, however, since the identifications have been made satisfactorily, using the line list calculated by Osterbrock and students for a normal chemical composition that is similar to the composition in our solar neighborhood, and the fact that these lines come up with the same predicted strengths, imply there are no strong differences in abundances from those in our own vicinity, which is somewhat surprising.

Since there is a great energy problem if the redshift is cosmological, the first alternative theory proposed was that z is gravitational in nature. Take a simple model for a QSO (Fig. 4.1): a sphere with the mass concentrated toward the center; energy production in the central region; a region of radius R producing continuum radiation, surrounded by a thinner region dR where line emission is produced. The difference in gravitational potential between points A and B leads to a broadening of the lines. Well, the lines are broad, but the line width over the redshifts sets a limit to dR/R which sets a limit to the volume into which one packs the emitting gas, depending on R, and on what mass you assign to the object. The number of hydrogen atoms needed is calculated from the strength of the Balmer lines, and one also knows that the electron density can't be too high, or you would see no forbidden lines. The argument in its strongest form, applied by Greenstein and Schmidt (1964), took R as that derived from collapse of the star, and putting in that number of hydrogen atoms leads to an abnormally

# Observational Problems of High Energy Astrophysics

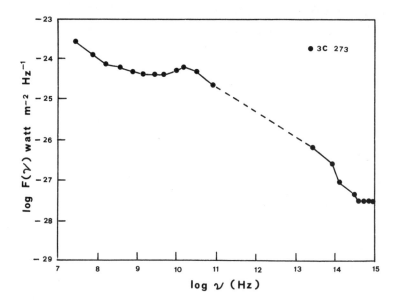

Figure 4.3

high density, $10^{10}$, too high, in fact, for forbidden lines to appear. The argument is less stringent if the object is of galactic mass and is outside the galaxy, but this is hard to satisfy particularly with the larger z values now known. Hoyle and Fowler (1967) proposed an alternative model, with gas not distributed outside the object but in a pool of matter in the gravitational well of a number of collapsed objects.

Consider next the form of the continuum radiation from QSO's. In the last lecture a little was mentioned about the form of the continuum radiation produced by synchrotron radiation in a non-thermal emitter. One has $F(\nu) \propto \nu^{\alpha}$, $\alpha$ for a typical radio galaxy being about -0.7, -0.8. Plotting log F vs. log $\nu$, one gets a straight line; note the higher flux from the lower frequencies (Fig. 4.2a). For the spectra already shown with a label of C, one gets a curved line, or variants on this slope, a much flatter slope (Fig. 4.2b). If one looks toward very low frequencies, eventually the curve turns over because of absorption by the interstellar plasma of our own galaxy. If one can see a turnover at higher frequencies, the turnover is produced by absorption in the source itself.

This is the spectrum of 3C 273 (Fig. 4.3), with frequencies

ranging from the low end of the radio to the optical and ultraviolet. 3C 273 is the strongest radio source QSO which enabled the mystery of the emission line spectrum to be deciphered. Following the spectrum one comes to the gap unobserved in the microwave and far infra-red, and one goes on to the observable infra-red, optical, ultra-violet, with the X-ray off the graph. So this is a complex spectrum, it doesn't have the simple power law form like the extended radio sources occuring about radio galaxies. It's the sort of thing characteristic of small diameter radio sources; it undoubtedly has components, and bursts of electrons; the hump in the spectrum is characteristic of absorption in the source.

Notice the rise in the infra-red flux over the optical and UV. This is a powerful way to identify QSO's; this method has been used by the Bologna group, particularly Braccesi. It has already been shown that to identify QSO's one may take a photograph in ultraviolet and a photograph in blue and pick out the UV images; one can do the same thing with photographs in the infrared, i.e., pick out strong infra-red images. So if one picks out the same object in the ultraviolet and the infra-red, except if the object is a strange double star, it is a non-thermal emitter, and the chances of getting a QSO identification increase. A lot of radio quiet objects may be picked out in this way.

There is a new class of object about which not much is known, although it has similarities to QSO's. The objects are radio emitters with a flat, or curved spectrum. This selection of objects was made by picking out those that had this type of radio spectrum; looking to see what kind of optical object would correspond, one found starlike objects, but not QSO's, because there are no spectral lines, therefore no redshifts. They have other properties however that can be determined and I shall return to these objects later.

Another property of QSO's important in understanding their physical nature is their variability, which occurs in every case in which one looks in sufficient detail. If one has a variable source of emission with a simple spherical shape of radius r, and one sees the light to vary on some time scale $\tau$, for an observer to see significant light changes in this object he has at least got to see light coming from a region of dimension R (supposing he's looking at the front hemisphere) in a period of time shorter than the period of light variation, i.e., the period of light variation should be longer than the light travel time from different parts of the object. This is always satisfied in all the variable stars seen. If one sees variability in QSO's, what does this say about the dimensions of these objects? A period of light variation $\tau$ sets a limit for $R \lesssim c\tau$, so that I get:

$\tau$ = year $\qquad R \sim 10^{18}$ cm $\sim$ 1/3 of a parsec

$\tau$ = month $\qquad R \sim 10^{17}$ cm

$\tau$ = day $\qquad R \sim 3 \times 10^{15}$ cm $\sim$ like size of solar system

So one has to get a vast energy from a region of these dimensions; or if there's a sudden cooling off, or a sudden switching off of high energy electrons producing synchrotron radiation, you still have to get it transmitted to the whole object and seen by the observer from the whole object to see any sign of variations.

One of the strong variable QSO's is 3C 446. It has a redshift of 1.4 and had a magnitude of 18.5; then in July 1966 it increased in magnitude by 3 mag, then there was a rapid decline of 1.5 mag in 10 days, then the object recovered, and there was a sharp peak in a few days, then a drop, a rise again. Some have attempted to fit periodicities to this object and to some of the other largely variable QSO's, but so far nothing has worked out satisfactorily. People think they have found the period, predict when the next brightening is going to occur, but the object does not always behave!

Next, let's turn to a subject which will take some time to go through, the spectra of QSO's which have been found to show absorption lines. As a consequence of problems like the 1). big scatter in z-mag relation, and 2). getting large energies out of a small volume (if one says the objects are extremely distant), there was a suggestion made in 1966 that QSO's are nearer, z is not cosmological in nature. There is a possible test as to whether z is cosmological of not. If it is cosmological, if there is enough mass in intergalactic space to give $q_0 \sim 1/2$ (i.e., $q \sim 2 \times 10^{-29}$ gm cm$^{-3}$), and if cold hydrogen provides that mass, it follows that one should see on the short wavelength side of the Ly $\alpha$ line a great absorption running up to the emission line and down to the limit due to our own atmosphere, by all the gas at successively smaller z. The same phenomenon is expected for QSO's which have an emission line of

H. The first QSO with z = 2 and showing Ly α was 3C 9, but there was no absorption. This set a severe upper limit of the amount of cold hydrogen lying there if z is cosmological: N $\sim 10^{-6}$ of what you would need to close the universe.

The next thing said was suppose there is that much of intergalactic matter there, but it's hot enough to be ionized so that less than one part in $10^6$ is neutral. This doesn't rule out the possibility of looking for clouds, or intervening galaxies; these would give discrete redshifts. So there would be absorption redshifts. Just about the same time the theoreticians were suggesting looking for these, the first of the spectra of QSO's with absorption lines, 3C 191, was taken by Roger Lynds with his image tube at Kitt Peak Observatory and by myself at Lick Observatory (Burbidge, Lynds, Burbidge, 1966). The spectrum has been very broadened and the object trailed over a very long length of slit; one sees the broad emission lines, the stronger ones of C IV and Ly α. The strongest thing one sees is the strong absorption (narrow compared to every emission line looked at, except in the low redshift and the forbidden lines) on the short wavelength side of the strong emission lines. All the resonance lines are broad, but the absorption lines are narrow, so narrow that, even on this low resolution spectrum, the Si IV pair of lines, one of the strong transitions in the Table 4.1, blended always in emission, is clearly resolved in absorption. One also sees doublets: every one is a permitted resonance transition, an absorption out of the ground level. If the line were not resonance, one would not expect to see absorption there; it's the sort of thing one expects to be produced strong, and which will be produced even if there's not much gas there. Well, the identifications were perfectly straightforward, there were many lines, no problem, everything from a spectroscopic point of view quite sensible. In Table 4.1 there are the identifications, the measured wavelengths running from 3407 to 5640 A°. Computing the mean redshift $\bar{z}$ from the absorption lines, one may compute the rest wavelength from $\bar{z}$, and $\lambda_{lab} - \lambda_{rest}$ is the error of the measurement of individual lines, which turned out to be $\sim$ 1-2 A°. (The errors are smaller than those involved in estimating the centers of emission lines because the absorption

## Table 4.1

### ABSORPTION LINES IN SPECTRUM OF 3C 191

| Ident. | $\lambda(\text{Å})$ | Ident. | $\lambda(\text{Å})$ |
|--------|---------------------|--------|---------------------|
| Si II  | 1190.4              | C II   | 1335.3              |
| Si II  | 1194.2              | Si IV  | 1393.8              |
| Si III | 1206.5              | Si IV  | 1402.8              |
| Ly α   | 1215.7              |        |                     |
| N V    | 1238.8              | Si II  | 1526.7              |
| N V    | 1242.8              | Si II  | 1533.4              |
| Si II  | 1260.4              | C IV   | 1548.2              |
| Si II  | 1264.8              | C IV   | 1550.8              |

lines are much narrower). It turns out that $z = 1.9523$ from the emission lines, and a slightly smaller value of $z = 1.9460$ from the absorption lines. Supposing this is due to Doppler shifts, the gas producing the absorption in front of the object, driven out by the object, has a velocity with respect to the object of 600 km/sec. There is nothing strange in gas coming out from a hot object; certain hot stars have gas coming out from their surfaces at fairly small velocities, seen by absorption lines shifted to the short wavelength region of the emission lines.

One can make an interesting comparison with the hot stars in our own galaxy, because the redshift takes the wavelength to the ultraviolet. In the spectrum of ε Orionis, in our own galaxy, a perfectly ordinary hot star (the spectrum was taken above the atmosphere with a rocket), the Ly α is interstellar; there is certainly cold gas in interstellar space which absorbs the ultraviolet light of stars, and the other **observed** absorption lines arise in an expanding shell. The interesting thing is that in the part of the spectrum, where the main lines coming from the star are in emission, there occur the same transitions as in QSO's, eg., C IV, Si IV, N V. If one considers the C IV emission lines with absorption lines on the short wavelength side and converts the wavelength difference to a velocity, one finds the star is surrounded by a shell moving out at 2000 km/sec. So if this sort of thing happens in an ordinary hot star, then surely it may happen in QSO's.

So there's no worry about 3C 191. It's fortunate that 3C 191 was one of the first absorption line QSO's to be found, because all subsequent ones have been hard to interpret; e.g., they have not been able to be interpreted at one value of z close to the emission line value of z.

The next link in the chain of deductions about the nature of the absorption lines was PHL 5200, originally known as a blue stellar object, but actually a 4C source, with redshift of $z = 1.98$.

In kind, one sees the same phenomenon occurring: emission with absorption on the short wavelength side. But, in this case, the absorption lines are not narrow, but extremely broad. The absorption in N V almost obscures the emission of the neighbouring Ly $\alpha$ and the absorption of Ly $\alpha$. Converting this width to a velocity, it corresponds to $\sim$ 10,000 km/sec, an order of magnitude greater than 3C 191. One can still use the same interpretation, likening this spectrum to things one sees again in neighbouring galaxies. It's possible to link this to a type II supernova explosion, whose spectrum, if you catch it before its light has decayed too much, has broad emission and broad absorption lines on the short wavelength side, assumed to be due to explosive ejection of thick clouds of gas, with high kinetic energy from the large energy release in the center of the supernova; to start with too, it's optically thick in absorption, and later on it breaks up and other things happen. The same thing is seen on a smaller scale in novae in our own galaxy. So it's a familiar phenomenon and ties in with what one would think offhand would be happening in these objects: a big energy release and the possibility of pushing out gas.

More puzzling things will be come upon in later lectures.

## Lecture 5

I will now deal with the spectra of QSO's which have many absorption lines; I will go over the identification of the absorption lines and give a possible explanation for them.

With the first QSO that was found with absorption lines, 3C 191, the identification was straightforward: the transitions seen were identified with transitions out of the ground level. During the intervals between absorbing quanta and then coming back down to the ground state, the gas does not get disturbed, so a radiative transition back to the ground state occurs.

PHL 5200 has very broad absorption lines, like a supernova, indicating a velocity width of $10^4$ km/sec. The spectrum of PHL 938 shows two strong emission lines with which there is no problem with identification; they are Ly $\alpha$ and C IV at a redshift of 1.9553. There are absorption lines on the short wavelength side of both of these emission features. When Kinman obtained the first low dispersion spectrogram of the object, the absorption lines near each other were blended, so he assumed he was seeing Ly $\alpha$ and C IV in absorption at somewhat lower wavelength than the emission lines. However, he noticed that the wavelengths didn't properly fit. As soon as a higher resolution spectrogram was taken, the absorption lines were seen to be split and more absorption lines were detected. However, the ratio of wavelengths of the pair of absorption lines near the C IV emission did not agree with the ratio of wavelengths of C IV. There were some lines near the Si IV emission, but on the other side of it. Fred Hoyle pointed out that wavelength ratio of the lines near the C IV emission was compatible with the ratio of the Mg II doublet. Roger Lynds, Stockton, and I (1968) tried to identify the other absorption lines in the spectrum assuming the redshift of the Mg II doublet and after a while Lynds realized that Fe II lines would work. Of the many lines in the Fe II multiplets, only some appear in the spectrum. The observed transitions come from the absolutely zero level only; the other five structure levels of the ground state are not populated. The identifications are given in Table 5.1. With the strong absorption near the Ly $\alpha$ emission there is nothing to identify but Ly $\alpha$ with z = 1.906. There

Table 5.1

IDENTIFIED ABSORPTION LINES

| $\lambda(\text{A}°)$ | Identification | z |
|---|---|---|
| 3533.2 | Ly α λ1215.7 | 1.9064 |
| 3644.2 | Fe II λ2260.1 | 0.6124 |
| 3779.1 | Fe II λ2343.5 | 0.6126 |
| 3818.6 | Fe II λ2366.9 | 0.6134 |
| 3827.1 | Fe II λ2373.7 | 0.6123 |
| 3841.6 | Fe II λ2382.0 | 0.6127 |
| 4171.2 | Fe II λ2585.9 | 0.6131 |
| 4193.0 | Fe II λ2599.4 | 0.6131 |
| 4509.1 | Mg II λ2795.5 | 0.6130 |
| 4520.1 | Mg II λ2802.7 | 0.6128 |

seems to be no doubt for the redshift of 0.613 for the iron and magnesium lines. They could be due to an object between us and the QSO at the cosmological distance corresponding to z = 0.613 or, if the absorbing object is associated with the QSO, it has a velocity of 160,000 km/sec relative to the QSO, about half the speed of light. The width of the lines is very narrow. It is instructive to look at the intensity ratio of the Mg II doublet; the theoretical value for the ratio is 2:1. But the lines look fairly equal, showing that they are quite strongly saturated. The width of the lines on our spectra is no more than the instrumental width. Stromgren developed a method for dealing with this sort of situation in a study of lines from gas in our galaxy. By taking into account the degree of saturation of the lines, one can get the velocity dispersion in the gas and the number of absorbing atoms or ions in the line of sight. For PHL 938 a velocity dispersion of 50 km/sec was obtained, which is smaller than typical rotation velocities found in galaxies, although it could be possible for the halo of a galaxy. The column densities were found to be

$N(Fe^+) = 9 \times 10^{14} \text{ cm}^{-2}$

$N(Mg^+) = 3 \times 10^{14}$ cm$^{-2}$

Since these have similar ionization potentials, they indicate the actual iron to magnesium ratio
$[N(Fe)/N(Mg)]_{PHL\ 938} \simeq 3$
while $[N(Fe)/N(Mg)]_\odot \simeq 1.6$.
Since there are many uncertainties in the transition probabilities for the iron lines, the agreement is pretty good.

Ton 1530 is another object which shows many absorption lines. It, as well as PHL 938 and 4C 25.5, is quite bright and has large redshift, so that it falls above the normal Hubble relation on a (m - z) diagram. The principal absorption lines in Ton 1530 is the C IV doublet, in which the line at lower wavelength is stronger than that at higher. The emission redshift is z = 2.046, while that of the C IV doublet in absorption is z = 1.936. There are a couple of weaker redshift systems, which are evidently due to Ly α and C IV absorption; they have redshifts z = 1.980 and 2.055. Note that the last mentioned system is slightly on the long wavelength side of the emission. Morton and his group have looked at Ton 1530 with higher resolution than previously achieved, over a short wavelength region. They found that each of the components of the strongest C IV doublet is resolved into three; there is not one redshift at z = 1.936, but three very close redshifts. My student, M. Chan, derived a velocity dispersion of 74 km/sec, but did not know the lines are triple; for a single line it may be something like 20 km/sec. The column density of C IV in the line of sight is $N(C^{+3}) = 1.6 \times 10^{15}$ cm$^{-2}$ for the z = 1.936 system.

PHL 938 and Ton 1530 are both radio quiet. They come from lists of blue stellar objects. On the other hand, PKS 0237-23 was found by the Parkes radio survey. Its declination is -23°, so it is not very easy to observe from the Northern hemisphere, but the large southern telescopes now in construction will probably start work on it early on. Its emission line redshift is $z_{em}$ = 2.223. The Ly α emission is very broad and there are three C IV doublets in absorption near it. There seem to be 7 absorption redshift systems, including $z_{abs}$ = 2.2017, 1.6744, 1.6715, 1.6564, and 1.3646. Two of the CIV doublets overlap. There seem to be close groupings of absorption redshifts. Lynds,

Stockton and I made the identifications mainly on the basis of the wavelength ratio argument. Bahcall, Sargent, and Greenstein (1968) made identifications by a computer technique and there was agreement on the redshifts mentioned above; they claim three more possible systems and we have two more. The z = = 1.3646 system is one of the best determined and the difference between it and the emission line redshift corresponds to a velocity difference of 90,000 km/sec or about one third the speed of light. This object seems to be strong evidence against the intergalactic interpretation of the absorption lines, since some objects have no absorption lines, while this one has so many systems. It is statistically implausible that the intervening objects would be strung out in certain directions. M. Chan has derived a velocity dispersion for the z = 1.656 system of 80 km/sec, if the lines are not multiple, and a column density $N(C^{+3}) = 6 \times 10^{14}$ cm$^{-2}$.

For all these three objects, if we assume that the carbon to hydrogen ratio is "normal" and thus do not have to worry about the ionization equilibrium, and that all the carbon is in C IV, which is reasonable since C II and C III lines are not seen, we find $N(H) \simeq 10^{19}$ cm$^{-2}$. This is small and could be produced by a filament or some such object near the source.

Two points to notice are (i) the close groupings in $z_{abs}$, and (ii) coincidences between strong resonance lines from different elements at different redshift, as we saw with PHL 938.

The emission line redshift of 4C 25.5 is $z_{em}$ = 2.358. The only absorption redshift that has yet been fitted is $z_{abs}$ = = 2.368. It should be noted that the Ly α absorption is not black at the center, that is, it is not absorbing all the radiation coming to us.

5C 2.56 is a faint QSO, which has an emission line redshift $z_{em}$ = 2.390 and $z_{abs}$ = 2.367. Its light varies between about magnitude 19.2 and 20 on a time-scale of years. It was the first of the QSO's with a large enough redshift for Ly β to come into view. It should be monitored spectrophotometrically to see whether the emission lines vary in absolute strength when the total light varies.

# Observational Problems of High Energy Astrophysics

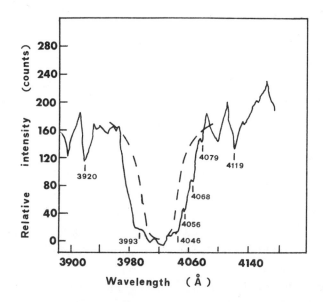

Figure 5.1  Ly α (z = 2.309) region of the spectrum of PHL 957 obtained with the digicon, showing possible Si II identifications and a line profile fit using the same velocity dispersion as the z = 2.309 C II line.

The object with largest known redshift is 4C 5.34, found by Lynds and Wills. It is a 4C radio source and has $z_{em}$ = 2.877. It has many absorption lines and, while one can pick out a few systems, there are many which lie shortwards of Ly α, which do not seem to have any other lines to fit with. Lynds has suggested that these might all be individual Ly α absorption. Ly β shows up and falls near some O VI lines.

PHL 957, a radio quiet object, was discovered by M. Schmidt. It has the second largest redshift, $z_{em}$ = 2.67, and has many absorption lines including a very broad feature thought to be Ly α; the Ly β absorption line corresponding to the Ly α is seen. By another strange coincidence, this Ly β absorption is very close to the Lyman limit in the emission system. PHL 957 was seen by the Princeton group using their TV camera with a resolution of 0.9A°. They collaborated with Palomar astronomers. The best of the absorption line systems are $z_{abs}$ = 2.662, 2.309, 2.225 and 2.206 and there are possibly 4 others. The lowest redshift is 1.824 which is a bit doubtful. Once again it seems implausible that the absorption systems are not connected with the QSO. The statistical methods developed by Bahcall show that only 1 in 10 identified absorption systems are chance coincidence. The Princeton-Palomar group noticed a C II ($\lambda_o$ = 1335A°) line which is quite narrow, $\sigma \leq$ 30 km/sec; the Ly $\alpha$ absorption is much broader; the Digicon was used to examine the profile of the broad Ly $\alpha$ absorption feature, which has a redshift of 2.309. If the sky is subtracted from the QSO spectrum, the center of the Ly $\alpha$ line seems to effectively go to zero (Fig. 5.1). The flux over the central 30 A° is 0.002 ± 0.006 of the continuum level. Since the object has magnitude 16.5, the $3\sigma$ level at the bottom of the line is about magnitude 21, showing the performance of the Digicon. Using the prototype Digicon, a 15th magnitude star would give 0.2 counts/A° sec; the prototype has a quantum efficiency of 4%. The dark current is less than $1 \times 10^{-3}$ counts/sec diode. A computed line profile was fit to the broad Ly $\alpha$ absorption line, but it did not seem to account for all the absorption. There are some identifications to be made over and above those made by the Princeton-Palomar group. Si II transitions out of excited fine structure levels of the ground state seem to be likely candidates. This implies that the lines originate from fairly near the QSO, where there would only be transitions from the absolute ground level. However, we can still not account for all the absorption in the broad feature, and we suggest there is Ly $\alpha$ absorption from systems with slightly different redshifts than the principal line. The only other way out of the puzzle appears to be that carbon is very underabundant in this object, $10^{-3}$ of the solar neighborhood value.

If we look at the distribution of redshifts of QSO's, there is a peak at $z \simeq 1.95$. The C IV doublet can only be found in objects with $z > 1.2$, and for lower redshift the Mg II lines are important. We are now in the process of measuring Mg II in high redshift objects where it appears in the infrared. We can then do statistical studies of the appearance of Mg II over a large range in redshift.

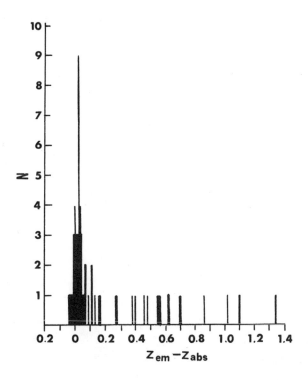

Figure 5.2

Looking at the distribution of $z_{em} - z_{abs}$ (Fig. 5.2) for the QSO's, there is a peak at small positive numbers which shows the majority of the absorption line redshifts are slightly less than the emission line redshifts. There are some with $z_{abs} > z_{em}$.

Our conclusion is that the absorption lines are produced by gas that is associated with the QSO, and one of the main questions we have to ask is how do the lines become so narrow, when there is so large a velocity difference compared to the emission lines. One hypothesis that has been proposed by a La Jolla group is that the gas giving rise to the absorption has been ejected and accelerated to high velocity by radiation pressure from the QSO. The assumption has been made that the QSO's are at cosmological distance. The hypothesis can be applied to ejection by Seyfert nuclei and also to the case of QSO's if the redshifts are not cosmological; it is only the scale of the masses that changes. A hot star, as we saw, has sufficient radiation pressure to push off gas at 2000 km/sec.

## Lecture 6

The problem of multiple absorption lines in QSO's is one of the most interesting and challenging in astrophysics, and if we could understand its nature, we would be a long way towards understanding the physics of QSO's.

We have seen before (Fig. 5.2) a histogram plotting the number of objects against the difference $z_{em} - z_{abs}$. The histogram has a big, fairly symmetrical, peak about the 0 value, and a spread of about 0.03 in redshift. The distribution for $z_{abs} < z_{em}$ has a long tail, extending up to the highest value, corresponding the PHL 938 which has $z_{em} = 1.955$, $z_{abs} = 0.613$, $z_{em} - z_{abs} = 1.342$, which corresponds to a velocity of about half the velocity of light, in the rest frame of the object.

We have discarded the argument that these absorption lines could be due to intergalactic matter or intervening galaxies, and concluded that these lines belong to the objects. We have thus to understand the production of large velocities in gas. This has bypassed the whole question of the nature of the redshift, because we have put in the assumption that the redshift is Doppler in origin. Of course, the shift could be due to other mechanisms, like gravitational redshift, in which case the gas is further away from the object, and is somehow stabilized under different gravitational potentials which would give a whole range of redshifts. But one cannot think of a sensible model to do that. So we made the hypothesis, and we have to take it as a working hypothesis and no more, that the differences in redshift are Doppler in origin; also suppose that the QSO's are at cosmological distances, although the whole argument can be scaled if they are not. Let's think of a mechanism that could give rise to an outward force on gas belonging to the QSO's. The one thing that comes to mind is radiation pressure. The QSO's, being put at cosmological distances, must have very high radiation output. Radiation pressure is therefore the obvious thing to think of for the outward force.

That was the idea behind the paper by Mushotzky, Solomon and Strittmatter (1972). They made the comparison between QSO's and hot stars, where gas streams are observed to be pushed out up to a velocity of about 2000 km/sec.

The radiative outward acceleration $g_R(i)$ due to species $i$ is given by:

$$g_R(i) = \frac{F_\nu}{c} \frac{N_i}{N_H} \frac{X}{m_H} \frac{\pi e^2}{m_e c} f, \qquad (6.1)$$

where $F_\nu$ is the continuum flux at the line center; $N_H$ and $N_i$ denote the number density of hydrogen and of the ionization state of the element; $X$ is the fraction by mass of hydrogen, and the remaining physical constants have their usual meaning. Considering the C IV $\lambda 1549$ transition, and assuming that $N_i/N_c \sim 1$ and $N_c/N_H = 3 \times 10^{-4}$, we obtain:

$$g_R \simeq 2 \times 10^{37} L_{46}/R^2$$

where $L_{46}$ is the total optical luminosity in units of $10^{46}$ ergs/sec. The gravitational acceleration is: $g = GM_S/R^2$. Both accelerations vary as $1/R^2$, and a stable situation can be established if other forces like magnetic forces can be ignored. The condition that gravity should exceed radiative acceleration can be expressed as

$$M_S \geq 1.5 \times 10^{11} L_{46} M_\odot \qquad (6.2)$$

This result was obtained neglecting the fact that, as one moves outwards from the object, the ionization level changes. If this is considered, a parameter $\beta$ of order unity has to be introduced: $M_S \geq 1.5 \times 10^{11} \beta L_{46} M_\odot$. The Schwarzschild parameter $z_G$ for the continuum-producing region satisfies then the condition:

$$z_G = \frac{2GM}{Rc^2} \gtrsim 0.02 \; \beta \; L_{46}/R_{17}, \qquad (6.3)$$

where $R_{17}$ is the radius of the continuum source in units of $10^{17}$ cm. Now, we know from light variation arguments that $R_{17} < 1$, for a spherically, symmetric model. So if one puts in enough mass to give outflow, the core of a QSO may be comparable to its Schwarschild radius. So we are in a region, where we cannot ignore gravitational effects. Next, we ask whether, if the mass is not quite enough to hold in the material ($M<M_s$), we can get high enough velocities to explain the observations. The flow velocity is approximately given by:

$$v^2 \simeq 2\int_R^{R+S} g_R \, ds \sim 2 \times 10^{37} \; \beta \; L_{46} \; S/R^2 \qquad S \ll R, \qquad (6.4)$$

$$\sim 2 \times 10^{37} \; \beta \; L_{46}/R \qquad S \gtrsim R,$$

where the acceleration takes place over a length scale S. Thus, we see that we can obtain quite large velocities. Let's go back to the emission lines. How sure are we that these emission lines are not produced in an outward moving atmosphere? Let's take the velocity that is given by the width of the emission line. A typical half-width of an emission line (e.g. C IV $\lambda 1550$) in the rest frame of the object is about $\Delta\lambda \sim 40$ A°; we then have:

$$\frac{v}{c} = \frac{\Delta\lambda}{\lambda} \lesssim 3 \times 10^{-2} \qquad (6.5)$$

Using $v^2 \sim 2 \times 10^{37} \; \beta \; L_{46} \; S/R^2$ we get:

$$R \gtrsim 2 \times 10^{19} \; \beta \; L_{46} \; (S/R) \; \text{cm} \qquad (6.6)$$

This condition has to be satisfied in order not to get a smearing of the emission lines which we do not see. Conversely, we could say that if we have gas producing emission lines which do not satisfy this condition, we would not see emission lines coming from it: they would be completely smeared out. In that region,

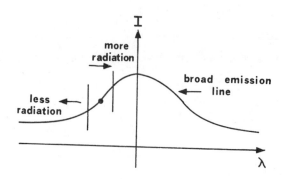

Figure 6.1

the Schwarzschild parameter satisfies (with $M \sim M_s$):

$$z_G \sim \frac{2GM}{Rc^2} < 10^{-3} \qquad (6.7)$$

So we don't have to worry about there being an appreciable gravitational redshift in the emission lines in this case ($M < M_s$).

Consider now what might happen in the absorption lines. We want some mechanism that would give us big shift, but no broadening. One such mechanism is what is called "Line locking". <u>We postulate now that the mass of the QSO is near the optical mass</u> $M_s \sim 10^{11} M_o$. We then get a pretty near balance between the gravitational forces and the radiation pressure. Consider a blob of gas that has achieved a velocity outwards by radiation pressure, such that, in the rest frame of the QSO, it sees radiation that is coming from a region where there is a strong gradient in intensity; this situation would occur naturally in the wing of an emission line (Fig. 6.1). Suppose we have a blob of gas going out that has a velocity so that it sees radiation at that point. If it acquires higher velocity, it shifts to the left (lower $\lambda$), and the radiation that it sees falls in intensity, and therefore it gets less of an acceleration. On the other hand, if its velocity decreases a bit, it would see stronger radiation, and that would give it an extra kick. We then can get some sort of a balance. This mechanism was suggested by the fact that we saw the Mg II absorption line sitting in the wings of a C IV emission line in PHL 938. One can show that, <u>if the acceleration occurs over a region of dimension S comparable with the radius R</u>, and <u>if the mass M is close to the initial mass $M_s$</u>, the following relation holds:

$$z_G = \frac{2GM_s}{Rc^2} \simeq \frac{v^2}{c^2} \tag{6.7}$$

So for $v/c \simeq 1/3$, $z_G$ is not any longer negligible in these conditions. Then the observed redshift would no longer be a true measure of the distance, even if the QSO's were at cosmological distances.

So, for this mechanism to work, we must have a near balance between the optical mass and the actual mass. This could then work as a way of limiting the velocities that you actually see, because the gas would tend to become grouped in velocity space, if one has a gradient in the radiation field. Something that would give an intensity gradient, besides an emission line, would be a continuum edge. In fact, it has already been suggested, before the line-locking mechanism was proposed, that the continuum edge produced by the Lyman limit might be such a case. Such a near coincidence is observed in PHL 938 and PHL 957. Another way of producing an intensity gradient is to have the following situation: an absorption line produced by one of the gas blobs, and the radiation countered by it is seen by a further out blob moving at a different velocity (quicker or smaller); we obtain then a gradient with the form of an absorption line. This effect might be an explanation for the rather close coincidence between an absorption line in one redshift system and an absorption line in another redshift system. For example, in PKS 0237-23, we have three sets of C IV. Two sets were very close together and one of them is actually multiple; in fact, it meant that one component of this doublet of C IV was almost superimposed on the other component of that doublet in the other redshift system. PHL 957 shows the same kind of phenomenon. This seems to be a strong evidence for the line-locking mechanism, although we have not been able to put the theory on a really firm or numerical basis: we should be able to find something that is locking the lines in any redshift system, and, so far, we have not been able to do it.

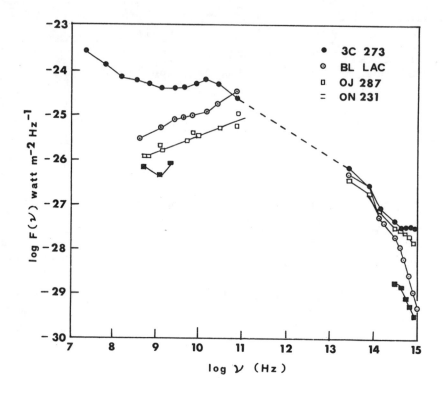

Figure 6.2

Could the same mechanism work in places other than QSO's? It turns out that it could very nicely work in Seyfert nuclei. In a typical Seyfert nucleus, the optical and X-ray luminosity is $\sim 10^{43}$-$10^{44}$ ergs/sec. Going through the same calculations as in the case of QSO's, we find that the critical mass is $M_s \sim$ $\sim 10^{8-9}$ $M_\odot$. This can be compared with the upper limit of $M \sim 3 \times 10^8$ $M_\odot$ obtained for the nucleus of the Seyfert galaxy NGC 1068 from the rotation curve. Anderson and Kraft (1969) have inferred the existence of outflow in the Seyfert galaxy NGC 4151 from the blue shift of HeI $\lambda 3888$ and H$\gamma$ absorption lines, which corresponds to a velocity of about $10^8$ cm/sec. The optical variation of the nucleus suggests a dimension $R \sim 10^8$ cm. The velocity can be shown to be :

$$v \sim 10^8 \, (\beta L_{43}/R_{18})^{1/2} \, \text{cm/sec}. \tag{6.8}$$

This agreement, while possibly fortuitous, lends additional weight to the hypothesis that radiation pressure against mass loss may play an important role, both in nuclei of Seyfert galaxies and QSO's.

An interesting observation has been made recently by Weedman. The object is a Markarian object, MK 231, a Seyfert-type galaxy, in which the usual emission line spectrum is seen with some very strong absorption lines blueshifted by $\sim$ 4000 km/sec (corresponding to a critical mass $M_s \sim 10^{8-9} \, M_\odot$). The absorption lines are again resonance lines, the strongest of which is Na.

Let us turn now to some very puzzling objects, objects of the type of BL Lac. I should describe how this prototype object came to be found. Fig. 6.2 shows the log of the flux as a function of the log of the frequency for 4 different objects. The top plot is that of 3C 273, the optically brightest quasar in the sky. 3C 273 is a double source: a compact radio source plus a large diameter source dominating at low frequencies. Its spectrum has a bump in the radio and rises in the infrared. A search was made in high frequency radio catalogues to pick out objects of a similar character, i.e., having a flat or **peaked** spectrum in the radio range, reflecting self absorption in a compact object. In this way, BL Lac was found in the Illinois catalogue. The Ohio radio group looked for selected objects that have unusual radio spectra (OJ, ON objects). The energy distribution of objects found in such a way is plotted also on Fig. 6.2. BL Lac was later found to coincide with a reasonably bright star ($m_v \sim 14m$), and it was realized that it was also present in the catalogue of variable stars as an irregular variable. The optical light was found to be highly polarized. Table 6.1 lists 5 of these objects, on which recent measurements at optical and infrared wavelengths have been made. From the variability of these objects, (see column 3 of Table 6.1) one would guess that they are related to QSO's. The distinguishing feature of all these objects is that they all show lineless spectra, and thus we don't have any idea about their distance. From their distribution on the sky, it is clear that they are not associated with the plane of our galaxy. 21-cm

Table 6.1

DESCRIPTIVE AND PHOTOMETRIC DATA OF COMPACT NONTHERMAL SOURCES

| Visual Object | Radio Source | Range of $m_v$ | $\alpha$ (1950) | $\delta$ (1950) | U | B | V | R | I |
|---|---|---|---|---|---|---|---|---|---|
| OJ 287 | | 12-15 | $08^h 51^m 57^s$ | 20°17'59" | 12.8 | 13.4 | 13.0 | 12.6 | 12.3 |
| | ON 325 B 1215+30 | | $12^h 15^m 21^s$ | 30°23'39" | 15.6 | 16.1 | 15.6 | | 14.8 |
| W Com. | ON 231 | 11.5-16.5 | $12^h 19^m 01^s$ | 28°30'36" | 16.9 | 17.2 | 16.7 | | 15.3 |
| AP Lib | PKS 1514-24 | 14.5-16 | $15^h 14^m 46^s$ | -24°11'21" | 15.8 | 16.0 | 15.4 | 14.5 | 13.9 |
| BL Lac | OY 401 VRO 42.22.01 | 12-15.5 | $22^h 00^m 38^s$ | 42°02'01" | | | | | |

Observational Problems of High Energy Astrophysics    169

Table 6.2

POLARIMETRY OF COMPACT NONTHERMAL SOURCES

| OBJECT | Date(UT) | P | $P^U$ | $P^B$ | $P^V$ | $P^R$ | $\Theta$ | $\Theta^U$ | $\Theta^B$ | $\Theta^V$ | $\Theta^R$ |
|---|---|---|---|---|---|---|---|---|---|---|---|
| OJ 287 | 2/18/72 | | | 2.9±0.1 | | | | | 39° | | |
| | 3/17/72 | | 5.2±1.0 | 5.0±0.3 | 5.0±0.4 | 7.0±0.6 | | 91° | 92° | 93° | 92° |
| | 4/16/72 | | 11.0±0.8 | 10.3±0.4 | 10.2±0.5 | 10.4±0.9 | | 83° | 87° | 87° | 83° |
| | 4/17/72 | | | 10.8±0.3 | 10.4±0.4 | 11.2±0.9 | | | 87° | 88° | 85° |
| ON 325 | 2/18/72 | 7.05±0.12 | | | | | | | | | |
| | 2/20/72 | | 8.2±3.1 | 7.2±0.4 | 6.6±0.7 | | 145° | 142° | 140° | 135° | |
| B 1215+30 | 3/17/72 | | | 4.4±1.4 | | | | | 146° | | |
| | 4/16/72 | | | 4.6±0.8 | | | | | 119° | | |
| ON 231 (W Com) | 2/18/72 | 3.3±0.7 | | | | | 95° | | | | |
| PKS 1514-24 (AP-Lib) | 3/17/72 | | | 5.6±0.9 | 3.0±1.9 | | | | 173° | 13° | |
| | 4/16/72 | | | 5.8±1.4 | 8.0±1.9 | | | | 139° | 149° | |

absorption studies in the direction BL Lac shows that the object is at a distance exceeding 200 pc.  The polarizations are large and tend also to be variable. This is shown in Table 6.2 where P and Θ refer respectively to the percentage and position angle of linear polarization and superscripts refer to the wavelengths used (no subscript means an unfiltered measurement).  It has been suggested that these objects might be blue-shifted QSO's: they have similar kinds of spectra (see Fig. 6.2).  They appear brighter than QSO's and they don't show any lines.  The absence of lines might be explained in two ways:  first, one might say that all the radiation is synchrotron radiation like in the jet of M 87, or one can say that there are lines, but they are too weak to be seen; the continuum might be brightened so much that it swamps the lines, or, alternatively, you might say that you are looking at a part of the spectrum where the lines are intrinsically weaker relative to the continuum.  We have seen that there is a steep rise in the infrared, and we might be looking in the IR which is blueshifted.  The strongest line one might see in a blueshifted object is the line Hα with rest wavelength λ6563, which, even if blueshifted with $z = 0.5$, would be observable, unless it was swamped by the continuum.

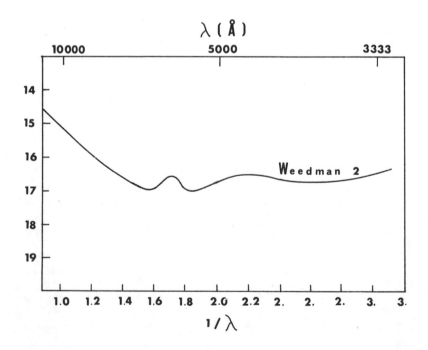

Figure 6.3

Are there objects similar to BL Lac associated with other galaxies? Weedman has made up a list of blue objects that show up in the same way in the Palomar Atlas as the QSO's, in the vicinity of bright galaxies. One such object is Weedman 2, near the double galaxy NGC 2992/3 (Arp 245 in the Atlas of Peculiar Galaxies). The spectrum is lineless, like the spectra of objects of the type of BL Lac. The object shows UV excess and is strong in the infrared (see Fig. 6.3), which are the characteristics of a non-thermal radiation spectrum. Weedman 2 has not been seen as a radio source. The emission lines in the galaxy are quite strong, indicating the presence of ionized gas, and the velocity field is very disturbed: it is not a simple rotational field.

Suppose we say that BL Lac is in some way connected with our Galaxy, at a distance comparable to that of the Magellanic clouds. Its apparent brightness is $m_v \sim 14$ mag. The distance modulus corresponding to the Magellanic clouds is: $m - M \sim +19$, which would give an absolute magnitude $M = -5$, which is a nova—like magnitude. For Weedman 2, $m \sim 17$ mag. If we take its distance to be the distance of the galaxy $\simeq 40$ Mpc, the distance modulus would be $m-M \simeq +33$ and so $M = -16$, which is in the supernova range.

## REFERENCES

Anderson, A., Kraft, R., 1969, Ap. J., 158, 859.

Bahcall, J. N., Greenstein, J. L., Sargent, W. L. W., 1968, Ap. J., 153, 689.

Burbidge, E. M., 1967, Ann. Rev. of Astro. and Ap., 5, 399.

Burbidge, G. R., 1970, Ann. Rev. of Astron. and Ap., 8, 369.

Burbidge, E. M., Burbidge, G. R., Sandage, A., 1963, Rev. Mod. Phys., 35, 947.

Burbidge, E. M., Lynds, C. R., Burbidge, G. R., 1966, Ap. J., 144, 447.

Burbidge, E. M., Burbidge, G. R., 1967, "Quasi-Stellar Objects", (Freeman and Co., San Francisco).

Burbidge, E. M., Lynds, C. R., Stockton, A. N., 1968, Ap. J., 152, 1077.

Greenstein, J. L., Schmidt, M., 1964, Ap. J., 140, 1.

Hoyle, R., Fowler, W. A., 1967, Nature, 213, 373.

Humason, M. L., Mayall, N. V., Sandage, A. R., 1956, A. J., 61, 97.

Lynds, C. R., 1968, Proc. Symp. on Beam-Foil Spectroscopy (New York: Gordon and Breach), p. 539.

Lynds, C. R., 1970, Ap. J. Letters, 159, L151.

Lynds, C. R., Sandage, A. R., 1963, Ap. J., 137, 1005.

Lynds, C. R., Wills, D., 1970, Nature, 226, 532.

Minkowski, R., 1957, I.A.U. Symp. No. 4 (Radio Astronomy), p. 107.

Mushotzky, R. F., Solomon, P. M., Strittmatter, P. A., 1972, Ap. J., 174, 7.

Schmidt, M., 1969, Ann. Rev. of Astron. and Ap., 7, 527.

Strittmater, P. A., Serkowski, K., Carswell, R., Stein, W. A., Merrill, K. M., Burbidge, E. M., 1972, Ap. J. Letters, 175, L7.

Theoretical Problems of High Energy Astrophysics

Lectures by

G. R. Burbidge

University of California, San Diego

Notes by

R. Epstein, K. Brecher, H. G. Hughes

# Theoretical Problems of High Energy Astrophysics

Lecture 1

In this lecture I shall attempt to summarize the various points of view presented by the observationalists concerning the main astrophysical problems before us, especially, the problems concerning QSO's.

This and subsequent lectures shall discuss the four basic cosmic components, intergalactic matter, stars and galaxies, QSO's and related peculiar objects, and relativistic particles. Of particular mystery is galactic formation, and the distances of the QSO's, which must be known to infer their brightness and density. The possible upper extremes of energy allowed by the present ambiguity in the distances of QSO's, may, if verified, call for some entirely new physics to explain them.

What, now, do we known about the universe in terms of the normal, popular, expanding Friedmann cosmologies and their highly compressed initial states? Also, what wisdom do the infinite, beginingless, uniform steady-state models have to offer?

The Friedmann models, shown to be solutions to Einstein's equation by Friedmann and Lemaître, predict expansion of the universe. The galactic redshifts observed some forty years ago by Hubble and Humason were simply interpreted as due to the expansion of the universe and were, at that time, thought to be confirmation of the expansion hypothesis. The Hubble constant inferred from these observations has evolved in time (see Fig. 1.1), as have the details of the interpretation, but the redshifts of normal galaxies and other objects still constitute one of the two observational pillars of an expanding Friedmann universe.

What is considered, by many, to be the second pillar of the hot big-bang universe is the 3 °K microwave background radiation, predicted by Gamow, Alpher, and Herman around 1950 and later partially observed by Penzias and Wilson (1965). Expanding Big-Bang models demand that the microwave background assume a blackbody spectrum. While this seems to be observed in the Rayleigh-Jeans portion of the spectrum, from which the 2.7 °K is inferred, rocket and balloon observations of the higher frequencies suggest a higher 6 °K temperature from the higher frequency portion of the spectrum. The most recent

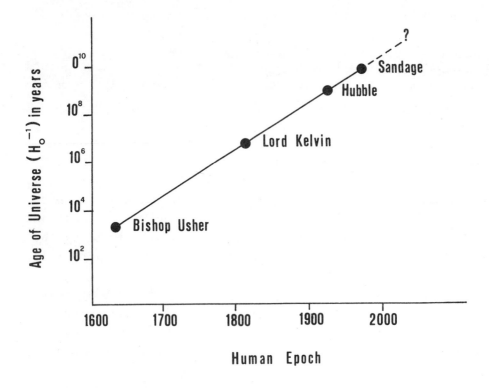

Figure 1.1

balloon observations of Muehlner and Weiss may not be incompatible with a 3 °K spectrum. It might be suggested that the 6 °K result is due to the detector having found an isolated peak superimposed on an otherwise 3 °K spectrum, but this would mean that most of the microwave background radiation from the sky is coming from that one peak. If the background radiation is to be explained as a feature of a steady-state cosmology, rather than as the result of one initial event, then the apparent smoothness of the spectrum must imply a large number of discrete sources ($\geq 10^{15}$).

The formation of galaxies is very mysterious in any cosmology. At the present state of research, it seems that everything of which a galaxy consists must be provided a priori. This includes the initial density fluctuation which, in the case of the expanding universe, must provide a deep enough gravitational well to overcome the local effects of cosmological expansion. In the Hoyle and Narlikar picture of the steady-state universe, galaxies, the nuclei in particular, serve an additional function as the sites of matter formation.

The work of Wagoner, Fowler, and Hoyle (1967) predicts the formation of helium in a big bang, and so the existence of genuine primordial helium is a crucial observational check on the big bang that is yet to be performed conclusively. The formation of the heavier elements in stars is relatively well-understood, independent of cosmological considerations.

There is, as yet, no explanation in any cosmology for the origin of primordial magnetic fields, which, in turn, are needed to explain the magnetic fields which presently pervade the cosmos.

Representative of observational tests for determining which cosmology we inhabit is the plot of the logarithm of the total number of sources N(S) whose observed flux exceed S versus the logarithm of S. The resulting log N - log S curve should, for example, yield a slope of -1.5 for a Euclidean universe. As yet, this scheme is not a very powerful cosmological test. It is not certain that the observed sources are a significant sample of deep space. Astronomers at Cambridge observe a slope of -1.8 at stronger flux levels which, they claim, supports the hypothesis that the radio objects are evolving in time and thereby undermines the steady-state theory. These results are still in dispute because of possible selection effects, among other reasons, and no results are recognized as conclusive at this time.

Another cosmological test is the luminosity-volume test for QSO's. The distribution of observed QSO redshifts takes the form $(1 + z)^\alpha$, where $\alpha = 6$, according to Schmidt (1972 a, b, c). This result is not predicted by a Friedmann model, and it can be explained only by a non-steady-state evolution scheme, such as a brief period of rapid QSO formation in the past that has been gradually slowing until the present. The above discussion, of course, assumes that the redshifts are cosmological Doppler shifts. The observations of Schmidt do not contradict a steady-state cosmology, if the QSO's are allowed to be local (Lynds and Wills, 1972). Thus we see that the meaning of the luminosity-volume test depends upon the answer to the even more fundamental question of the locality of the QSO's and of the causes of their redshifts.

An observational verification of a steady-state universe would be the observation of $q_o = 1$, the characteristic steady state value of the deceleration parameter. These observations

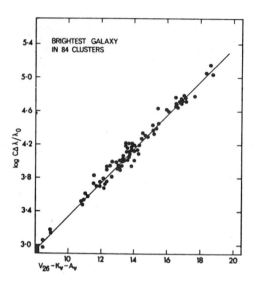

Figure 1.2

are very difficult, however, and none have been conclusive.

As suggested above, QSO's present problems and big, fundamental physical questions unassociated with any cosmology in particular. Especially important is the dilemma of quasars in the compelling apparent proximity of companion objects with redshifts differing from those of the companions by more than can be explained by any reasonable application of the Doppler effect. This dilemma is most strongly stated as the Arp Hypothesis, that these redshift discrepancies cannot be explained by cosmology and the expansion of the universe.

QSO's, furthermore, do not yet seem to fall on any particular distance-redshift distribution. In the Fig. 1.2, it is seen that galaxies fall along a line in the log z - magnitude plot, that suggests that they are objects of very roughly the same luminosity, whose redshifts are due to the expansion of the universe. The next Figure 1.3, shows that once you go dim enough to see QSO's, their redshifts are generally large, but, apart from that, there is no identifiable corelation between their redshifts and brightnesses that would give any cosmological information.

Looking at the distribution of QSO redshifts (291 red-

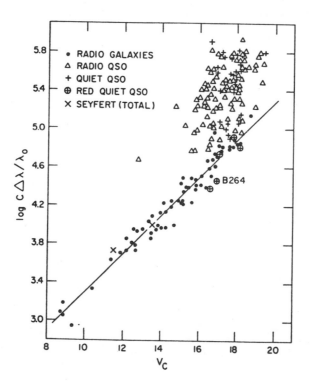

Figure 1.3

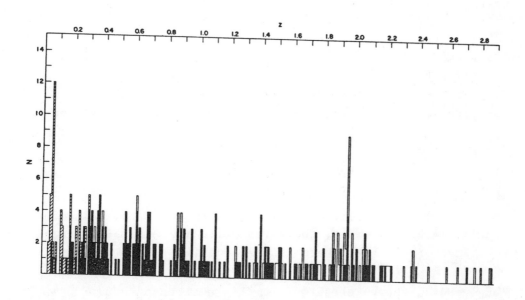

Figure 1.4

shifts from emission and absorption lines of 235 objects), one sees a peak at z = 1.95 (Fig. 1.4). There is yet to be any published statistical support for the reality of this peak. The peak does suggest some event significant to QSO activity at that time, perhaps, for example, a pause in the expansion as in a Lemaitre universe. Adding to this histogram redshifts from 39 related objects, some periodicity becomes apparent. By "related object", we mean bright objects with continuum spectra with no absorption features or any other evidence suggesting stars. The two single large peaks (the big peak being at z = .06) are of weak statistical significance. O'Dell (Burbidge and O'Dell, 1972) finds a periodicity of 0.031 in this histogram that is significant to a 97.5% level. What cosmological, or even numerological meaning there is in this periodicity is not at all apparent. The existence of any sort of fine structure at all is especially mysterious when one considers that the dispersion of redshifts due to random velocities, possible slight variations in intrinsic redshifts, and the variations in cosmological redshift, simply due to arbitrariness of locating an object in a particular spot in time, would be bound to smear away any sort of fine structure, such as periodicity in such a histogram.

Problems associated with the small sizes of QSO's and the superlight velocities are connected with the relation of large distances with large redshifts. Specifically, are large redshifts cosmological? In this course we will see how the assumption of the purely cosmological origin of redshifts causes difficulty.

Given that a QSO varies significantly ($\sim$ 1m) in luminosity on a periodic time scale, $\tau$, its radius can be inferred to be $R < c\tau$. Simply by presuming that to vary in any coherent manner, the extremeties of the object must be within communication distance of each other, sufficient for mechanical communication, for example, on the time scale of observed systematic variations. From observed $\tau$'s, $R \lesssim 10^{17}$ cm. Assuming QSO's are at a distance implied by purely cosmological redshifts, the observed luminosity, $L \sim 10^{46} - 10^{47}$ ergs/sec, implies a huge photon energy density from the relation $\rho_E = L/\pi R^2 c \sim$ $\sim 100$ ergs/cm$^3$. If, as suggested by Hoyle, Sargent, and

Table 1.1

| Name | Type | z | v/c |
|---|---|---|---|
| NGC 1275 | Seyf.-Perseus Cluster | 0.016 | 0.2 |
| 3C 120 | N-System | 0.033 | 2 |
| 3C 273 | QSO | 0.158 | 3-4 |
| 3C 279 | QSO | 0.54 | 6-10 |

myself (1966), the QSO radiation is incoherent synchrotron radiation, there appears a paradox. Compton scattering would dominate the synchrotron process because of the very high radiation energy density demanded by the above considerations. QSO models at non-cosmological distances, however, would not be prevented from emitting synchrotron radiation by their large energy densities.

With the advent of very long Baseline (VLB) interferometry, resolution on the order of $5 \times 10^{-4}$ sec. of arc has become feasable. With this technique, some radio sources have been observed to consist of components that change in time. The data can be successfully fit to a model consisting of two blobs receding from a common point. Assuming that they are at the cosmological distances indicated by their redshifts, very high velocities are inferred from their apparent angular speeds. The results for four objects are listed in Table 1.1.

The uncertainties in v/c are primarily due to uncertainties in $q_0$. It must be emphasized, that the superlight velocities here are artifacts of our assumption of cosmological redshifts. This apparent paradox is resolved if you allow some noncosmological origin of the redshifts that would bring the objects closer to us.

## Lecture 2

I would like to discuss further the association between quasi-stellar objects and galaxies. The fundamental problem is that if there are two objects, A and B, there are only two methods of making a distance measurement. One uses the redshift, assuming it is cosmological; the other says I have reason to believe they are physically associated, and thus at the same distance: given the distance of one, the other is fixed. What happens if these two methods disagree? How does one test physical association?

The method used by Gunn (1971) is the following: Galaxies can only be seen on Palomar plates with redshifts $z \lesssim 0.3$. If we restrict ourselves to looking for QSO's with $z \lesssim 0.3$, and look for galaxies associated with them then, since galaxies with $z \gtrsim 0.3$ cannot be seen, we have already prejudiced the result we can find. It is thus not good to also limit $z_{QSO} \lesssim 0.3$, since then one can never see $z_{QSO} \gg z_{gal}$!! Gunn has tried to prove in this way, by finding QSO's in clusters at the same redshift, that the $z_{QSO}$ is indeed cosmological in origin. The most famous case is that of the QSO PKS 2251+11 ($z \simeq 0.33$). On the plate one sees several very faint galaxies near it. Originally Gunn reported several emission lines in the galaxies, standing out above the continuum, indicating that $z_{gal} \simeq z_{QSO}$. When others tried to repeat this observation these lines could not be found. It appears that Gunn was only seeing light from the QSO which got into the slit while measuring the galaxy. It is therefore not surprising that he found $z_{QSO} = z_{gal}$. Later, Wampler and Robinson (1972) found the H and K lines of calcium in <u>absorption</u> in the galaxies, with $z_{gal} \simeq z_{QSO}$. However, it should be noted, it is exceedingly difficult to measure redshifts of point objects. Taken at face value, however, this is taken as proof that quasi-stellar objects have cosmological redshifts. There is one other cluster with a QSO in it (3C 61.1) with $z \simeq 0.3$. One must wonder whether the QSO is really in a proper cluster. Further, what is the chance that a QSO will appear to be in a cluster? Unfortunately, at 22m little is known about the distribution of faint galaxies, so the probability of chance association is difficult to estimate.

Another way to test the association hypothesis, is to look for correlations between galaxy and QSO positions independent of magnitude. We have taken a well defined sample of QSO's (all 40 3C QSO's), and looked for associations with galaxies in another well defined catalogue (de Vaucouleur's), which contains all galaxies with $m \lesssim 13$, with about 2500 galaxies. Do the QSO's show any tendency to be associated with these galaxies? We found by counting the number of galaxies around each QSO, that 4 QSO's lay very close to spiral galaxies, 3C 232 being associated with NGC 3067, 3C 268.4 with NGC 4138, 3C 275.1 with NGC 4651, and 3C 309.1 with NGC 5832. Each of these QSO's lies within about 7' arc of the associated galaxy. This is very close indeed. Statistically, we found (Burbidge, Burbidge, Solomon, Strittmatter, 1971) that the chance of finding such a close association was $10^{-2} - 10^{-3}$. This then is a demonstration of association with discrepant redshifts ($z_{gal} < 0.1$, $z_{QSO} > 0.5$). A similar test, however, done with QSO's chosen from the Parkes' catalogue by Bahcall, McKee and Bahcall (1972) lead to the result that these QSO's were randomly distributed with respect to the galaxies. Nonetheless, one can make arguments based on selection effects which would indicate that in going to a fainter sample of objects, associations are more difficult to find (Burbidge, O'Dell, Strittmatter, 1972). For example, if one searches for QSO's by looking for radio emitting objects, one has the following situation. In the 3C catalogue, radio sources fall into two groups: QSR's and radio galaxies. Identification of one of these or the other was made on the basis of radio positions with an accuracy of minutes of arc. If an error box contained a galaxy, then that was called the identification (for example, 3C 455). When the identifications were made, QSO's were not even known. Now when an error box also contained lots of stars, even if one was a QSO the galaxy was still identified as the radio source. But if there was a QSO in the box, one would have selected against finding it. One such case is 3C 455. It is a galaxy with a small redshift. As the radio position was refined, it moved off the galaxy, and on to a stellar object only 22" away from the center of the galaxy,

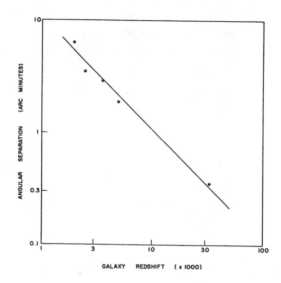

Figure 2.1

with a redshift $z \simeq 0.5$. One should really say that there are in fact 5 cases now with 3C QSO's near galaxies. Even more remarkable, if one plots the angular separation between the galaxy and the QSO vs. $z_{gal}$, one find a straight line on a log $\Theta$ - log $z$ plot (cf Fig. 2.1). Thus the linear separation between the galaxy and the QSO is constant. (Note that variations in $\Delta\Theta$ are partly due to defining the center of the galaxy). This result adds weight, or addition to the statistics, that there is a real association between QSO's and nearby galaxies. I conclude that there is evidence that there are objects physically associated with discrepant redshifts. (As a final remark, let me note that if a radio QSO is too close to a galaxy, it will not be identified as such. It may well be that the nuclei of galaxies themselves are QSO's).

What kind of explanations have been offered for this kind of phenomena?
1. First, is the suggestion by Terrel (1964), and Burbidge and Hoyle (1966) that QSO's are ejected from galaxies in all directions, and the redshifts are simply Doppler shifts, with the QSO's moving at relativistic speeds. Note that $z = 2$ implies $v \simeq 0.8c$. Terrell had all QSO's coming out of our

galaxy, and since they had already passed they are all redshifted. This leads to problems with the number of QSO's, energetic difficulties, etc. The main difficulty is that if they are radiating spherically symmetrically in there own rest frame, then, if they are ejected from other galaxies, one expects to measure more blueshifted objects than redshifted in the ratio

$$N_{blue}/N_{red} \simeq (1 + z)^4.$$

This is because objects moving toward you would be brighter. If one surveys to a given limiting brightness, one will tend to find more blueshifted than red shifted objects in the sample. With $z = 2$, one expects $N_{blue}/N_{red} \simeq 81$. In fact $N_{blue} = 0$.

Thus since no blueshifts have been found, one must look for some <u>intrinsic</u> source of redshift. Gravitational redshifts have been suggested for QSO's. In the original discussion of 3C 48 and 3C 273, Greenstein and Schmidt (1964) analyzed what is to be expected if a QSO has a strong gravitational field, and the radiation is emitted from a shell surrounding a central object. There will be a strong gradient of the field throughout the radiating region, which will give rise to broad lines. No sensible model could be constructed of this type. Especially difficult to explain, in addition, on this basis are objects with $z \simeq 2$. Hoyle and Fowler suggested that one put the gas in the middle of an object (at the bottom of a deep gravitational well), say filled with neutron stars or black holes, then one can eliminate the problem of $\Delta z$, the gradient in the field. However, it is still difficult to get large redshifts, $z > 1$. Note also that if the suggestion that apparently normal galaxies have anomalous redshifts is true, this explanation would not apply to them.

3. Finally, one can try to construct a theory to change the Rydberg constant. This would change the effective wavelength of all the atomic transitions, and could explain the redshift.

The electron charge, $e^2$, cannot be changed, that is $e^2/\hbar c$ is known to be normal from the observed splittings of resonant transitions in distant objects (e.g., Mg II). Thus the only variable is the effective electron mass, $m_e$, which can vary (which is also equivalent to allowing G to vary). But so far, except for Hoyle and Narlikar (see Hoyle's lectures), there has not been offered a theory to explain variable electron masses. Under this heading, one can also add quark-atoms as a suggested way of changing the energy levels of an atom. Beside the question of the existence of bare quarks, one is not sure of what transitions to expect in complicated quark atoms.

# Lecture 3

What I should like to do in this lecture is to sketch for you some of the properties of the matter and radiation in the universe, the material between the galaxies and the peculiar objects that we observe, and then, in the latter part of the lecture, I shall indicate the kinds of general arguments that are being used to account for the releases of matter and radiation from the nuclei of galaxies and from QSO's.

Let me write a few formulae as follows:

$$\Lambda + 3H_o^2 q_o = 4\pi G(\rho_o + 3p/c^2)$$

In the evolving universe of the classical Friedmann type we find this relation among the cosmological constant, the Hubble constant, the deceleration constant, and the mean density of matter in the universe. In the steady state universe we find instead:

$$\rho_{ss} = \frac{3H_o^2}{8\pi G}$$

If we drop the cosmological constant, and put $q_o = \frac{1}{2}$ (remember that the universe is closed for $q_o \geq \frac{1}{2}$), then we find that the critical density in the universe in this case is equal to the value for the steady state theory and is

$$\rho_{crit} = 1.1 \times 10^{-29} \text{ gm/cm}^3.$$

(for $H_o$ = 75 km/sec per megapc.). This is the so-called critical density of the universe. Now we have heard in other lectures that from counting galaxies, estimating their space density, and knowing their individual masses, or from using the mass-to-light ratio, one finds various values of $\rho$ on the order of

$$\rho_g = 5 \times 10^{-31} \text{ gm/cm}^3,$$

and the discrepancy between these two numbers is called the

"missing mass problem". It is, of course, entirely possible that there is no more mass than this, so that $q_o$ has a quite small value, and the universe is open. At the same time, this is in principle an argument against the steady-state cosmology.

On the other hand, if we are going to try to find the remaining 95% of the mass needed to close the universe or to give the steady-state universe, what sort of attempt can be made? Let me now describe the observations that have been attempted, based on the idea that the missing mass exists in the form of diffuse gas. This may or may not be a reasonable hypothesis, depending on your ideas about the formation of the universe. If we live in the classical "hot big bang" universe and if galaxies form from condensation of diffuse matter, then there is no reason not to expect that some matter should be left over. Not all the matter would condense into galaxies. If this matter is indeed primordial and has never been in galaxies, then because of the expansion of the universe we should expect it to be very cold, and largely in the form of hydrogen. We know that the heavier elements cannot form in the early stages of the evolution of the universe, but must be synthesized in stars. If this matter has never been in stars, then it should be hydrogen.

Thus we come to the conclusion that it is reasonable to look for cold hydrogen. People have done this in several ways. The first way was to use the 21 cm line in radio astronomy, a line that is due to a spin-spin transition in atomic hydrogen. One could in principle detect the 21 cm line as absorption in the intergalactic medium (Fig. 3.1). If there is cold atomic hydrogen in the line of sight between a continuum source and the observer, then one could expect absorption of the 21 cm line to occur due to the presence of the intergalactic medium.

There have been a number of attempts to see this effect, and they have all led to null results. The present limit set

Figure 3.1

by this method is:

$$\rho_H \lesssim 5 \times 10^{-30} \text{ gm/cm}^3,$$

a number which is significantly below the closure density, or the steady-state density $1.1 \times 10^{-29}$ gm/cm$^3$.

A much more sensitive method to detect atomic hydrogen is to use the strongest transition of hydrogen, the Lyman α transition. Since the wavelength of the Ly α transition, 1216 A°, puts it in the far ultraviolet, which cannot penetrate the atmosphere, you must use one of two methods to observe it. Either you must be above the atmosphere, or you must find Ly α with a large enough redshift to put it into the terrestrially observable region. Direct observations of galaxies from above the atmosphere have not yet been made with sufficient resolution in the far ultraviolet. What one would hope to see from such an observation is indicated in Fig. 3.2.

One should see a Ly α emission line in the object itself and absorption of the continuum radiation by the Ly α transition in the intervening hydrogen, which is at smaller redshift because it is closer to us than the object. If the intervening hydrogen is uniformly distributed, we should see a trough extending all the way from the redshifted Ly α emission line of the object being observed, down to the z = 0 position of Ly α at 1216 A°. This would represent absorption all along the path of the light from the object to our local neighbourhood. We should be able to measure the lowering of the continuum on the blue side of the emission line, and from a knowledge of the distance of the object, estimate the amount of the intervening hydrogen.

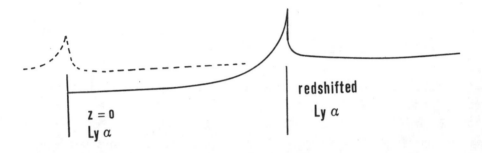

Figure 3.2

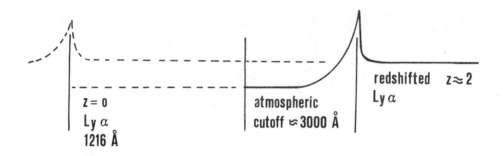

Figure 3.3

The other method, of course, is to use the QSO's. They show the Ly α line in the observable region because of their very high redshifts. You can use exactly the same method to look for absorption, except that now there will be an atmospheric cutoff (Fig. 3.3).

In principle you should see this same trough, provided you are at z at least = 2 to get above the atmospheric cutoff at around 3000 A°.

When this method was first applied to the quasi-stellar 3C 9, which has z = 2.012, Gunn and Peterson (1965) thought that they saw such a trough, and they concluded that the interstellar hydrogen had a density of

$$\rho \simeq 10^{-34} \text{ gm/cm}^3,$$

a very low value. However, later observations did not find this effect at all, and the conclusion now is that the upper limit is

$$\rho \simeq 10^{-35} \text{ gm/cm}^3.$$

There are, however, two possible explanations for this null result. Firstly, one may accept it at face value, and conclude that there is no intergalactic hydrogen. But if the QSO's are not at cosmological distance, then you would not expect to see any absorption, since in this case the intergalactic hydrogen would not have a sufficiently high redshift to put any part of the absorption trough above the atmospheric cutoff.

Now if one wants to maintain the existence of the intergalactic hydrogen in the face of these null results, one must make

# Theoretical Problems of High Energy Astrophysics

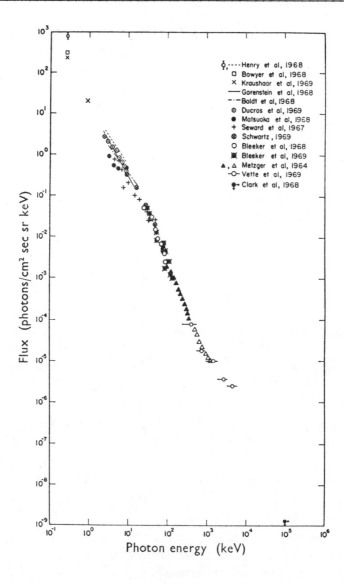

Figure 3.4

some change of hypothesis, and the first possibility is that the hydrogen is not cold, but hot. If less than one part in a million of the gas is cold enough to be in a neutral form so that it can absorb, then the two numbers $10^{-29}$ and $10^{-35}$ can be reconciled. From this, one can compute the necessary temperature, which turns out to be on the order of

$$T = 5 \times 10^6 \,°K.$$

There had been theoretical suggestions concerning a hot intergalactic gas before this. Hoyle and Gold (1958) proposed a variation of the steady-state universe in which matter was created in the form of neutrons. As the neutrons decay, the resulting hydrogen gas is, of course, quite hot. But that point of view has not held up.

Now if we accept this hot gas (remembering that there is one major hole in the argument, namely the assumption that the QSO's are at cosmological distance) we must expect that this gas should emit thermal bremsstrahlung in the form of X-rays at energies around a few kilovolts. Therefore we should see background X-rays of thermal origin. Of course, we do see background X-rays, and people began to think about whether they could come from this source.

Fig. 3.4 shows the form of the background X-rays. You can see that there is some sort of power law behavior, not the form you would expect from thermal bremsstrahlung, which would be the normal exponential type of behavior. This is a compilation made about two years ago by Brecher and Burbidge (1970) from observational data on the X-ray background.

The next Fig. 3.5 is a very recent one made by Dr. Brecher, whose point here is to show that the observed X-ray spectrum is not inconsistent with the radio spectra of two randomly chosen radio galaxies. There is quite a good case to be made from this type of argument that the X-ray background is generated by discrete sources of X-rays, which are most likely to be associated with radio sources, and that the X-rays are produced by the inverse Compton process, in which the radio electrons, the relativistic electrons present in the radio sources, scatter on the microwave background radiation and produce X-rays. This is a perfectly good explanation, from what we know at the moment, for the X-ray background, and it seems to me that this kind of data does not support the idea that you are looking at a hot intergalactic gas.

There was another argument made. The group at NRL maintained that the few points at the soft end of the spectrum on Fig. 3.4 (fractions of a keV) did not fit with the rest of the curve, and that while the bulk of the data fitted inverse Compton processes, these last few points indicated something else going on and might suggest the effects of the hot gas after all.

It seems to me that, based on the lack of evidence for hot

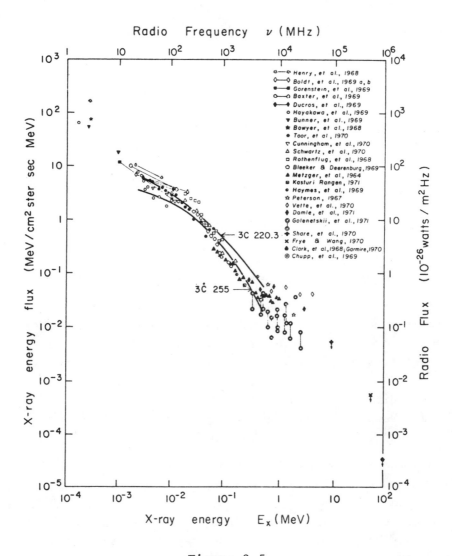

Figure 3.5

gas and on the null results for cold gas, that there is no indication of any appreciable amount of intergalactic gas at all, certainly not enough to close the universe. If you want to make the best case you can, I think you must say that the QSO's are local, so that you get around the null Lyman α results, and then, from the 21 cm measurements, you can take the upper limit, so that there is ten times as much hydrogen as galaxies; nevertheless, this is still not enough to close the universe, or give us the steady-state density.

I should also mention some indirect arguments used by people who want to argue in favor of the existence of intergalactic gas. We have seen a good bit of evidence that galaxies do eject matter, so there should be some intergalactic matter from that source. But clearly, it would take a rather remarkable picture of the evolution of galaxies to have them now in a stage where they have already ejected about twenty times the mass that they now possess. But certainly galaxies do eject matter, and also they seem to eject radio sources. The radio sources have a characteristic form that several people have already spoken on. You have a galaxy, or in some cases a QSO, and two blobs, that Ken Brecher calls by the incredibly coarse name of "double bubbles" (Fig. 3.6). It is these lobes that contain the relativistic particles and magnetic fields, and it is here that we find the colossal amounts of energy associated with these sources. When we do the so-called minimum energy calculation, we find typically numbers on the order of $10^{60}$ ergs, for minimum energies. The question is how can one get these double features and how can these tremendous energies be confined. Some people have argued that the only two ways of combining the lobes are to assume enough inertia to keep them together during ejection, or to accept some material outside which can act on the plasma, if that is what the lobes are, and keep them combined. Another possibility is that what is ejected consists of coherent objects that have enough mass to hold the system together over time scales of $10^7 - 10^8$ years.

Since most of the radio sources are associated with clusters of galaxies, you may argue, if you believe that the confinement is due to an intergalactic medium, that at least in clusters there must be a considerable amount of gas. Since galaxies do eject gas, and there may be gas left over from the formation of the clusters, I see no particular objection to this view, but the point I want to stress is that there is no reason to think

Figure 3.6

# Theoretical Problems of High Energy Astrophysics

that there might be enough gas in clusters to bring us up to the critical density.

Now let me say something about the radiation in the universe. To begin with, we have the so-called 3 °K black-body radiation, the microwave background. Now this has an energy density of about 1 eV/cm$^3$ if we have a 3 °K black body. This is about $10^{-12}$ ergs, which amounts to $10^{-33}$ gm/cm$^3$, so this is of no significance compared to the critical density. The radio background, which comes from the integrated effect of all the galaxies, and the X-ray background are very small indeed. They are about $10^{-14}$ or $10^{-15}$ ergs for radio and $10^{-16}$ ergs for X-rays. We have not yet said anything about cosmic rays in this meeting. Let me just say that the most energetic assumption you can make is that the rays are universal at our local energy density, and this again gives a very small value.

I might also mention neutrinos and antineutrinos. It cannot be ruled out that there might be a high density of neutrinos and antineutrinos. If we assume a completely degenerate Fermi gas of neutrinos, we have

$$\rho = 1.64 \times 10^{12} E_F^4,$$

where the Fermi energy that would give the critical density is

$$E_F = .01 \text{ eV},$$

so that we are dealing with a sea of very low energy neutrinos and antineutrinos. Now we can only get an estimate of the observational limits from the β-decay arguments. The β-decay will be cut off at an energy below the maximum energy available, and if we look at the results for $T^3$, we find $E_F \leq 60$ eV, so you are a factor of 6000 above what you need, and since the energy density goes as $E_F^4$, this sets limits at a factor of $(6000)^4$ above the critical density. So there is nothing to stop your saying that the universe is absolutely filled with neutrinos, but of course there is no understanding at all of why we should be living in a totally neutrinos-dominated universe, especially since they cannot be primordial. Steven Weinberg showed a few years ago

what the neutrinos density should be for a big-bang universe and
found that they should have a black-body spectrum at about 1°K
rather than the 3° for the microwave radiation.

Then what is left? One can argue that there is a large fraction of the mass of the universe tied up in dark, collapsed objects. These can be evolved galaxies, galaxies that have gone
all the way to the white dwarf or neutron star stage, or they
could be black holes. Black holes into which matter and radiation
may come, are certainly not excluded by any of the considerations
we have discussed here, but of course there is as yet no direct
evidence for any such hypothesis.

Now I want to talk very briefly about the nuclei of galaxies
and about QSO's. In these objects one sees luminosities from
$10^{38}$ to $10^{46}$ ergs/sec, ejection of masses up to $10^8 M_\odot$, ejection
velocities on the order of $10^4$ km/sec, large outbursts of nonthermal radiation, energy reservoirs of the order of $10^{60}$ ergs,
etc. What sort of processes can produce such large outbursts
of energy? This problem has been wrestled with ever since the
strong radio sources were discovered in the 1950's. I should
emphasize that it is not a problem that came only with the QSO's.

We know that this energy comes from very small regions.
From light-transit time in the case of variability, and from
angular size measurements in some cases, we know that these
violent regions are quite small, ranging down to one light
year or in some cases to sizes as small as the solar system
($\simeq 10^{15}$ cm).

There are basically two known sources of great energy:
thermo-nuclear reactions and gravitational energy. Now thermonuclear reactions are known to be normally nonexplosive. In
normal stars the reactions are very slow; at the end of stellar
evolutions, however, this can change. As stars exhaust their
thermonuclear fuel supply and begin burning their carbon,
oxygen, etc., they can reach stages of explosive nucleosynthesis;
this, together with some release of gravitational energy, is
thought to explain supernovae. Therefore an obvious suggestion
to make is that the cores of galaxies are violently active because of multiple supernovae going off within a very short
space of time. A number of people have discussed this idea,
but I think that it does not work well. It is not clear why

many stars should become supernovae at the same time. It is hard to get large-scale coherent ejections outside the object. And there are other problems as well.

In order to make anything on this scale work on a galactic scale, you need very large concentrations of matter. Now conventionally one thinks of galaxies as forming from primordial turbulence during the expansion of the universe. The direction of evolution of the galaxy is a shrinking to higher densities together with ejection of stars. The tail of the velocity distribution tends to escape as the system condenses, and you gradually go to higher densities of stars in the center of the galaxy. Now the time scales for this process are very long, longer than the Hubble time, if you start with normal densities. The densities of stars in the centers of galaxies, as we measure them, are no greater than 100-1000 stars/$pc^3$ which is very low from the point of view of anything violent occurring. The condensation will go on until the stars are close enough to interact and even collide. Then the radiation emitted by one star will have some effect on the surrounding stars, and this situation can lead to chain-reaction supernovae; but you need a very high star density ($10^6$-$10^7$ stars/$pc^3$) for this to happen.

It is also possible to form, from many stellar collisions, a single very massive object that can gravitationally collapse. In analogy to pulsars, rotating neutron stars, we can imagine a rotating massive body, from which you might hope to get large energy outputs.

The alternative point of view, not yet well developed, is that the centers of galaxies and QSO's are the places where matter is coming out in essence from its initial state. This makes the centers of galaxies the places of matter creation, an idea with which we must be careful, since we see matter of apparently normal composition being ejected from these cores. However, the idea may lend itself more naturally to the questions of ejection of coherent objects, and perhaps to the discrepant redshift problem, which presumably requires an idea of some new kind of process, than would the classical view. Remember that the classical scheme would require that the galaxies be in a late stage of evolution to be doing these things, and yet the classical approach to QSO's, etc., would say that these are objects back in time, early in the universe, in early stages of their development. So all these things are very unclear.

I (Burbidge, 1970) have made a recent discussion of all these arguments including the properties of nuclei of galaxies and of QSO's and a fairly extensive discussion of the various theoretical schemes that have been proposed, and an extensive list of references.

I might in conclusion mention one rather amusing idea, which is that if the initial state of matter is massive bare quarks, then there is a large energy release in the production of baryons, and this energy will come out in essentially the form that you want in order to explain high-energy phenomena, because you want very energetic particles. Of course we have no evidence that quarks exist, or that they would occur in such conditions as the initial state of matter, but if you think of it this way, that the quarks appear and then form baryons, then the formation of the elements from the quarks is the really primordial nucleosynthesis.

## REFERENCES

Bahcall, J. N., Robinson, L. B., 1972, Ap. J. Letters, $\underline{171}$, L83.

Brecher, K., Burbidge, G. R., 1970, Comments Astrophys. Space Phys., $\underline{2}$, 75.

Burbidge, G. R., 1970, Ann. Rev. of Astr. and Ap., $\underline{8}$, 369.

Burbidge, G. R., Hoyle, F., 1966, Ap. J., $\underline{144}$, 534.

Burbidge, G. R., Hoyle, R., Sargent, W., 1966, Nature, $\underline{209}$, 751.

Burbidge, G. R., Burbidge, E. M., Solomon, P. M., Strittmatter, P. A., 1971, Ap. J., $\underline{170}$, 233.

Burbidge, G. R., O'Dell, S. L., Strittmatter, P. A., 1972, Ap. J., $\underline{175}$, 601.

Burbidge, G. R., O'Dell, S. L., 1972, Ap. J., $\underline{178}$, 583.

Greenstein, J. L., Schmidt, M., 1964, Ap. J., $\underline{140}$, 1.

Gunn, J. E., 1971, Ap. J. Letters, $\underline{164}$, L113.

Gunn, J. E., Peterson, B. A., 1965, Ap. J., $\underline{142}$, 1633.

Lynds, R., Wills, D., 1972, Ap. J., $\underline{172}$, 531.

Penzias, A. A., Wilson, R. W., 1965, Ap. J., $\underline{142}$, 419.

Schmidt, M., 1972a, Ap. J., $\underline{176}$, 273.

Schmidt, M., 1972b, Ap. J., $\underline{176}$, 289.

Schmidt, M., 1972c, Ap. J., $\underline{176}$, 303.

Terrell, J., 1964, Science, $\underline{145}$, 918.

Wagoner, R. V., Fowler, W. A., Hoyle, F., 1967, Ap. J., $\underline{148}$, 3.

Wampler, E. J., Robinson, L. B., 1972, Ap. J. Letters, $\underline{164}$, L113.

X-Ray Astronomy

Lectures by

R. Giacconi

Harvard College Observatory

Notes by

R. Bland, G. Palumbo

# Lecture 1

In this lecture we are going to give a sketch of what is known about X-ray emission from celestial objects; the subject of X-ray astronomy has grown so big in the past few years that a complete summary requires much more than a few hours.

We will be mainly concerned with a general review of aspects of galactic and extragalactic sources. But to begin with we should ask ourselves: why X-ray astronomy at all?

X-ray astronomy has been of interest mainly to physicists up to the present time, but now the subject has become a branch of astronomy in its own right. This change has been brought about thanks to the tremendous amount of work that has been done with balloons, rockets and satellites and in particular with Uhuru, the first orbiting observatory, an X-ray satellite. Historically 1962 is the year in which the first extra solar source was detected, SCO-X-1, which has an equivalent X-ray flux at the Earth of $10^{-7}$ ergs sec$^{-1}$ cm$^{-2}$ in the energy range 1-10 KeV. In 1972 we can detect sources $\sim 10^{-4}$ times weaker and there are plans to gain another factor $10^{-4}$ in the next 10 years.

With these accomplishments X-ray astronomy can be on the same footing with other more traditional branches of astronomy (optical, radio). The main interest of X-ray astronomy comes from:
1) The point of view that more information can be gained studying a wider range of the electromagnetic spectrum rather than a restricted range in greater detail;
2) The study of explosive phenomena in which X-rays are likely to be produced.

With X-ray astronomy it becomes therefore possible to study explosive events taking place in our own galaxy, particularly for stellar objects at the end of their evolution, and in external galaxies. Let us first get some idea about the energy output of X-ray sources. The best known is the Sun with intrinsic X-ray luminosity in the 1-10 keV energy range $L_x \sim 10^{28}$ ergs sec$^{-1}$ to be compared with its luminosity in the visible region of the spectrum $L_v \sim 10^{33}$ ergs sec$^{-1}$. If the ratio $L_x/L_v$ were the same for all sources there wouldn't be any chance of seeing

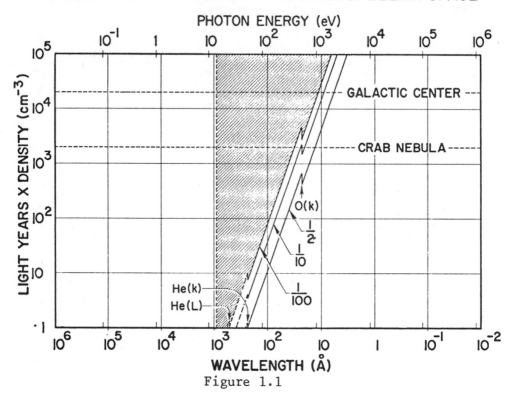

Figure 1.1

any other star; however, sources have been observed with intrinsic luminosities $L_x \sim 10^{35} - 10^{39}$ ergs sec$^{-1}$ to be compared with optical luminosities which are again $L_v \sim 10^{33}$ ergs sec$^{-1}$.

These objects are therefore intrinsically very bright X-ray emitters and the ratio $L_x/L_v$ between 8 and 11 orders of magnitude greater than for the Sun. Such sources do not appear to be so uncommon. There have been observations of X-ray sources in massive binary stars and it is likely that short period binaries ($\sim 40\%$ of all binaries) at some point in their evolution are violent X-ray emitters.

The reason why the study of X-ray astronomy between 1-10 KeV could not have been attempted before space flight is due to atmospheric absorption. Interstellar gas is also responsible for some absorption and its influence can be understood by inspecting Fig. 1.1, which shows the attenuation due to interstellar gas for a given frequency at a given distance.

It might be inferred that more information could be gained by going into the field of γ-ray astronomy, since photons of this energy travel larger distances in the interstellar gas. The results are however disappointing because of the downward-sloping spectra of sources which attain a very low intensity into the γ range.

Most information can therefore be gained at energies close to the absorption cut off. The same cut off can of course give some information about the distance of the emitting object. The energy range, between 1-10 KeV, will form the subject of these lectures; this is precisely the range at which the Uhuru satellite operates. From the $L_x/L_v$ ratio we can deduce that X-ray sources are ∿ 99.9% efficient. Such an efficiency could not be achieved in a laboratory by bremsstrahlung of electrons impinging on a cold material. Therefore some other mechanism must be at work; one possible candidate is thermal bremsstrahlung to which we know most of solar X-radiation is due. We would expect to see continuum radiation temperatures of ∿ $50 \times 10^6$ °K down to $10^6$ °K. At lower temperatures for a thin plasma, lines should be superimposed on the continuum.

Black body radiation is also a possible mechanism but no sources so far have shown its characteristic spectrum. There are two more mechanisms, making use of high energy ($E_e \gtrsim 10^{14}$ eV) electrons impinging on magnetic or radiation fields. In the first case we have synchrotron radiation, in the second inverse Compton effect.

The latter is particularly interesting because electrons producing radio or IR radiation by the synchrotron mechanism might then Compton scatter on the radiation field and produce X-rays. We are thus able to get limits on various parameters of the source. The main features of the radiation mechanisms so far considered are summarized in table 1.1.

The rough type of measurements that are done on X-rays give pulse height analysis of spectra with a resolution of $\Delta E/E \sim 20\%$. The best that can be done is to fit the data to an a priori assumed law (power law or exponential). To complete this very sketchy summary: in 1962 the first extra solar source was detected, while the Sun was known as an X-ray star since 1948. X-rays are generally detected with thin window proportional counters filled with gas; an X-ray that passes through the window is absorbed in the gas and produces a photoelectron, a cascade is started and an electrical pulse is received, at the

Table 1.1

**THERMAL BREMSSTRAHLUNG**

optically thin   $f_\nu \sim \exp(-h\nu/KT) + \text{lines}$

optically thick  $f_\nu \sim \dfrac{\nu^3}{\exp(h\nu/KT) - 1}$

intensity  $\sim h^2 V(T)^{1/2}$

**SYNCHROTRON**

power law electrons  $f_\nu \sim h\nu^{-\alpha}$

intensity  $\sim H^2 E^2$

$h\nu \sim HE^2$

**INVERSE COMPTON**

power law electrons  $f_\nu \sim h\nu^{-\alpha}$

intensity  $\sim nh\nu' E^2$

$h\nu \sim h\nu' E^2$

---

anode whose height is proportional to the energy of the incoming photon.

It takes about 30eV to ionize the gas so for a 1KeV photon about 30 electrons are produced. The statistics here is $N \sim 30$ electrons $\pm (30)^{1/2}$ and this shows where the 20% energy resolution comes from. So one can measure energy with this kind of precision and determine spectra only making use of some assumptions.

In 1963 there was the first identification of an X-ray source with a body previously known and studied: The Crab Nebula. The identification was made in a very elegant experiment by Friedman and his collaborators at NRL. In 1966 came the optical identification of SCO X-1 with a blue faint star which has been much studied ever since. We have some idea of the distance of

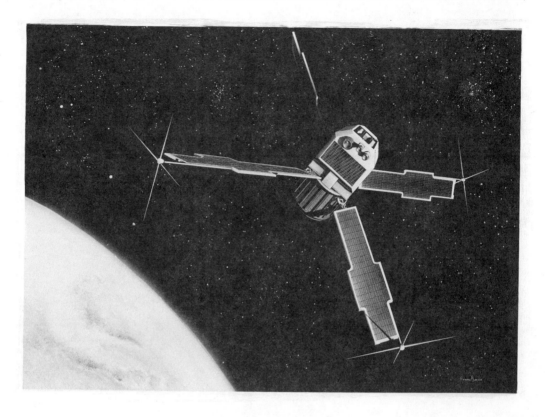

Figure 1.2

such sources, and it can be shown that this star is very different from the supernova remnant type of source like the Crab, and therefore it is deduced that some other production process must be at work. In 1968 the first extra-galactic source was detected and in March 1970 Uhuru was launched, which marks a different approach to X-ray astronomy. Most of what will be said here has to do with data from Uhuru and therefore it may be justified to point out very briefly what this observatory looks like. It was called Uhuru because it was launched from Kenya on the anniversary of Kenya independence (Uhuru means freedom in Swahili). Fig. 1.2 shows the satellite, which is about 6 feet tall and 10 feet in diameter. Data are tape recorded and stored and transmitted to the ground after each orbit. The orbit is equatorial, the satellite is below the Van Allen radiation belts, so its counters have very little background.

The primary elements of Uhuru are sketched in Fig. 1.3.
Let's first talk about the proportional counters detecting

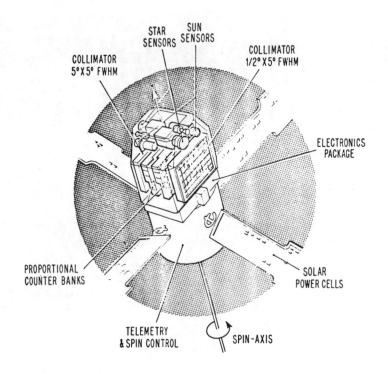

Figure 1.3

the incoming photons. It must be pointed out that the strongest source known (SCO X-1) gives $\sim$ 50 photons $cm^{-2}$ $sec^{-1}$ at the Earth. The detectors need to be large and in Uhuru they cover an area of $\sim 1000$ $cm^2$ and have very thin window. The fields of view are defined by slit collimators in two directions. When the detector passes by a source a characteristic triangular response is observed. Star sensors are used to give the position of the X-ray source with respect to visible light stars. The satellite is slowly rotating about its axis, and in order to maintain stability it has a rotor which contains all of the angular momentum. It rotates about 1 rotation/10 minutes, but it can be controlled from the ground and made to rotate slower than that, e.g., 1 rotation/100 minutes or perhaps stop altogether. The spin axis direction can be arranged to point anywhere in the sky; coils are provided which torque the satellite with respect to the Earth's magnetic field.

The two fields of view are different: $x_1$ = 5° x 5° FWHM and

X-Ray Astronomy                                                                 209

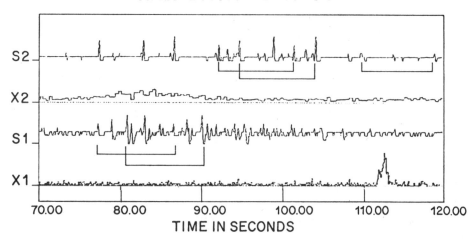

Figure 1.4

$x_2 = \frac{1}{2}° \times 5°$ FWHM. The latter is to separate sources in the sky, and the former to give longer observing time on isolated sources. There are 4 output channels. ($S_1$ and $S_2$ are star sensors). The satellite can respond quite fast (in a few hours) to commands to change its axis orientation. The aspect of the data from the 4 channels are shown in Fig. 1.4. Residual background is $\sim 10^{-2}$ counts $cm^{-2}$ $sec^{-1}$. One detector can be used to study X-ray background and another to detect residual Cosmic Rays which are not anti-coincided out. Azimuth and elevation of any observed object can be computed from the signals of the star sensors. The visible light from a star passes through an N shaped slit. By measuring the time between successive signals both azimuth and elevation can be computed. One can therefore superimpose the data from many passages in a particularly interesting region of the sky and find results statistically significant. A map of the sky can be constructed in many passes during many days. Fig. 1.5 shows a map of $\sim 70$ days of observation. The traces are plotted in galactic coordinates.

Before Uhuru was flown some 50-60 sources were known and only

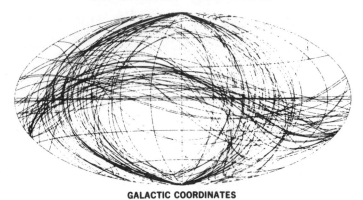

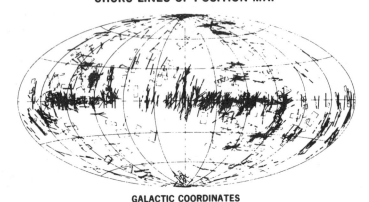

Figure 1.5

few of them were identified. There are now ~ 130 sources known. Fig. 1.5 shows in galactic coordinates where they have been found in the sky; the empty part of the map is a part of the sky which has not yet been observed. The concentration of sources above the galactic equator simply shows that many of them are of galactic origin. The interesting fact is that some 50 sources have been detected at galactic latitude $|b| > 20°$ which means that probably most of them are of extra-galactic origin, though some may be very close nearby stars. To confirm this point of view we may add that those among the 50 that have been identified correspond to extra-galactic objects.

Let us first consider the sources with $|b| < 20°$. The position of very strong sources has been determined within ~ 1' arc,

# X-Ray Astronomy

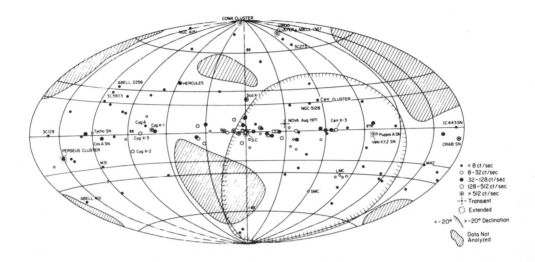

Figure 1.6

and it becomes as bad as ½ x ½ degree for the very extended sources. Spectra are obtained, in the energy range 2-20 keV. The shortest integration time constant is $\Delta\tau \sim 0.1$ sec. (Unfortunately when Uhuru was planned and designed pulsars were not yet known and shorter integration time constants were not considered).

Galactic sources are estimated to emit $\sim 10^{35}$ to $10^{38}$ ergs, but this is very hard to determine precisely since it is not known how far away the sources are. For 4 objects in the Large Magellanic Cloud, though, the distance is known and it turns out that they are very luminous at about $10^{38}$ ergs, much more than anticipated. This means that when one studies the distribution of galactic sources one can take the view that they are as far as the galactic center (or even twice that distance and put them in the spiral arm opposite to ours) without obtaining impossibly high luminosity.

Fig. 1.7 shows a plot of the number N(S) of sources whose intensity is greater than S versus S (S is in counts $sec^{-1}$). The plot is divided in galactic latitude $|b| < 20°$ and $|b| > 20°$.

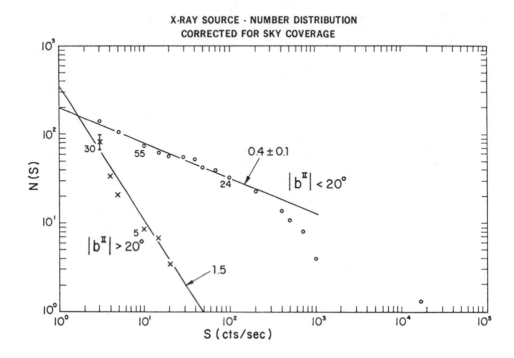

Figure 1.7

For $|b| > 20°$ the distribution is as one would expect for a population of celestial objects with the same intrinsic luminosity uniformly distributed in space.

For $|b| < 20°$ we believe that most of the sources are stars in our galaxy. For constant-luminosity sources distributed uniformly in the disk of our galaxy one would expect a slope of 1. The fact that we find a different slope, with structure in the slope, can be interpreted in two different ways. One is that we are seeing sources all the way to the edge of the galactic disk, and so that in fact we are just seeing the intensity distribution of the sources. The other is that the particular slope found may also be due to a different class of objects, one with high intrinsic luminosity, placed mostly at the center of our galaxy, the other of lower luminosity distributed in the spiral arms.

None of the data collected up to now exclude the possibility of a very large number of objects in our own galaxy with very

## SPACE DISTRIBUTION    X-RAY SOURCES
## 21cm (WEAVER)

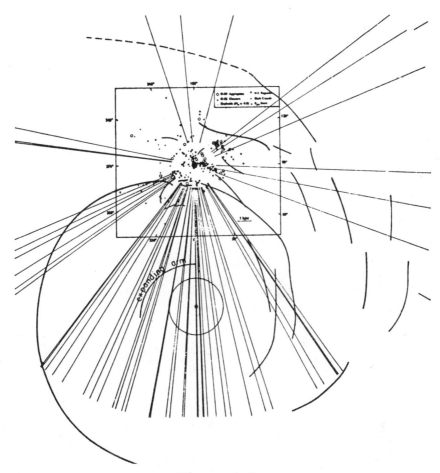

Figure 1.8

faint intrinsic luminosity ($10^{32} - 10^{34}$ ergs sec$^{-1}$). The only limit one can place comes from observing an excess of radiation from unrevealed sources in the direction of the galactic equator. Such an excess has not been observed and this allows us to put a limit of $10^{37}$ ergs sec$^{-1}$ as a total luminosity output of our galaxy for this kind of sources. These sources therefore play a small role in the total luminosity of our galaxy as it would be seen from outside. We can try to find out more about the distribution of these objects. Some attempts have been made to relate the direction in which we see them with the spiral arm

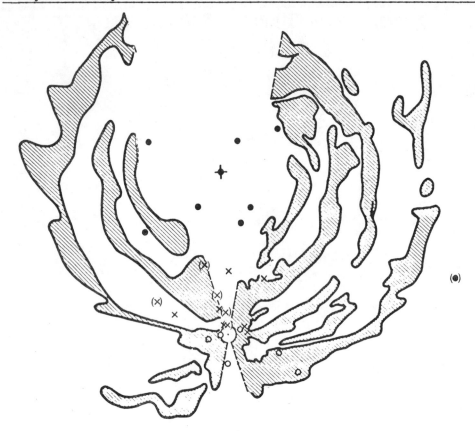

Figure 1.9

structure of the galaxy as observed at 21 cm. Fig. 1.8 shows that there is a concentration towards the center of the galaxy. Some sources seem to follow the spiral structure but are by no means fully in agreement with it. It appears that there might be two kinds of sources, those connected with the spiral arms and those connected with the galactic center. There has been an attempt by Seward (1972) to associate not only directions (which are well known) but distances (which are more uncertain). This was done by observing the emission of all sources at longer wavelengths, and trying to use a cut off in the spectrum at the energies observed and turning it to a scale of distances. Fig. 1.9 by Seward and co-workers shows the results of such an attempt, which is quite daring since there are large clouds of gas present about the sources. The sources at the center have an energy output of $\sim 10^{38}$ ergs $sec^{-1}$ and alone could explain the X-ray emission from our galaxy when seen from the outside. This is very much like in the Large Magellanic Cloud, where the integrated contribution X from weak sources is < 10% of the total.

Let us summarize the main features of galactic objects before moving to extra-galactic ones.

X-ray stars, which are the most powerful objects:
SCO X-1 has been optically identified and in addition has an unusual radio source associated with it. It is a triplet source, and its physical meaning has not been fully understood. Comparing the radiation emitted in X-rays, optically and in the radio band of the spectrum has also proven not very illuminating. In all SCO X-1 type of objects (there is a whole class of them) an absorption has been observed in IR and this leads one to believe that one is looking at sources which are quite dense. The total extent of the source might be $\sim 10^8$ cm, with a density of $\sim 10^{16}$ atoms $cm^{-3}$; then immediately one thinks of an object smaller than a white dwarf, possibly a neutron star. One could expect to see emission lines, if the emission is due to thermal bremsstrahlung, but no lines have been observed and therefore a limit for line width can be placed at $\sim$1eV. A reasonable model for SCO X-1, and perhaps the most accepted view, is that it is part of a binary system. However, there is very little evidence for it. One can nonetheless imagine that SCO X-1 is in a binary system and therefore argue that the energy released is simply due to the infalling material from a companion star onto the X-ray star, which must be a small object. Another possible model is that there is a rotating neutron star or white dwarf, which is surrounded by gas. The rotational energy is spent in heating the gas; we would not be able to see the rotating object, but we would just see the thermal radiation from the heated gas.
Supernova Remnants: we know that there is a neutron star in the Crab Nebula, the very well known pulsar NP0532, from which pulsations have been observed at optical, radio and X-rays frequencies. These pulsations are correlated among them and it is understood that the energy for the pulsating emission and for the extended emission is entirely provided by the rotational energy of the neutron star. It may be that all other supernova remnants have the same energy source; there seems though to be indications that this is not the case. The energy source then could be the expanding shell itself giving up its kinetic energy.
X-Ray Novae are observed to abruptly change their emission by a

X-Ray Astronomy                                                              216

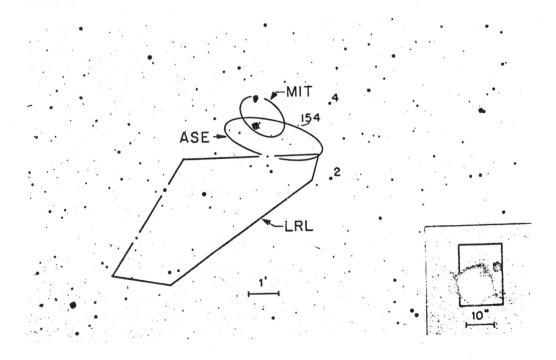

Figure 1.10

factor between 10 and 100.  This occurs in a few days, then the emission decays in a period of months.

Some of the brightest sources known (as bright as SCO, like CEN X R-2 and CEN X R-4 discovered in 1967) have been seen by Uhuru.  None of them had been identified with an optical object until recently.  There is indication that there is a decrease in temperature as time goes on, but this process is not yet very well understood and even less the abrupt changes and increases of emission.  Recently however the identification with an optical nova in Norma has occurred and from this we should be able to have some information about the distance of the object and therefore find out which of the possible models of production is at work.

Binary X-Ray Sources:  Cyg X-1 is one of the most interesting objects discovered by Uhuru.  There is now evidence that stars are pulsating, some regularly, some irregularly.  The time of the pulsations seems to be connected with binary system and gives us a great deal of information about such systems.  Cyg X-1 is one of the most interesting sources from this point of view. It was discovered in 1966 with a large    error   box connected to it.  In Fig. 1.10 is shown how the situation has improved since

# X-Ray Astronomy

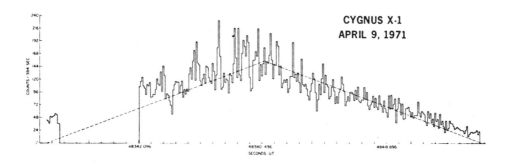

Figure 1.11

thanks to new and more precise measurements. The position for the X-ray source seems to indicate that it lies on the star (or stars) shown in the expanded view (10" of arc) in the right hand side of Fig. 1.10. The most recent position seems to coincide with the type B0 star. This star has been identified by Bolton (1972), Webster and Murdin (1972) as a visible star with a binary invisible companion. Spectroscopic observations of the velocity dispersion of the lines emitted allow one to determine the reduced mass and some of the parameters of the system. Assuming that the visible star has a mass $M_v \sim 12 M_\odot$ (which seems to be an underestimate), the unseen companion, the X-emitting object, should have a mass $M_x \gtrsim 3 M_\odot$. Since this object is pulsating on a time scale down to 5 msec, it has been assumed one was discussing a compact object (radius $< 10^8$ cm). Therefore $M_x \gtrsim 3 M_\odot$ gives rise to the problem that either this is a neutron star with mass larger than expected, or it is a new object (Black Hole), which is to be expected at the end of stellar evolution. There has been a great deal of controversy on this subject. The data presented in Fig. 1.11 show the pulsating nature of the object. Statistical errors are very small; the triangular shape of the data is simply due to the detector and collimator response. The data have been represented on different time scales, 1/10 sec, 3/10 sec, 1 sec, many seconds. Variations occur at all time

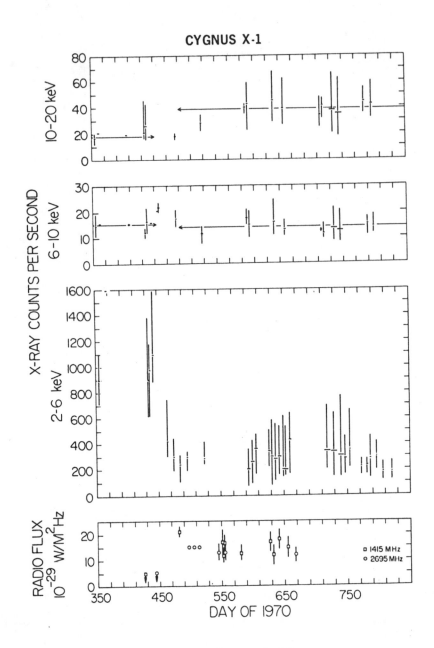

Figure 1.12

scales. Fourier analysis of the data shows characteristic peaks at each passing in the power spectrum. The peaks do not occur at the same place, but wander about; this was first interpreted as an effect caused by mismatch between telemetry and the real time constant oscillator. Now it is known that the pulsation is not truly periodic and there are two general approaches to explain what happens. One is by Oda (1971), who says that there is a train of pulses which is emitted, followed by silence. Terrell (1972) argues that the observed pulses are just random noise pulses. Anyway there are pulsations present and this implies a very small size source. If the source is indeed connected with the binary system one can ask oneself if it is possible to see any connection with 4, 5, 6 days cycle. Here Uhuru is extremely useful since it can observe the source for months. Fig. 1.12 shows variations during days: all kinds of variations are present. If one tries to analyze the data folding the intensity modules of 3 days, 5, 6 and 6.2 days (which is the period of the observed binary system), there is no further significant effect. The dispersion seems to be consistent with the fact that the source is variable, but does not support the identification with a binary. This is in apparent conflict with conclusions at high energies. One can then ask whether the behaviour is different at different energies. Such a difference should give information about different emitting regions of pulsation. The data show there must be a coherent region and the answer must be looked for somewhere else. There is, however, a piece of evidence connecting radio emission and X-ray emission, and consequently, through radio and optical, connecting X-ray and optical emission.

This can be seen studying the average emission over one year. An abrupt ($\sim$ 1 month) decrease of the emission is observed at low energies (4-6 KeV), while an increase of the emission is observed at high energies (10-20 KeV). This abrupt change in spectral shape occurs in coincidence with an increase in radio emission. Thus, if this link can be confirmed, we have to conclude that X-ray emission comes from the unseen companion in the 5-6 days binary, and Cyg X-1 is quite probably a very unusual object (Black Hole?).

# X-Ray Astronomy

## Lecture 2

In the first lecture I described the X-ray sky in general terms; many galactic sources and, distinct from these, a large number of novae, supernovae, and X-ray stars such as SCO X-1, CYG X-2, and others. Then we considered in detail one irregularly pulsating variable, CYG X-1. It was this source which originally suggested the very short-time variability of X-ray sources. This turned out to be so general that now only the Crab Nebula is thought to be constant enough to be used for calibration (and, of course, one averages over its 33 ms pulsations). Most of the others vary on one time scale or another; novae, for instance, with rise times of weeks and decay times of months; SCO X-1, with irregular pulsations, as short as 20 sec; CYG X-1 changing by as much as a factor of 2 over as little as 5 ms. CYG X-1 thus suggested a slow-spin mode of operation of the satellite to search for periodic pulsations, and some such sources were found.

Let us now discuss CEN X-3, the first regular pulsating source which we saw, near the galactic equator, and thought from the beginning to be a pulsating star. Data from one slow pass (∿ 100sec) are shown in Fig. 2.1. The pulsations are quite regular, with period about 4.8 sec, in contrast to the irregular CYG X-1 pulsations that we have already seen. Next look at the intensity, averaged over the pulsation cycle for one entire day of observation, shown in Fig. 2.2. There is first a low-level, but positive, signal, with no detectable pulsation; we can put a limit of 20% on the pulsating component. Then the signal rises, and

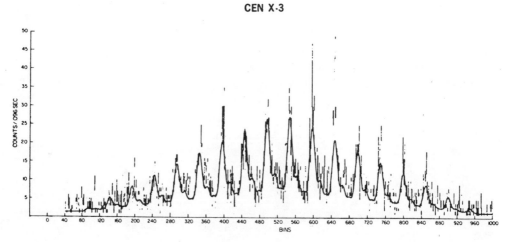

Figure 2.1

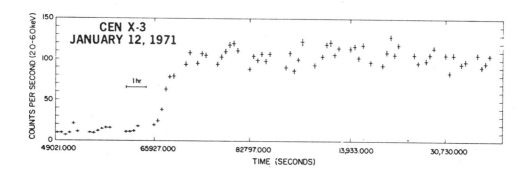

Figure 2.2

note that the rise is not instantaneous, to a much higher level, where the pulsation is observed. We know now that CEN X-3 is a binary system, and that the rise corresponds to the source coming out of eclipse.

To study the period of the pulsations over a longer period of time, we changed to a fast-spin mode of operation of the satellite, corresponding to seeing fewer pulsations at any one sighting, but having a shorter interval between sightings. These data were interpreted by calculating the time difference $\Delta t_n = t_n - n\tau^*$, where $t_n$ is the time of the $n^{th}$ pulsation, and $\tau^*$ is a trial value

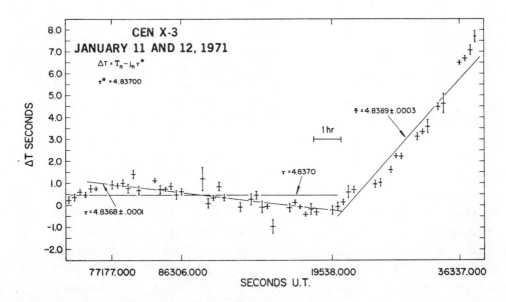

Figure 2.3

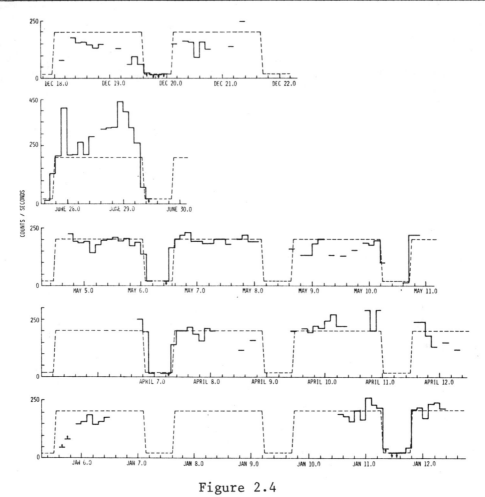

Figure 2.4

of the period. This period must already be well-enough known to correctly assign n for pulses in different sightings. When $\Delta t_n$ is plotted against $t_n$, a constant period gives a straight line (a horizontal line if $\tau^*$ is exactly correct), and changes in period show up as breaks in the slope. Data are shown in Fig. 2.3 for the same interval of time as for Fig. 2.2. Our first interpretation was that this corresponded to two slopes, and the length of the first interval of constant period, $\sim 20,000$ sec, gave us confidence in interpreting the phenomenon as truly periodic. (In fact, however, a sine curve can be fitted through the data.) The first interpretation was that this was either a pulsating or a rotating star. It was surprising, however, that the period of a rotating body could change so much without enormous releases of energy, and that the change in period did not correspond in time with the change in intensity.

# X-Ray Astronomy

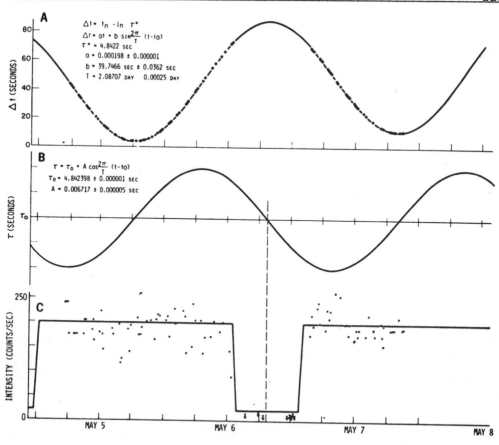

Figure 2.5

The large value of the change in intensity also bothered us.

We subsequently took data over a much longer period, about a year. The intensity as a function of time is shown in Fig. 2.4. The curve corresponds to the hypothesis of regular eclipse in a binary system, and agrees with the data in the following sense: the source was never up when it should have been occulted, though sometimes it was not up when it should have been. In fact the width of the eclipse is not always the same, varying by as much as one hour in 24, and suggesting a change in the occulting disk. The period of occultation is about 2.08 days.

We then returned to a study of the period as a function of time. $\Delta t$ vs $t$ is shown in Fig. 2.5 over several periods of occultation. We now see a clear sinusoidal variation of the period which persisted over all the time when we were observing. (Looking back on our first observations, over less than a day, we were lucky to have taken them at a time when the period was nearly constant over a fairly long time interval, which encouraged us to continue studying CEN X-3, rather than discarding it as another wild variable like CYG X-1). Using the data for the full time of observation, the period of the occultation cycle can be determined very accurately from the fitted curve for $\Delta t$; we find $\tau_{occult} = 2.08707$ days, equal within errors to the period derived from the intensity variation. Taking this to indicate that these two phenomena are related, it does not seem too extravagant to assume that the variation in period is due to the Doppler effect, as the X-ray source approaches us or recedes from us. This interpretation is confirmed by the fact that the zeroes of the sinusoidal period variation correspond with the maxima and minima of intensity. From the maximum and minimum values of $\Delta t$ we then get directly the projected radius of the orbit in light-seconds. The variations of the data from a sinusoidal form give information about the eccentricity; we find $\varepsilon \leq 5\%$. Then, using the standard theory of binary systems, we calculate its parameters. All of these results are summarized in table 2.1. It is amusing to note that the value of $v \sin i$

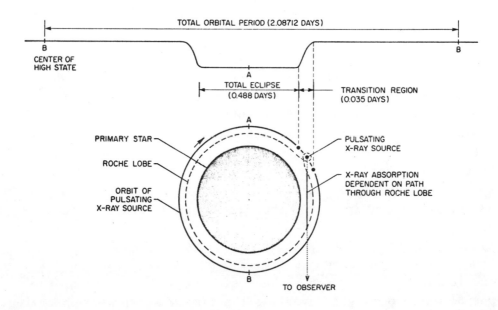

Figure 2.6

## Table 2.1
## PARAMETERS OF THE BINARY SYSTEMS CENTAURUS X-3 AND HERCULES X-1

| | CEN X-3 | Hercules X-1 |
|---|---|---|
| $\tau_{pulsation}$ (sec) | 4.84 | 1.24 |
| $\tau_{orbital}$ (days) | 2.08707 | $1.7002 \pm 0.0006$ |
| eccentricity $\epsilon$ | $< 0.05$ | $< 0.1$ |
| $v \sin i$ (a) (km/sec) | $415 \pm 0.4$ | $169.2 \pm 0.4$ |
| $r \sin i$ (cm) | $(1.191 \pm 0.001) \times 10^{12}$ | $3.95 \times 10^{11}$ |
| reduced mass $\equiv \dfrac{m_2^3 \sin^3 i}{(m_1 + m_2)^2}$ (b) (gm) | $(3.074 \pm 0.008) \times 10^{34}$ ($\simeq 15\, M_\odot$) | $(1.69 \pm 0.01) \times 10^{33}$ ($\simeq 0.85\, M_\odot$) |

(a) $i$ is the angle between the axis of rotation and our line of sight.
(b) $m_1$ and $m_2$ are the masses of the emitting and occulting bodies respectively.

---

is as precise as any astronomical measurement at any frequency of a radial velocity.

The binary system envisaged is shown schematically in Fig. 2.6. We assume that the X-ray source, being a white dwarf or something even smaller, is the smaller member of the system. From the ratio of the occulted and "visible" intervals we determine the radius of the occulting region (the star and its atmosphere out to a radius where the density is small enough to transmit X-rays). We find this to be about 0.8 of the separation. From the slope of the light curve just before or after occultation we can determine the scale height of the atmosphere of the massive object. We find roughly $10^{10}$ cm, while the commonly accepted value is about $10^9$ cm, a discrepancy that we do not consider significant

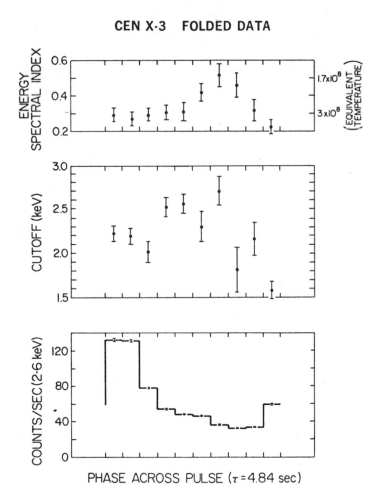

Figure 2.7

so far. The variability of the time of eclipse is evidently also telling us something about instabilities in the upper atmosphere of this star. The important point is that we are observing properties of the atmosphere of the star, and that in the future we may be able to get a lot of information by observing the absorption edges corresponding to the high -Z elements. Note that while this model explains most of the observed properties of this X-ray source, it still does not give us the energy source or the clock mechanism.

The first candidate for the emitter, suggested by Tucker and others (1972), was a white dwarf, for which the pulsation period of 4.8 seconds would be a reasonable vibration period.

X-Ray Astronomy                                                                    227

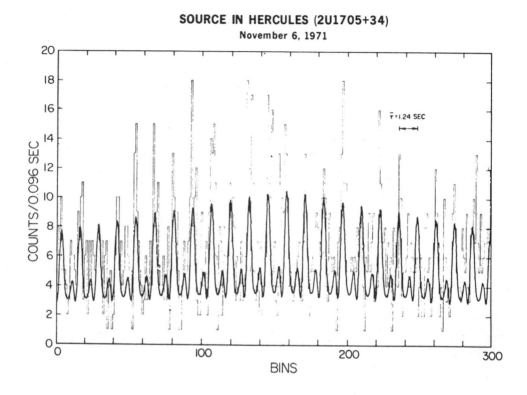

Figure 2.8

As a consequence of this model they predicted the temperature profile of the pulse. Since all detailed models will make such predictions, let us look at the detailed properties of the pulse. (Fig. 2.7) Note the long duty cycle, and the smoothness of the curve. The smoothness is real, not due to the resolution of the instrument, and the pulse cannot be represented just as two sharp peaks. The temperature (as deduced from the spectrum assuming thermal bremsstrahlung emission) rises rapidly at the beginning of the pulse, then falls off during the pulse.

There have been a number of papers in Nature on the optical identification of CEN X-3, but no positive identification has been made. This identification is of course very important, since we could then measure the masses very accurately as with double spectroscopic binaries.

Later another similar object was found, Hercules X-1. The pulsations of this object are shown in Fig. 2.8. The interpulse seen here is a real effect, with a varying intensity, sometimes as large as the main pulse. There is a fairly large duty cycle, and most of the energy ($\geq$ 90%) is pulsed. From the beginning we interpreted this as the same phenomenon as CEN X-3, and the same analysis was applied, giving the parameters in Table 2.1.

Now, what is the occulting object? We had originally assumed it to be the relatively low-density atmosphere of a star. It was pointed out by Wilson (1972) and by many others that to explain both the sharp rise in intensity and the duration of occultation that we see, one must suppose that all of the atmosphere of the star is contained within its Roche lobe. (The Roche lobe is the zero-velocity surface of the binary system, as seen in Fig. 2.6.) The size of the Roche lobe is known from the ratio of the occulted to visible intervals on the light curve, and from this Wilson calculates an upper limit on the ratio of the masses. He finds a very small upper limit, $\simeq$ 1/100 for CEN X-3, and Van den Heuvel (1972) finds $\simeq$ 1/16. This gives a very small upper limit on the mass of the X-ray emitting body, $\simeq$ 0.1 - 0.25 $M_\odot$.

The mechanism of mass transfer as a source of energy in closed

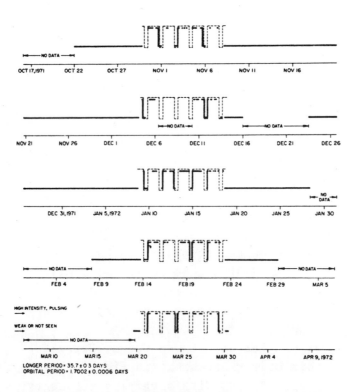

Figure 2.9

X-Ray Astronomy                                                         229

## CEN X-3

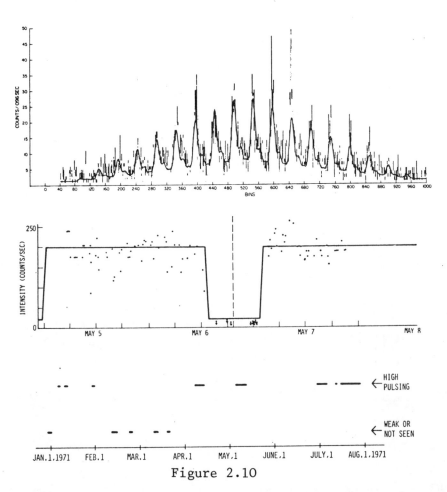

Figure 2.10

binary systems can now be invoked.  The mechanism involves transfer from the (filled) Roche lobe of the larger star onto the smaller star.  The energy for X-ray production comes from the gravitational energy of the mass falling onto the smaller star. The clock mechanism would be given either by a magnetic rotator or by some other peculiarity of the (rotating) X-ray emitting region.

Some other features of the X-ray source complicate the picture. Let us look at the light curve of Hercules X-1.  (Fig. 2.9)  It is not always "on" when it should be.  There seems to be a second, longer period of about 35.7 ± 0.3 days: "on" for 7 days, "off" for 28.  One suggested explanation is a precession of the emitter. Look, however, at the light curve for CEN X-3, shown in Fig. 2.10 on different time scales.  We see the   5-second pulsation, the

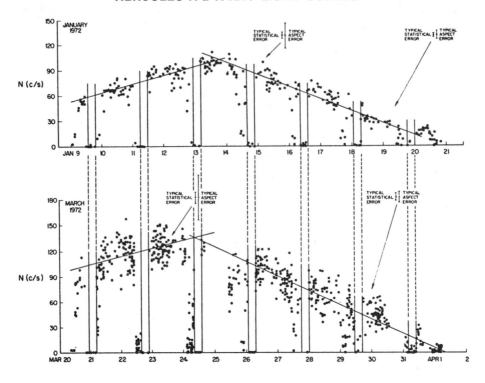

Figure 2.11

2.087-day period, and finally an effect on a longer time scale, like that for Hercules X-1, but so erratic as to be difficult to explain by a precession, unless the period is longer than 200 days, with much substructure in one cycle of precession. We will in the future put much more effort into studying the spectrum, and in particular into the question of whether it is valid to average over many pulses.

A new fact has come to light recently and is still preliminary. Take the 8-day cycle in the Hercules X-1 system. One sees a smooth systematic variation of intensity during the 8-day cycle. This is just what one would expect from the precession mechanism. It now seems increasingly difficult to interpret Hercules X-1 and CEN X-3 as the same phenomenon. After all, the reduced mass is smaller for Hercules X-1, and being at high galactic latitude it may be much closer and intrinsically weaker than CEN X-3. There is an additional effect in Hercules X-1 (see Fig. 2.11). We see an additional dip in the light curve which marches across the light curve, being slightly out of phase with the main pulsation. The spectrum at the dip is more severely cut off. We have no explanation, but could imagine a third body in the system, or a disk around the massive companion, or an accretion disk around the X-ray source itself. It would be nice if one could show that this has to do with the 36-day period. This could also be due to an instability of the atmosphere of the occulting star near the zero-velocity surface. Calculations of such instabilities have not been carried out, but the changes in spectrum seem to suggest that this is an absorption effect and corresponds to some matter interposing itself between us and the source.

Another source that we have seen is shown in Fig. 2.12. We can't yet see the pulsations, but it shows the occultation indicating a binary system of 3.6-day period. It also shows the on-off aspect of Hercules X-1. Since this source is in the SMC we can estimate the intrinsic $L_x$ which is $\sim 10^{38}$ erg sec$^{-1}$

Now, why do these particular binary systems emit X-rays? One is now beginning to understand that evolution is affected by the fact that stars belong to binary systems. Evolution occurs much more rapidly in more massive stars. The evolution of both members of a binary pair may thus be altered by the transfer of matter from one Roche lobe to the other, at the point in the star's evolution when it expands to fill its Roche lobe. It is believed that in the evolution of a star it eventually expands, and that whether or not this atmosphere leaves the Roche

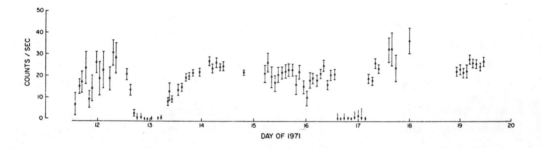

Figure 2.12

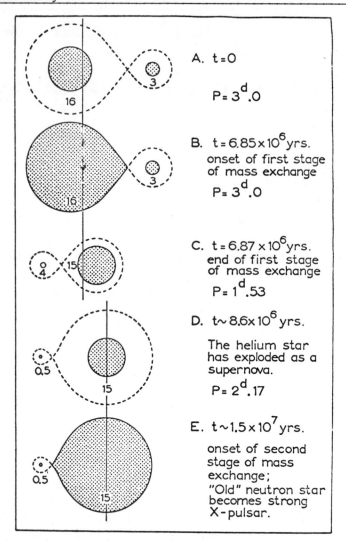

Figure 2.13

lobe has to do with the period. It may reach the Roche lobe before helium ignition.

A possible history of a system such as we are studying, due to Van den Heuvel (1972), is illustrated in Fig. 2.13. At the start the heavier star evolves faster, and at some point fills its Roche lobe and spills over into the other star, leaving finally (by design) 4 $M_\odot$ in the originally heavier star. This 4 $M_\odot$ star then evolves rapidly to a supernova explosion. It can be shown that the system remains bound after the explosion, and the authors assume that about 0.5 $M_\odot$ remains. This becomes a neutron star. Then, as the other star goes through its evolution, it fills its Roche lobe and transfers matter to the neutron star, finally giving the X-ray emitter as we see it.

Now, if, assuming this mechanism, one predicts the number of X-ray stars that should be seen, one gets a number too large by a factor of 25 to 100. Even if the X-ray emitting mechanism is active only part of the star's life, the problem remains.

From sources in the Small Magellanic Clouds we estimate the intensity at $\sim 10^{38}$ to $10^{39}$ ergs/sec. This is just at the limit due to radiation pressure: at a certain level of energy emission the radiation pressure stops the mass transfer. The fact that many binary-star X-ray sources are close to this limit is telling us something about the mechanism.

Let us ask next if all X-ray sources can be binary stars. The Crab pulsar is not, but it might be true for a very large number of X-ray stars, including the ones near the galactic bulge, emitting at $10^{38}$ ergs/sec, Cygnus X-1, SCO, etc.

## Lecture 3

This last lecture is a series of short topics in extra-galactic X-ray astronomy. We seem to find that all galaxies emit X-rays at some level or another, going from single galaxies from about $10^{39}$ ergs/sec to about $10^{45}$-$10^{46}$ ergs/sec in intrinsic luminosity. This means that in some of the galaxies the X-ray emission exceeds the total of optical, radio and IR emission. Now in this very large range of intrinsic luminosity there is a great variety of phenomena taking place. We will start with normal galaxies, similar to our own, and will then turn to individual galaxies which emit at a much higher level. I will show from observations of the Local Cluster of galaxies that in galaxies such as our own the total emission which an outside observer would see is simply the total of the emission from the various stars that I have been discussing. I will then show that in some other galaxies the X-ray emission is much greater than what could be explained by the integral contribution of stars and that it must be due to processes which seem in most cases to take place in the nuclei of galaxies. Then I will describe a different kind of emission which takes place in clusters, and which may have to do with the gas or radiation field within the cluster. In conclusion I will try to show that the collective emission from very distant extra-galactic bodies might in fact explain the diffuse background of X-rays which we have observed. I will then make a few remarks about future directions for X-ray astronomy.

To start with our own galaxy, let us look again at Fig. 1.6, the sky map of discrete sources. At low galactic latitudes are the galactic stars, over whose contribution we will integrate to find out what our galaxy emits as a whole. At high galactic latitudes are the extra-galactic sources, on which I will spend most of my time. A number of them, indicated on the map, are identified with previously known objects, and many more are unidentified. Now, the total contribution of our galaxy consists of a number of components, either observed or for which we can set upper limits. First, there is the emission from the objects believed to cluster around the center of our galaxy, emitting at about $10^{38}$ ergs/sec each. About 10 of these objects can explain the whole emission from our galaxy. (Similarly in the

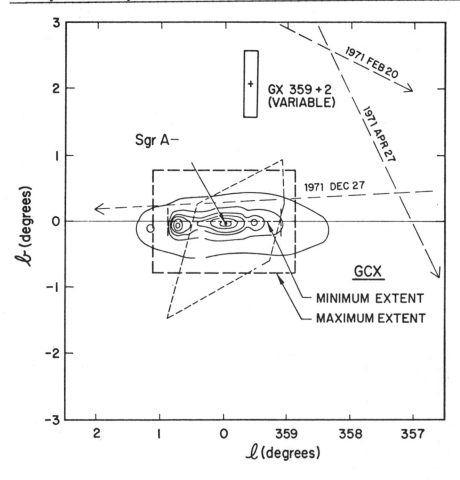

Figure 3.1

Large Magellanic Cloud the total integrated emission comes mainly from a few bright sources.) Then, a tentative observation shows an extended source in our own galactic nucleus, as seen in Fig. 3.1. The observation is very rough, consisting of two scans through this region. The source could be as large as the blob indicated, or as small as a line of the same length as the blob, along the galactic equator. This result could also be due to four or five unresolved sources. If, though, one assumes an extended source, the correspondence of this source with the IR contours given by Hoffmann (1971), shown in Fig. 3.1, suggests that we are seeing the inverse Compton effect from the electrons producing the radio waves in our own galaxy, against the IR radiation. This effect could not take place in a diffuse region, simply because the radiation field would be too diffuse to give a sufficient X-ray flux. We would then have to fall back on a large number of discrete sources, where the radiation field is intense: discrete radio sources, discrete IR sources, discrete electron sources. In any case this whole complex, if we assume that it is at the galactic center, is radiating at about $10^{37}$-$10^{38}$ ergs/sec, and therefore contributes much less to the total than the very strong sources clustering about the center. Another contribution could be given by sources, of which there must be millions, of X-ray luminosity between $10^{28}$ (as for our sun) and $10^{36}$ ergs/sec. Clearly the coronas of all stars emit X-rays. Large amounts of absorbtion would be present. To see this one would have to look for soft X-ray emission, and I won't go into this at all. We are not in a position to see such

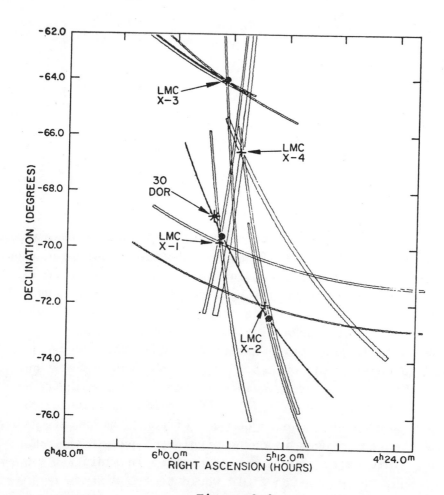

Figure 3.2

sources in the energy interval 2-20 KeV, and we can only say that, if they exist in very large numbers, then, as we cross the galactic equator, we would see a large bulge. This bulge is not observed, and the integral contribution of the entire galaxy from this kind of sources can be evaluated at $\sim 10^{37}$ ergs/sec. This could be due to $10^9$ stars like the Sun, or to between 10 and $10^9$ with intermediate luminosities. So, the conclusion is that most of the X-ray flux from our galaxy is due to a few very bright sources clustered near the center.

This is confirmed by our observations of the Large Magellanic Clouds (Fig. 3.2), where emission occurs in discrete sources. They are unresolved by our measurements, limiting their angular size to about 15' of arc, and we know that they are stellar-sized objects because one of them is variable in a time of the order of days. The total contribution of the rest of the cluster does not exceed 10% of the contribution of these discrete sources. So, the picture is similar to that for our own galaxy. The total emission of the galaxy, $10^{39}$ ergs/sec, is also confirmed by the observations of M31. Here we cannot resolve the individual sources, due to the weak signal, but the total flux is again consistent with this value. The ratio of X-ray emission to visible light for these ordinary galaxies is thus about $10^{-4}$.

Let's pass now to some abnormal galaxies, like NGC 4151, a Seyfert galaxy. We find that its emission is about $10^{42}$ ergs/sec, a larger part of the total emission of the galaxy. Its spectrum is cut off at low energies, indicating that the X-rays that reach us have traversed a considerable amount of matter. There remains the very interesting possibility of determining whether or not the X-rays are emitted in a small nuclear region, 1 to 10 kpc in diameter, where much of the activity is taking place.

The quasar 3C 273 also shows X-ray emission, with a low-energy cutoff. We cannot seem to interpret this as absorbtion in intergalactic matter. If this is occuring in the object, it measures the total amount of matter in the line of sight. Since no absorption is seen in the radio, this sets a lower limit on the size of the cloud that must surround 3C 273. This cloud is very large, consistent however with its usual cosmological distance.

## Cen A REGION

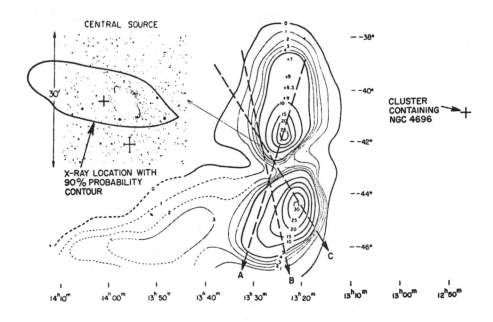

Figure 3.3

The X-ray luminosity is about $10^{45}$-$10^{46}$ ergs/sec, more even than NGC 4151 (accepting the cosmological distance). Some other individual sources, such as Cygnus A, Hercules A, and Centaurus A, have luminosities between $10^{42}$ and $10^{46}$ ergs/sec.

For Centaurus A we have somewhat more detailed data, as seen in Fig. 3.3. Centaurus A is one of the brightest radio sources in the sky. Two large radio lobes are observed. Bowyer, et al. (1972) originally thought that they had observed X-ray emission from these lobes, presumably due to the inverse Compton scattering on the black-body radiation by the electrons producing the radio emission. Uhuru observes a source at CEN A, and another nearby, possibly associated with the Centaurus Cluster, but they have nothing to do with the lobes. Our upper limit on emission from the lobes is 1/10 to 1/20 that from the center; the emission from Centaurus A is entirely from the central region. In the blow-up of the central region one sees that we do not have sufficient resolution to distinguish between the central galaxy obscured by dust in Fig. 3.3, and the two small radio sources nearby. However,

## Table 3.1

### PREDICTED X-RAY FLUXES FROM CENTAURUS A

| B (gauss) | $\rho = 0.25$ ev cm$^{-3}$, $3°K$ | $\rho = 6$ ev cm$^{-3}$, $7°K$ | $\rho = 13$ ev cm$^{-3}$, $8°K$ |
|---|---|---|---|
| $10^{-6}$ | 0.046 | 1.03 | 2.30 |
| $4 \times 10^{-6}$ | 0.004 | 0.09 | 0.20 |
| $10^{-5}$ | 0.001 | 0.02 | 0.04 |

X-ray Fluxes in KeV/cm$^2$ sec from 1 to 10 keV.

Uhuru Upper Limit on X-ray Flux from Centaurus A

$2\sigma$   0.035   KeV/cm$^2$ sec from 1 to 10 keV

---

a very large absorbtion, of the order of 3 to 4 keV, suggests that the emission is from the nucleus of the central galaxy, as the X-rays must have traversed a very dense region of matter.

Now, we are still trying to evaluate the possibility that the X-rays that we see might be produced by inverse Compton scattering between photons of the infrared or radio emission and the same electrons that produce the radio and infrared emission by bremsstrahlung. Our calculations on this process are not yet complete. However, in table 3.1 we make use of this line of argument to limit possible values of the magnetic field and electron energy. The observed radio flux gives a relationship between these two quantities, assuming inverse Compton scattering. Then, if we see no X-rays, this sets a bound on the product of the number of electrons present and the square of the magnetic field, or on the temperature of the black body assumed. This leads to discarding certain possibilities for the magnetice field and electron density; as seen in table 3.1, we must have $4 \times 10^{-6} < B < 10^{-5}$ Gauss and $T_{black\ body} < 7°$. These values are not surprising, but they show how the X-ray emission now permits us to tie in the infra-red and the high-energy electrons emitting

X-Ray Astronomy                                                            240

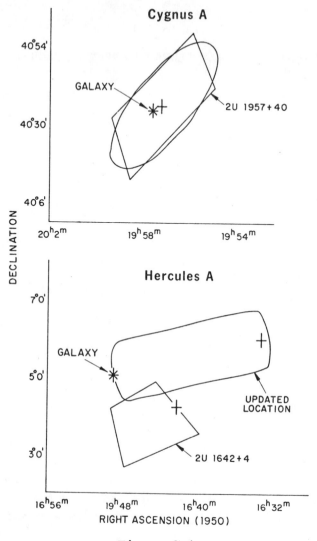

Figure 3.4

the radio waves, and to get some quantitative results about it.

Hercules A and Cygnus A are found to be X-ray sources (see Fig. 3.4). For Cygnus A the position is quite good. For Hercules A the position is rather uncertain, and in fact we cannot decide between emission from a single galaxy, from a cluster, perhaps not yet recognized, or from several sources. As single objects the luminosities are $10^{42}$ ergs/sec for Cygnus A and about $10^{46}$ ergs/sec for Hercules A, a tremendously large flux. Whether indeed this is emission from a single galaxy or not we cannot yet say.

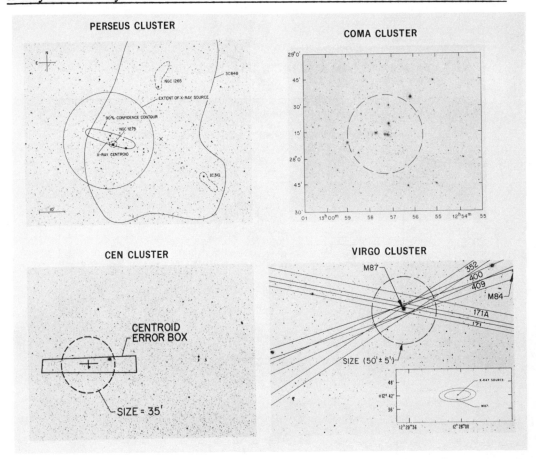

Figure 3.5

One of the most interesting results from Uhuru is the detection of X-rays from clusters of galaxies. Fig. 3.5 summarizes the results obtained. The Virgo Cluster containing the galaxy M 87 was first detected by the NRL group and observed later on by the MIT group, in both cases with a rocket flight. The source of X-ray emission observed by Uhuru is extended with an angular size $\Delta\theta \simeq 1°$. At first the location was quite uncertain, placed between M 87 and M 84; now that more data have been accumulated the center of the source can be firmly placed on M 87. The emission detected could be due to a number of individual galaxies, but later on reasons will be given to support the belief that this source is a different phenomenon from the one observed in the region of the nucleus of other galaxies. During the scan of the sky other extended sources ($\sim$ 30 of them) have been discovered, among which Perseus Cluster, Coma Cluster and Centaurus Cluster.

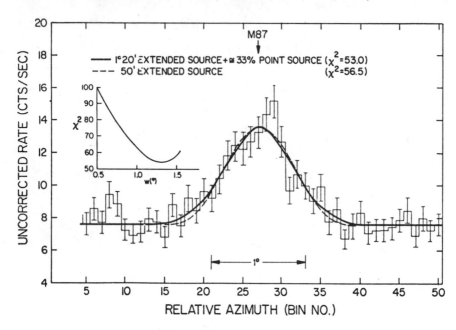

Figure 3.6

The following will be an account of significant details: Fig. 3.6 gives a sketch of how the data appear when the satellite scans across M 87. To see if it is a point source or not we first have to find the collimator response to a point source, and then ask ourselves if we are observing an extended region of uniform distribution, with the angular size that best fits our data. The results were obtained in this way. There is no reason to believe the chosen distribution is the correct one. In fact one can try all kinds of different distributions, e.g. a point source plus an extended region. For an extended region we get a distribution of the type sketched below

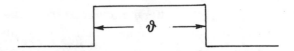

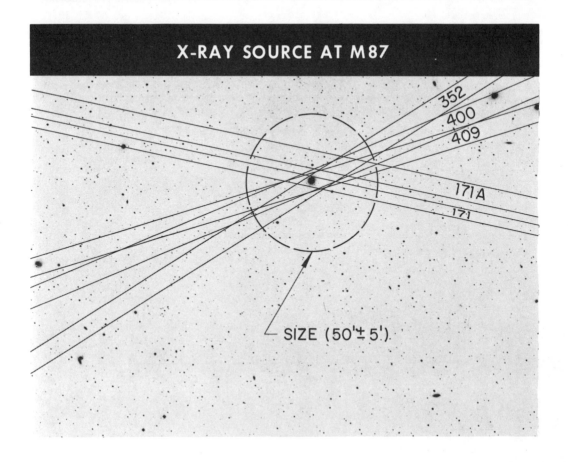

Figure 3.7

while for a point source superimposed to it would appear as shown:

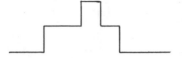

After examining the data, the only reasonable conclusion one can draw is that, if there is only a diffused source, it has to be an extended source of about 50'. If there is a point source plus an extended source superimposed the region of diffused emission must be wider ($\sim 1°\ 20'$) and in this case the point source allowed can contain $\sim \frac{1}{3}$ of the total emission. A recent measurement seems to support the view that $\sim 30\%$ of the emission comes from a point source, and the remaining flux from a

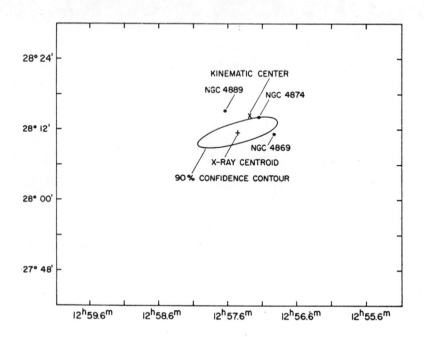

Figure 3.8

more extended region. As far as the location of the X-ray source in relation to M 87, the error box shows, in Fig. 3.7, how it is placed respect to the galaxy.

Fig. 3.8 shows the location of the X-ray source in the Coma Cluster. The + indicates the center of the X-ray position and some of the brighter galaxies in the field are shown. The x indicates the kinematic center of the Cluster. Most of the X-ray emitting clusters have been studied in the radio band and radio contours are available. The presence of radio emission can be taken as evidence for some kind of action taking place in the clusters.

Fig.3.9 shows Perseus Cluster, in which similar information has been obtained. The centroid of the X-ray position here appears to coincide with most active galaxy in the Cluster, NGC 1275, and the type of emission observed seems to be quite different from the one observed from NGC 4151. The region of sky involved, as it can be seen from the scale in the figure, is very large.

## Table 3.2

### SIZES OF EXTRAGALACTIC X-RAY SOURCES

|  | SIZE | | $L_x$ |
|---|---|---|---|
|  | angular | kpc | (erg/sec) |
| ABELL 2256 | 35' ± 15' | 2800 | $5 \times 10^{44}$ |
| Perseus - NGC 1275 | 35' ± 3' | 740 | $3 \times 10^{44}$ |
| COMA | 36' ± 4' | 1050 | $2 \times 10^{44}$ |
| CEN - NGC 4696 | 37' ± 8' | 500 | $2 \times 10^{43}$ |
| VIRGO - M87 | 50' ± 5' | 200 | $7 \times 10^{42}$ |
| NGC 4151 | $\leq 15'$ | $\leq 60$ | $1 \times 10^{42}$ |
| NGC 5128 | $\leq 10'$ | $\leq 20$ | $6 \times 10^{41}$ |

**EXTENDED X-RAY SOURCE IN PERSEUS**

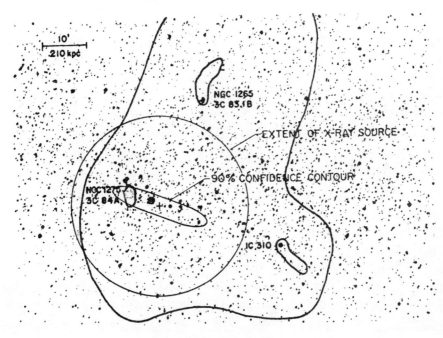

Figure 3.9

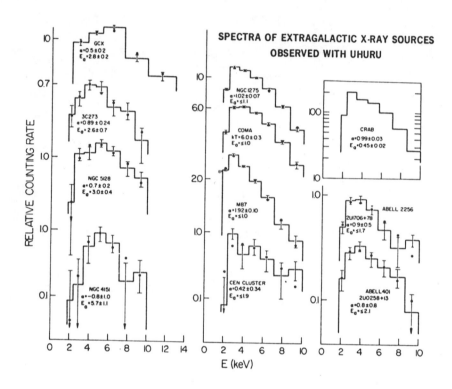

Figure 3.10

Now one can ask the question: are all great clusters X-ray emitters? Do all of them show extended sources? Table 3.2 shows a summary of the results obtained on clusters. For all clusters for which we can measure the angular size an extended source seems to be present. The angular size obtained for the two single galaxies NGC 4151 and NGC 5128 seems to indicate the different nature of these objects compared to clusters. The range of luminosities is very wide and this leads us to wonder whether Hercules and Cygnus A are, in fact, single sources or if they could also be clusters. A further question arises: is the emission from clusters due to the contribution of single galaxies in the cluster? To answer this question spectra have been measured for most of the observed sources.

Fig. 3.10 shows the spectra obtained from individual galaxies, spectra from clusters and the spectrum of an object chosen as a test object: the Crab Nebula. These results have a very high statistical significance. For instance, errors on the Crab Nebula are mostly due to errors in calibration rather than statistical errors. The Crab Nebula was used as a test object since there is a high degree of confidence that it has a power law spectrum, with a value of the power law index determined with high precision (its value being 0.99 ± 0.03) and an absorption energy (energy absorbed by a factor $\frac{1}{e}$) of 0.45 ± 0.02. These values find good agreement among observers.

Let's first examine the emission from GCX, the nucleus of our own galaxy. If we assume a power law spectrum, we find a fairly flat spectrum of 0.5, and a very high absorption energy of 2.8 KeV, which agrees fairly well with the absorption energy measured from observation of stars clustered around the galactic center (2.2-2.5 KeV), which implies a column density of $\sim 10^{22}$ H atoms $cm^{-2}$. In 3C 273 we find a cut off energy of 2.6 KeV which can be used to set a lower limit on the size of the absorption region. In NGC 5128 we find also a fairly large absorption energy of 3.0 KeV, and NGC 4151 has the highest absorption energy value of 5.7 KeV, which requires $\sim 10^{23}$ H atoms $cm^{-2}$. One can therefore speculate that there is a central source within 50 pc around the Galactic Nucleus. It would be very interesting to observe time variations on this region over a period of months or perhaps years. One can then conclude that the main common feature for this class of objects is their large absorption energy.

If the emission from clusters were of the same type we could expect the same absorption from all of them, but this is not so. We find that for all clusters observed there is no evidence for low energy absorbtion, thus it is unlikely that emission from individual galaxies integrated over the whole cluster gives rise to the X-ray detected. Therefore we must think of some other kind of mechanism. Two working hypothesis have been proposed:
1) If we observe galaxies in a cluster the velocities in it are not all the same (velocity dispersion). Considering the galaxies as a gas, according to the Virial Theorem, a Virial mass must be present to maintain closure. This mass exceeds by a factor of $\sim 10$ the sum of the masses of all galaxies seen

in the cluster. Therefore some gas must be there. If there is such a gas heated, either by galactic collisions, or from emission of particles by galaxies, one expects a spectrum of thermal bremsstrahlung. The mass required to produce it is only 1/10 of what predicted by the Virial theorem. What we have seen therefore is not the missing mass. There is though some relation between the missing mass and X-ray emission. It has been found that there is a strong dependence of the emission on the galaxy velocity dispersion. The astronomical estimates for velocity dispersion are still very crude and incomplete. For three clusters for which there are data, and for one for which it can be guessed, it is found that emission $\propto$ (velocity dispersion)$^4$.
2) The X-ray flux observed may be due to inverse Compton emission from high energy electrons impinging on photons. Several arguments have been raised in favour of this hypothesis.

From the cosmological point of view an interesting connection has been made by Gunn and Gott III: if the clusters have been there for a long time and one assumes a certain density of intergalactic medium, one expects that the intergalactic gas would fall into the galaxies and emit X-rays. From this point of view one predicts much larger X-ray fluxes than observed. The

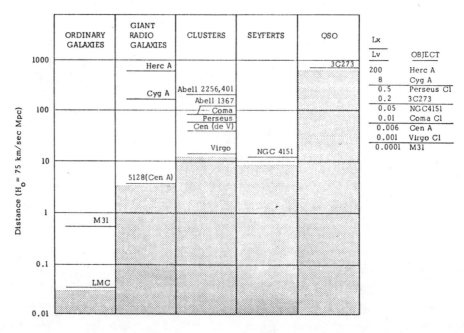

Figure 3.11

## SUMMARY OF DIFFUSE X-RAY MEASUREMENTS

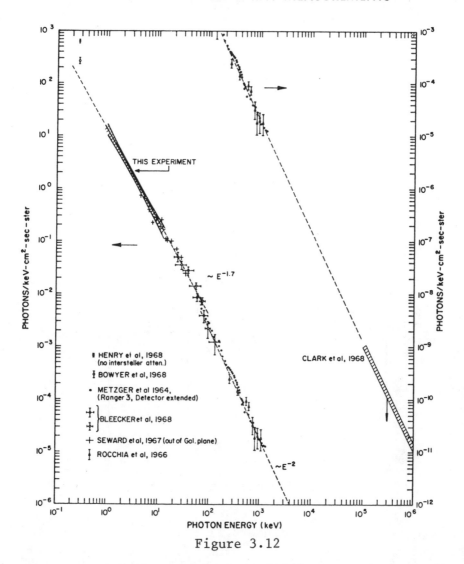

Figure 3.12

amount of gas required would be sufficient to close up the Universe. The measurements therefore put an upper limit on the amount of gas outside a cluster.

Fig. 3.11 summarizes the results on extragalactic sources. It can be concluded that Uhuru sees emission from all objects close to us and increasing sensitivity will allow us to see more distant objects.

Could the integrated effect of all these objects give rise to the X-ray background which has been observed since 1962? Fig. 3.12 shows the results so far obtained on the background. First of all one must note the large range of energy over which

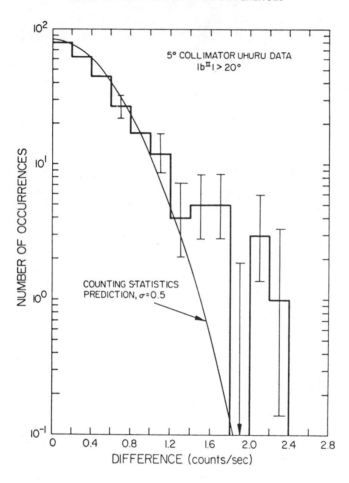

Figure 3.13

the isotropic background has been observed. Many kinks and breaks, it has been claimed, are present and a lot of discussions about their meaning have taken place. Such discussions were perhaps premature, since there is no evidence that there is a contribution from a diffuse source. If normal galaxies emit $2 \cdot 10^{39}$ erg/sec and are spaced in the sky with a density of $\sim 0.03$ Gal/Mpc$^3$, we find that the contribution they can give to the total background of $\sim 3 \cdot 10^{-8}$ ergs/cm$^2$ sec ster is very small. The contribution from radio galaxies is also very small. Seyfert Galaxies could give a significant amount, and they are very numerous, but so far only one has been detected (NGC 4151).

The others that should have been detected were not observed and perhaps it is not legitimate to assume they contribute substantially to the backgrounds. QSO's can give a small contribution and Clusters of Galaxies offer a substantial amount of energy. Clusters are very abundant and very luminous, but there is still great uncertainty about which objects contribute to the total background, even though $\sim 80\%$ of it can perhaps be accounted for.

From Fig. 1.7, the Log N - Log S plot, we can see that only the most intense of the extra-galatic sources have been seen. The great majority of them ($\sim 30-50$) are the weakest ones and have not been identified. The Log N - Log S curve could be extrapolated to fainter and fainter sources to see if they could justify the background. If one assumes that all these sources are extra-galactic, one comes to the conclusion that it is necessary to go $\sim 30$ times further out or 900 times fainter in apparent luminosity. One would need $\sim 10^6$ sources.

Fluctuations on the background are shown in Fig. 3.13, where it is indicated that one can detect even 1 count/sec. The fact that the curve is quite smooth indicates that there must be $\geq 10^6$ sources, the same figure obtained above. This seems to be a fairly consistent picture to explain the background. To be more precise, one should take into account cosmological effects, which has not yet been done.

The question arises: are these sources really extragalactic? A distribution in galactic coordinates shows no clustering towards the galactic plane. Furthermore looking towards the center of our galaxy we see $\sim 5$ times a larger volume than looking away from it. A ratio of 5:1 is expected for galactic sources, but the ratio observed is 1:1. The conclusion is these sources either are very local or extragalactic. If they are extragalactic one wonders what type of sources are they? Certainly not conventional galaxies; in fact for normal galaxies X-rays are a minute part of the electromagnetic spectrum. For NGC 4151 the ratio $L_x/L_o$ increases, and for Hercules A $L_x/L_o \sim 200$. If we look in the error box associated with the unidentified objects, we see only faint galaxies. Therefore an upper limit on visual magnitude can be set, and we can derive the apparent X-ray magnitude, and we get a $L_x/L_o$ reasonably large. These objects must be peculiar objects and could be unidentified QSO's as recently suggested by Setti and Woltier.

A few concluding remarks on the future of X-ray astronomy:
1) X-ray astronomy has become relevant in the study of stars at the end of their evolution; it may be a way to detect black holes. It contributes to the understanding of cosmological problems and to Extra-galactic Research in general. Therefore X-ray astronomy is here to stay.
2) What will be achieved in the near future? Measurements of angular sizes, spectra, locations, etc., will be improved. Energy resolution will reach $\Delta E/E \sim 10^{-2} \rightarrow 10^{-4}$. Polarization measurements should become $\sim 200$ times better than now. (At present polarization can be measured only from the Crab, with 3 $\sigma$ result). One should expect to see objects $\sim 10^{4}$ times fainter than the ones observed so far. Most of these things will probably be achieved, if the High Energy Astronomical Observatory will be launched by NASA as planned for 1975-1980.

## REFERENCES

Bolton, C. T., 1972, Nature, 235, 271

Bowyer, S., Lampton, M., Margon, B., Mahoney, W., and Anderson, K., 1972, Ap. J. Letters, 171, L 45

Gunn, J. E., and Gott III, J. R., 1972, Ap. J., 176, 1

Hoffmann, W. F., Frederick, C. L., and Emery, R. J., 1971, Ap. J. Letters, 164, L 23

Oda, M., Gorenstein, P., Gursky, H., Kellogg, E., Schreier, E., Tananbaum, H., and Giacconi, R., 1971, Ap. J. Letters, 165, L 1

Setti, G., and Woltier, L., 1972, Cospar I. A. U. Symp. No. 55 on "Non-solar X-ray and gamma ray Astronomy"

Seward, F. D., Burginyon, G. A., Grader, R. J., Hill, R. W., Palmieri, T. M., 1972, Ap. J., 178, 131

Terrell Jr., N. J., 1972, Ap. J. Letters, 174, L 35

Tucker, W. H., Blumentahl, G. R., Cavaliere, A., Rose, W. K., 1972, Ap. J., 173, 213

Van den Heuvel, E. P. J., and Heise, J., 1972, Nature Phys. Science, 239, 67

Webster, B. L., and Murdin, P., 1972, Nature, 235, 37

Wilson, R. E., 1972, Ap. J. Letters, 174, L 27

Statistical Thermodynamics of Strong Interactions

Lectures by

R. Hagedorn

CERN European Organization for Nuclear Research

Notes by
R.K.P. Zia, R. Bland

# Statistical Thermodynamics of Strong Interactions

Lecture 1

The subject of these lectures will be the statistical thermodynamics of strong interactions at high temperature and high density. They are intended to be qualitative rather than quantitative, with the details to be found in the guide to the literature included in the references at the end of these notes.

## Introduction

Until recently in cosmology the thermodynamics used has been that of non-interacting particles; this is, of course, inherently wrong. The properties of such a thermodynamics are rather trivial: 1) always in equilibrium; 2) as $t \to o$ (in a big-bang cosmology), $T \to \infty$; 3) the particle number density $n \propto T^3$; and 4) the fluctuations, proportional to $n^{-\frac{1}{2}}$, vanish at $t = o$. Now, at densities of the order of nuclear densities ($10^{14}$ g cm$^{-3}$) strong interactions cannot be neglected. (Neither can electro-magnetic interactions, but they cancel due to the presence of equal numbers of positive and negative charges.) The problem is to make a thermodynamics which includes the strong interactions. The thermodynamics of strong interactions was in fact first developed to calculate production rates and angular distributions of particles in high-energy physics, in order to set up experiments at the big accelerators, and the cosmological applications came out as a by-product.

Let us just anticipate the results, to stimulate the reader's curiosity:

1) In this thermodynamic there is a good chance of including all of the strong interactions, even though the formalism looks rather like that for non-interacting particles.

2) The strongly interacting particles (hadrons) seem to be composed of each other; there are no really elementary particles, but all particles are composite states of other particles, and each one is as elementary as any other one. This relates to a long-standing problem of philosophy, dating back several thousand years, namely, when you divide matter, what happens? Do you come to final particles which can be called atoms, or elementary particles, or do you not? The atom can be further divided, as we learned about 100 years ago. Then you come to the nucleus; it can be divided further. Now we are at the elementary particles. But the problem is, that whenever you have a particle which has a certain size, you can at least <u>think of</u> splitting it into two

halves. What are the two halves? This is an old dilemma, because if finally you come to truly elementary particles, then you have to build up all the rest from them. Now, if these elementary particles still have structure and properties, the question is how do these properties arise? Then you try again to split it up and to explain the properties of this elementary particle by the more fundamental properties of more fundamental particles. That never ends. Now if, however, all particles are composed of each other, then there is no most elementary particle, all of them are elementary, and each one is a composite state of all the others. This seems to be a result of the model that I am discussing.

The number of different kinds of hadrons is infinite. The number of particles in the mass interval between m and m + dm is $\rho(m)$, the mass spectrum, which grows exponentially with the mass:

$$\rho(m) \propto c a^{-3} e^{m/T_o}. \tag{1.1}$$

A trivial consequence of this form for the mass spectrum is that the constant $T_o$, which turns out to be about one pion mass, is a universal maximum temperature:

$T_o \simeq 140 \text{ MeV} \simeq 1.6 \times 10^{12} \text{ °K}.$

No matter whatsoever in equilibrium can have a higher temperature.

The resulting thermodynamics has the following structure: the energy density is

$$\varepsilon \sim \frac{T_o^2}{(T_o - T)^{1/2}} ; \tag{1.2}$$

the energy density fluctuations are

$$\frac{\Delta \varepsilon}{\varepsilon} \simeq \text{const.} \times (\varepsilon/V)^{1/2} \tag{1.3}$$

where V is the volume. Thus, if the volume is kept constant and $T \to T_o$, the fluctuations become infinite. Of course, if the volume is allowed to increase in proportion to the energy density, the fluctuations remain constant. In the limit $\varepsilon \to \infty$, $P \to$ constant and <u>not</u> $P \to \infty$. The number density of particles goes to a constant,

$$n \xrightarrow[T \to T_o]{} \sim \frac{1}{V_o}, \text{ where } V_o = \frac{4\pi}{3} m_\pi^{-3}.$$

I am using units such that $\hbar = c = k$ (Boltzmann's constant) $= 1$, so $m_\pi^{-1}$ is the pion Compton wavelength, about $10^{-13}$ cm.

For $T \to 0$ and baryon number $B \neq 0$, and large energy density, as in a neutron star, $P \sim \varepsilon/\ln(\varepsilon/\varepsilon_o)$, where $\varepsilon_o$ is some normalizing energy density. This pressure is a little lower than that which one finds in most of the other models. For instance, it is not proportional to the energy density, as in ordinary blackbody radiation. The factor $\ln^{-1}\left(\frac{\varepsilon}{\varepsilon_o}\right)$ will just remove the troubles which one had with causality, namely that in very dense matter the velocity of sound could be greater than the velocity of light.

A final result, as yet unpublished, by Carlitz, Frautschi, and Nahm, concerns the early state of a big bang universe. Firstly, they find that one does not have thermal equilibrium until $t \simeq 10^{-4}$ sec, in sharp contrast with conventional theories. Secondly, the fluctuations are big enough for formation of galaxies, even starting from the most homogeneous possible initial condition. Thirdly, however, there is no mechanism for matter-anti-matter separation. If you suppose that the universe started with baryon number $B = 0$, one must invoke the method of Omnes or something else to explain why matter and antimatter separated.

The early universe is matter dominated and has great density, so that strong interactions dominate. We shall neglect electromagnetic and weak interactions as well as gravity for the following reasons:

1) electromagnetic forces vanish on the average in a neutral plasma; thus only the presence of quasi-free $e^+$, $e^-$, $\mu^+$, $\mu^-$, and $\gamma$ actually enters the partition function. But their density is (because of the limited temperature) low compared to that of hadrons; one has only to consider them as a perturbation in quantitative discussions.

2) the same holds for the neutrino density.

3) gravity is always attractive and long range and, of course, determines the global structure of the universe. Locally, however, it only can matter for a thermodynamics of strong interactions, if its strength becomes comparable to strong forces. A good measure for this to happen is when hadrons become so heavy that they enter their own Schwarzschild radius. All hadrons have the same size, given by the range of strong interactions, of about $m_\pi^{-1}$. The maximum mass is then given by

$$2GM_o = m_\pi^{-1} \longrightarrow M_o \simeq .386 \times 10^{40} \, m_\pi \simeq 10^{15} \, g.$$

As long as hadrons are lighter than this, no problem with gravity arises. We shall come very near to this limit, however.

Strong forces, however, seem to be attractive on the average and no cancellation occurs. We therefore should aim at including them as fully as possible in our thermodynamics.

# Statistical Thermodynamics of Strong Interactions

Lecture 2

## PHENOMENOLOGY OF STRONG INTERACTIONS

Let us briefly review the main features of strong interactions which we have learned from the big accelerator in the past few years.

The first thing is that strong interactions are mediated by the exchange of field quanta of some mass. Yukawa postulated a meson to explain the short range of nuclear forces, which is of the order of $10^{-13}$ cm. ($m_\pi^{-1}$). In high energy physics, we can no longer deal with nuclear structure. We have to learn from collision experiments, where particles are smashed to pieces.

Let us consider the energy ($\Delta E$) used for smashing an object, of mass M, to pieces in several different processes:

Mechanical (Dropping something on the floor) $\quad \Delta E \sim 10^{-16} M$

Chemical (NaCl $\to$ Na + Cl) $\quad \Delta E \sim 10^{-9} M$

Nuclear (Na $\to$ 23N) $\quad \Delta E \sim 10^{-2} M$

Subnuclear (p $\to$ ?) $\quad \Delta E > 30 M$

In this table, we start with an example of dropping, say, a piece of crystal on the floor to smash it. The energy needed for it is of the order of $10^{-16}$ the rest mass of the crystal. But this will never dissociate it into its chemical elements. In the second example, we use higher energy and succeed in breaking a molecule up. But to break up the nucleus of the elements, we need still higher energies. Finally, we tried to split the proton in high energy collision experiments, with energies up to 30 times the mass of the proton itself, but we failed. Instead new stuff is created in the collision and the protons that emerge from the collision are as good as new!

A new situation starts below the decomposition of nuclei. Above this level, the original object disappears. Below this level, the original is not destroyed but new particles are created. The number and sorts of new particles created differ from event to event. This new situation could be summarized in a definition of elementary particles: elementary particles are particles which cannot be decomposed in a unique reproducible way into elementary particles. In short, elementary particles consist of

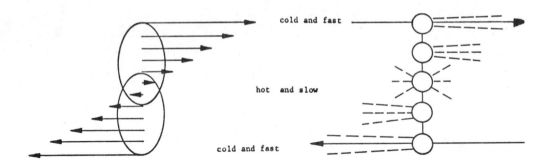

Figure 1  A hadron-hadron collision seen from the centre-of-momentum frame. Left: visualized as a continuum of hadronic matter; right: in a multi-peripheral picture.

elementary particles. Of course we don't know if this is true, but the content of these lectures will show that there is a good argument in favor of this interpretation.

The main features of high energy scattering of hadrons can be summarized as follows. In the center of mass frame, the two colliding particles are Lorentz contracted. Most of the stuff will continue to go in the original direction, but in a small region of overlap there will be high excitation, because it is brought to a sudden stop. The picture underlying these lectures is that the overlap region is hot, and the other region, cold. (Fig. 1)

The cross sections seem to tend to a constant as the energy grows. This might have been <u>expected</u> since, if we believe that these particles have a diameter given by $m_\pi^{-1}$, then as the energy increases, nothing much happens, so that the cross section becomes of the order of the geometric cross section:

$$\sigma \sim \pi m_\pi^{-2} \sim 63 \text{ mb}.$$

In particular, $\sigma(pp) \to \sim 30$ mb. (Fig. 2)

Another feature is that many particles are created. This is

# Statistical Thermodynamics of Strong Interactions

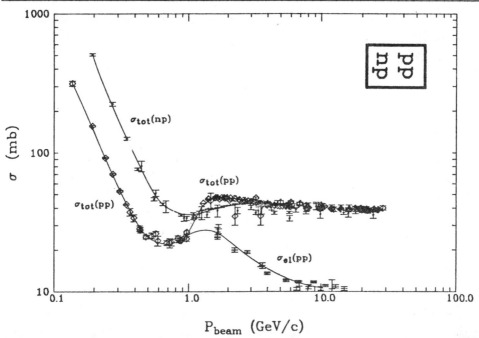

Figure 2 The proton-proton and proton-neutron collision cross sections at high energy. The curves end at 30 GeV/c primary momentum. The value of $\sigma_{tot}$ at 1500 GeV/c has been found to be 37 ± 1.5 mb (CERN-ISR) which shows that between 1 and 1500 GeV/c the total cross section is nearly constant and of the order of the geometrical one.

also <u>expected</u> just on the basis that mass and energy are equivalent. Kinetic energy which is destroyed can show up only as newly created particles. The multiplicity (= number of new particles produced) grows slowly with energy. This is also <u>expected</u> if we look at Fig. 1. Most of the energy is preserved in the kinetic form, and not available for production.

Next, the longitudinal momentum ($p_\parallel^*$) grows as $E_{CM} \equiv (s)^{1/2}$. This is again <u>expected</u> if we look at Fig. 1. A lot of particles will still have a good deal of the incoming velocity so that the longitudinal momentum will be proportional to the primary energy. Following Feynman, we define

$$x \equiv 2p_\parallel^*/(s)^{1/2},$$

which lies between -1 and +1. If we plot the momentum distribu-

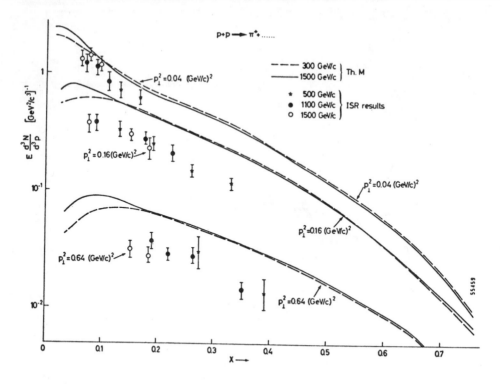

Figure 3 $\pi^+$ spectra as function of x with $p_\perp^2$ as parameter for 300 and 1500 GeV/c. For $x \gtrsim 0.2$, scaling is well approached, at low x it is not yet.

tion in this variable, we should get something roughly independent of the primary energy. (Fig. 3)

Next, the transverse momentum ($p_\perp$) distribution is Boltzmann-like: $\exp(-p_\perp/T)$. Phenomenologically, T lies between 120 and 160 MeV for CM energies between 2 and 1000 GeV[1]. (This feature is totally <u>unexpected</u>.) The upper limit for T is reached very rapidly; for energies in the ISR (50 GeV) T is very close to 160 MeV. There seems to be no reasonable classical explanation for such a behavior.

The last feature is also <u>unexpected</u>: There seems to be a very rich spectrum of particles. At present there are about 1000 different particle states. In this number is included the factor

---

[1] 1000 GeV events are cosmic ray events

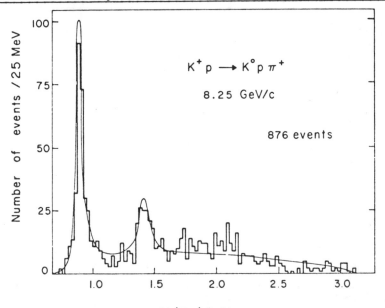

Figure 4  The $K°\pi^+$ mass spectrum for $K^+p \to K°\pi^+p$ at 8.25 GeV/c.

$(2J + 1)(2I + 1)2^\alpha$ (J = spin, I = isospin, $\alpha$ = 1 if particle ≠ anti-particle, $\alpha$ = 0 if particle = anti-particle) for the proper counting of distinct states. Most of these are <u>resonances</u>.

Let us briefly review the phenomenology of resonances. Resonances are unstable particles with a lifetime of the order of $10^{-23}$ sec. Note that this lifetime compared to a millisecond is the same as a millisecond compared to the age of the universe. Although at first sight it seems hard to study these particles, the experimentalists are able to identify their mass, lifetime, spin, parity, isospin, etc.

To identify these objects, we plot the distribution of events against $M_{ik}^2 \equiv (p_i + p_k)^2$, the invariant mass squared for any pair of particles (i and k) that emerge from a collision. The expectation from phase space considerations is also plotted. An observed peak over this expectation is then interpreted as a strong correlation between the pair i-k, so that we can think of particles i and k as being stuck together for a brief time in the form of a new particle which then decayed with a lifetime roughly equal to the inverse of the width of the peak. (Fig. 4)

Next, such resonances also show up in scattering experiments, whether we measure the cross section or the scattering phase shift.

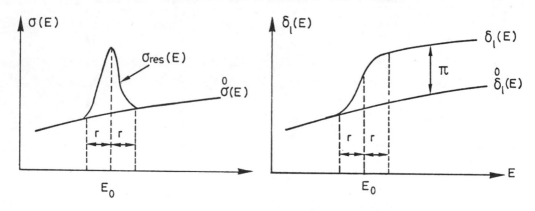

Figure 5 Phase shift and cross section at resonance.

The cross section will have a bump with a width that is proportional to the inverse of the lifetime. The phase shift has the property that as we go through a resonance, it suddenly jumps by the amount $\pi$. This rise also occurs within the same width (Fig. 5). If we were to know all the phase shifts, then we would know the entire interaction. (because phase shifts are eigenvalues of the S-matrix). So, if we incorporate all the phase shifts in our thermodynamic treatment then we have incorporated all of the interactions. This is an idea due to Beth and Uhlenbeck (1937).

## A SPECULATIVE ATTEMPT TO EXPLAIN A LIMITED $P_\perp$

Many years ago, the following picture was proposed to explain these features of high energy interaction. Two particles come in, overlap in a region, heating it up. After that, two pieces of highly excited hadronic matter, called fire balls, go out. Immediately, we get the result that the distribution in $p_\perp$ will be the thermal distribution in the rest frame of the fire ball, whereas the large $p_\parallel^*$ will be due to the large fire ball velocity. In particular, $p_\parallel^*$ will grow with the primary energy, and $p_\perp$ will grow slowly.

However, there is a problem. The transverse momentum seems to be <u>limited</u> and not only growing slowly. If we take such a fireball, the number of particles which will come out will be

# Statistical Thermodynamics of Strong Interactions

very large so that we might guess a statistical distribution. Fermi (1950) was almost the first to do this: Consider the fireball to be black-body radiation confined to $V_0$ imbedded in a temperature bath. If the temperature (T) is low, we would find only $\gamma$, $\nu$, $\bar{\nu}$. When T reaches 1 MeV, we get $e^+e^-$ pairs. As T rises to ~100 MeV and beyond, we will get $\pi$ mesons and hadrons. But, if there is a finite number of species, and if T >> largest mass, then we are back to a Stefan-Boltzmann law, where the masses can be neglected and the kinetic energy becomes larger than the rest masses. The result is that T will grow as the fourth root of the energy contained in this box. So, T grows slowly, but it grows. Only if the mass spectrum never ends, can T be kept bounded.

To do this, let us speculate and assume a mass specturm $\rho(m)$ such that $\rho(m)dm$ is the number of states with mass between m and m + dm. First let us take $\rho$ to be an unknown function of m. How many particles will we find in our box, which is in the temperature bath? The rate of production of a single species of particle goes as exp (-m/T) because this is just the barometric formula. There is an algebraic factor f(m) in front, but the essential feature is the exponential. To get the total number, when there are many different kinds of particles allowed, we must sum over all of them:

$$N(T) = \int_0^\infty \rho(m) dm\, f(m)\, \exp(-m/T). \tag{2.1}$$

Looking at this formula: if and only if $\rho(m) \sim \exp(m/T_0)$, will N(T) diverge for $T > T_0$, i.e. is the temperature limited, the limit being $T_0$. Such a $\rho(m)$ is, in fact, realized in the limited range of particles (and resonances) we know today.

At the same time, this $T_0$ should be also the temperature governing the momentum distribution of the particles which get out of the box if we open it. Therefore the growth of $\rho(m)$ and the momentum distribution in high energy collisions should be intimately connected just by this fact. $T_0$ becomes a universal upper limit of temperature for all matter in equilibrium. Even if we start with a neutrino gas with $T > T_0$, the neutrinos would collide to make, eventually, hadronic matter. The hadronic matter, then, will limit the temperature to $T_0$.

# Statistical Thermodynamics of Strong Interactions

Exactly this, the exponentially rising mass spectrum, is the result if one tries to construct a consistent statistical thermodynamics including all strong interactions. But it came as a surprise from particles physics, from trying to calculate the production rate of particles in an accelerator not from an attempt to explain the limited transverse momentum.

Before continuing further, let us raise two questions which will be answered later. One is the justification of including resonances in $\rho(m)$. What is the meaning of using these extremely short lived particles in equilibrium thermodynamics? We will see that in fact they play the most important role in a thermodynamics which includes strong interactions. The other question has to do with quarks! What if we found that all particles are composed of quarks? Wouldn't we be reduced to three fundamental particles suddenly? As we will see, the existence of quarks will not have an effect on our approach to the problem. There will only be three more lines in $\rho(m)$.

## RELATION BETWEEN QUANTUM MECHANICAL PRODUCTION PROBABILITY AND STATISTICAL THERMODYNAMICS

Fermi (1950) proposed a statistical model to explain the production of particles. Consider a completely central head-on collision in which everything joins into one big fireball. Then the probability to get N particles out of this fireball is given by a simple quantum mechanical formula:

$$P(N) = \int |S|^2 \, \delta^3\left(\sum_{i=1}^{N} \vec{p}_i\right) \delta\left[M - \sum_{i=1}^{N} (p_i^2 + m_i^2)^{1/2}\right] \prod_{i=1}^{N} d^3 p_i \quad (2.2a)$$

$$= \bar{S}(M) \left\{ \left[\frac{V_0}{(2\pi)^3}\right]^N \frac{1}{N!} \int \delta^3\left(\sum_{i=1}^{N} \vec{p}_i\right) \times \right. \quad (2.2b)$$

$$\left. \times \delta\left[M - \sum_{i=1}^{N} (p_i^2 + m_i^2)^{1/2}\right] \prod_{i=1}^{N} d^3 p_i \right\}$$

# Statistical Thermodynamics of Strong Interactions

In (2.2a), S is the scattering matrix; M is the mass of the fireball so that the $\delta$ functions simply express energy-momentum conservation. To obtain (2.2b), Fermi argued that, in order for all these particles to come out of a small volume, $V_o$, they have to be all within that $V_o$ for a brief period of time. The probability that N particles are found in such a volume is $V_o^N$. The $(2\pi)^3$ comes in as a normalization by $h^3 = (2\pi)^3$. These factors must be in S, otherwise they wouldn't have interacted. So we write these factors explicitly, and include $\bar{S}(M)$ as a slowly varying function of mass. These factors represent an average of the S-matrix taken out of the integral. The N! is supplied to avoid double counting. The rest is phase space.

Nox, if we take the curly bracket of (2.2b) and define

$$\sigma_N(M, V_o, \vec{P}=o) \equiv \left[\frac{V_o}{(2\pi)^3}\right]^N \frac{1}{N!} \int \delta^3\left(\sum_{i=1}^{N}\vec{p}_i\right) \times \quad (2.3)$$

$$\times \delta\left[M - \sum_{i=1}^{N}(p_i^2 + m_i^2)^{1/2}\right] \prod_{i=1}^{N} d^3p_i ,$$

we get the density of states of a system of N particles of total mass M and momentum zero, confined to volume $V_o$. We can neglect $\vec{P}=o$ in the case where $N \gg 1$. Then this is the same formula that appears in statistical mechanics for the density of states.

Neglecting $\vec{P}=o$, we can integrate over all $\vec{P}$ and consider a Laplace trasform of $\sigma$:

$$\int \exp(-M/T)\sigma(M,V_o,\vec{P}) dM d^3P = \left[\frac{V_o}{(2\pi)^3}\right]^N \frac{1}{N!} \times \quad (2.4)$$

$$\times \int \exp\left[-\sum_{i=1}^{N}(p_i^2 + m_i^2)^{1/2}/T\right] \prod_{i=1}^{N} d^3p_i = \frac{1}{N!} z^N$$

where $z \equiv \dfrac{V_o}{(2\pi)^3} \int \exp\left[-(p^2 + m^2)^{1/2}/T\right] d^3p$ (2.5)

Expression (2.4) is nothing but the partition function for N particles enclosed in a volume $V_o$, at temperature T. z is just the single particle partition function.

Since in collisions N is not fixed, we sum over N and arrive at the grand partition function:

$$Z(V_o,T) = \sum \frac{1}{N!} z^N = \exp\left\{\frac{V_o}{(2\pi)^3} \int \exp\left[-(p^2 + m^2)^{1/2}/T\right] d^3p\right\}$$ (2.6)

So, we see that, in a few lines, we can transform the description of particle production from quantum transition probabilities to a statistical mechanical ensemble. In ordinary statistical thermodynamics, we have the notion of relaxation times and equilibrium. Here, there is no such concept. The formulae express a formal equivalence rather than an actual thermodynamic system.

INCLUSION OF RESONANCES (for a more detailed justification see Hagedorn, 1971 a, b; or 1970 b)

In the Fermi model, we must include the interaction which causes resonances in the S-matrix to get a description that agrees with experiment. If we can further show that these interactions can be taken out of the S-matrix and put into the phase space as resonances or bound states, then it follows that we can do the same in thermodynamics. Expression (2.6) as it stands, takes into account only one kind of particles with mass m. We will need the entire mass spectrum to give a full account of strong interactions.

To see this in more detail, let us consider an analogy. Suppose we are to study the thermodynamics of a $He^4$ gas. What do we do? We start with a certain large number ($10^{23}$, e.g.) of free $He^4$ atoms, write down the partition function and make various statements about the thermodynamic properties of the system.

But, we also know that $He^4$ atoms are made of 2 protons, 2 neutrons and 2 electrons; so perhaps we should start with 2N protons, 2N neutrons and 2N electrons <u>and</u> take into account the strong and electromagnetic interactions that are responsible for making $He^4$ atoms. Yet we do <u>not</u> do that, by starting with postulating the existence of $He^4$ atoms we <u>get rid of all those interactions</u>. This is precisely the idea with strong interactions: by starting with <u>all</u> the particles and resonances, we can get rid of strong interactions between the hadrons. We begin by postulating the existence of a mass spectrum that is generated by the strong interactions. After that, we can treat these particles <u>as if</u> they are not interacting.

Now at high densities or temperature, this method does not work for the $He^4$ gas. However, in the case of hadrons, it still may work, because we allow all kinds of reactions, the number of particles being not fixed. Since we must build in the transmutations of all kinds of particles and the non-constancy of number of particles, the essential features of strong interactions are contained in it. So, there is a good chance that it will work.

But why does this work? It works because the density of states of a certain system depends on the number of particles. Therefore, interactions of, say, a pair of particles that causes a bound state could be "loaded" into the change in the density of states, as is the case in the $He^4$ gas. The fact that we may generalize this result to resonances was shown by Beth and Uhlenbeck (1937) long ago. The first time that it has been used in particles physics was by Belenky (1956)(see also Dashen, Ma, Bernstein (1969) for generalization to relativistic particles). Further analysis shows that it doesn't matter how long the resonance lives and that a resonance could be treated as if it were a free particle by itself.

## THE STATISTICAL THERMODYNAMICS DESCRIPTION

The program is the following:
i) Count <u>all</u> the particles and resonances in $\rho(m)$. This is one part of the interaction, the part that accounts for the existence of the resonances.

ii) Allow an indetermined volume of all sorts of particles. This is another part, one that allows the reactions between the particles.

iii) Deal with ideal fireballs. This reminds us that in collisions the asymmetry in the momentum distributions ($p_\perp$ vs $p_\parallel^*$) is a kinematical effect, so that there are really two or more fireballs going in opposite directions at the end of the collision.

Ideal fireballs can be obtained in high energy $e^+e^-$ collisions, for example.

Since this program takes into account the major features of strong interactions through the first two points, it is reasonable to expect that the partition function will describe strong interactions from a statistical point of view.

Given $\rho(m)$, we generalize (2.6):

$$\sigma(M,V_o) = \sum_{N=1}^{\infty} \frac{1}{N!} \int \sigma_N(M,V_o,m_1,\ldots m_N) \prod_{i=1}^{N} \rho(m_i) dm_i \qquad (2.7)$$

Here we start with a density of states with N particles in it with various masses $m_i$. Give each of these particles the possibility of running through the wide spectrum $\left[\prod_{i=1}^{N} \rho(m_i)\right]$. Sum over all mass values and divide by N! to avoid double counting. Finally, sum over all possible N. This gives us the total number of states.

Note the difference between $\sigma$ and $\rho$: $\rho$ is the number of different species of hadrons that lie between m and m + dm. On the other hand, $\sigma$ is the total number of states of the whole composed object with mass between M and M + dM. The latter is the function we need in thermodynamics.

Continuing as before, we take the Laplace transform and obtain the partition function:

$$Z(V_o,T) = \exp\left\{\frac{V_o}{(2\pi)^3} \int \rho(m) \exp\left[-(p^2 + m^2)^{1/2}/T\right] dm\, d^3p\right\}. \qquad (2.8)$$

Taking into account the difference between Bosons and Fermions

# Statistical Thermodynamics of Strong Interactions

we get the correct formula:

$$Z(V_0, T) = \exp\left\{\frac{V_0 T}{(2\pi)^2} \sum_{n=1}^{\infty} \frac{1}{n^2} \int \rho(m,n) m^2 K_2\left(\frac{mn}{T}\right) dm\right\}. \quad (2.9)$$

Here, $\rho(m,n)$ is defined to be $\left[\rho_{Boson}(m) - (-)^n \rho_{Fermion}(m)\right]$ and $K_2$ is a modified Hankel function. Taking only the first term, $\rho(m,1)$ is just the total hadronic spectrum $\rho(m)$. This is a sufficient approximation in most cases. Further approximating $K_2$, we have

$$Z(V_0 T) = \exp\left\{V_0 \left[\frac{T}{2\pi}\right]^{3/2} \int m^{3/2} \rho(m) \exp(-m/T) dm\right\}. \quad (2.10)$$

(For a derivation of (2.9) and (2.10), see for instance Hagedorn, 1971 a, b).

From this Z, if $\rho$ were known, we can calculate the mean particle number, the pressure, the energy density, etc., in the usual manner. For example:

$$<N> \sim \ln Z$$
$$<\varepsilon> = (T^2/V) \frac{\partial \ln Z}{\partial T}$$

But we need $\rho(m)$. We wish to re-emphasize that we must put <u>all</u> particles and resonances into $\rho(m)$, including the undiscovered ones.

## SELF CONSISTENCY AND THE DETERMINATION OF $\rho(m)$

We do not have the complete $\rho(m)$, and, further, it appears improbable that we will learn much from high energy collision experiments because the number of states becomes too dense. So, let us search for a self-consistency requirement. It turns out that this requirement tells us a lot about the functional form of $\rho(m)$.

The self-consistency argument runs as follows:
i) Suppose a thermodynamic description of strong interactions does exist, i.e. <u>that fireballs exist</u>.
ii) Such fireballs may have any mass and their level density,

$\sigma(M,V_o)$, may be calculated once $\rho(m)$ is known. Given $\rho$, Z can be calculated. Performing an inverse Laplace transform, we obtain $\sigma$ as a functional of $\rho$.

iii) Consider the continuum of fireball masses. For low masses, decay channels become fewer, and we discover that fireballs become resonances. For high masses, up to several GeV, these resonances find an increasing number of decay channels. These are just what we have been calling fireballs. So, in fact, resonances ≡ fireballs.

iv) Then a description which includes all resonances in $\rho(m)$ must also include fireballs themselves.

v) So, $\rho(m)$ cannot stop with resonances but also must include $\sigma$. So, at large mass, $\rho$ and $\sigma$ must become the same function. Detailed analysis in the canonical language leads to try the condition

$$\frac{m^\alpha \rho(m)}{\sigma(m,V_o)} \to 1 \text{ for } m \to \infty \qquad (2.11)$$

In fact, in a micro-canonical ensemble, a stronger condition is possible: $\alpha$ in (2.11) must be zero (Frautschi, 1971).

vi) The requirement (2.11) with $\alpha = 0$ leads to a unique asymptotic solution. (Frautschi, Hamer, 1971; Nahm, 1972):

$$\rho(m) \to \text{const.} \times m^{-3} \exp(m/T_o) \qquad (2.12)$$

vii) $\rho(m)$ is now known since its low mass behaviour can be obtained from particles data and its high mass behaviour is fixed by the self-consistency requirement.

viii) If $\rho$ is known, the thermodynamics is completely determined, and we have closed the circle of the self-consistency argument.

## COMPARISON OF $\rho$ FROM THE SELF-CONSISTENCY SOLUTION WITH EXPERIMENT

Frautschi and Hamer (1971) solved the problem by iteration using a computer. Referring to Cerulus (1970), the constant turns out to be $\simeq 8\frac{m^2}{\pi}$ and $T_o \simeq 150\text{MeV}$. Nahm (1972) proved analytically that the exponent of m is exactly -3. This is also confirmed in the computer solution up to the accuracy inherent in computational methods.

It is worthwhile to note that (2.12) is also the result of Veneziano-type dual models (Fubini, Gordon, Veneziano, 1969;

Fubini, Veneziano, 1969; Huang, Weinberg, 1970).

Fig. 6 shows the comparison between the predicted mass spectrum and the present experimental spectrum. The solid line is the predicted spectrum while the dotted lines are the various experimental spectra dating from 1964 till now. As we see, the experimental line fits the predicted one better and better as new states of higher masses are discovered. The curving off of the dotted lines indicates that the states are becoming so numerous and so much overlapping that it is impossible to resolve all of them and a direct further experimental confirmation seems to be excluded beyond m ≃ 2GeV.

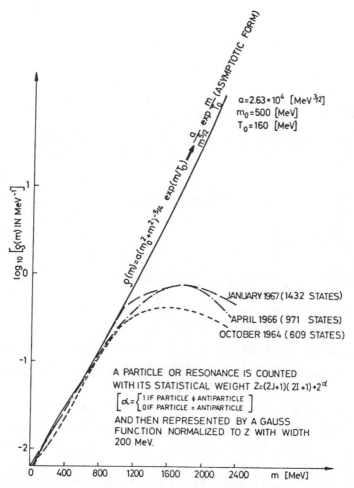

Figure 6  The hadron mass spectrum compared to the asymptotic form $am^{-5/2} \exp(m/T_o)$. Today the power $-5/2$ is replaced by $-3$. On the figure the effect would be invisible.

# Statistical Thermodynamics of Strong Interactions

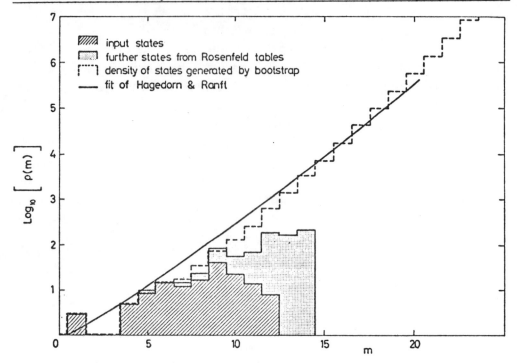

Figure 7  The hadronic mass spectrum generated by computer iteration of the bootstrap equation.  (Frautschi and Hamer)

On the other hand, the transverse momenta are still limited by $T \sim 160$ MeV in highest energy cosmic ray events. This is a kind of indirect comparison with the exponential spectrum. From the latter evidence, the spectrum could be shown to be exponential up to $M \simeq 15$ GeV. But <u>consistency</u> requires that it does not stop there, but go on to infinity, at least up to the mass $\sim 10^{40} m_\pi$ where gravitation comes in.

In Fig. 7, we show the results of an iterative calculation by Frautschi and Hamer. They put into the spectrum the physical one (shaded region) and, using the iterative method, found that the theoretical result converges to the dotted curve. It compares favorably to our first fit. (Hagedorn and Ranft, 1968).

In Fig. 8, we have a plot of the predicted relationship between the mean transverse momentum in a collision and the mass of the particles considered. The points are experimental points. The curves are ones of different T, corresponding to different collision energies up to $10^5$ GeV:

# Statistical Thermodynamics of Strong Interactions

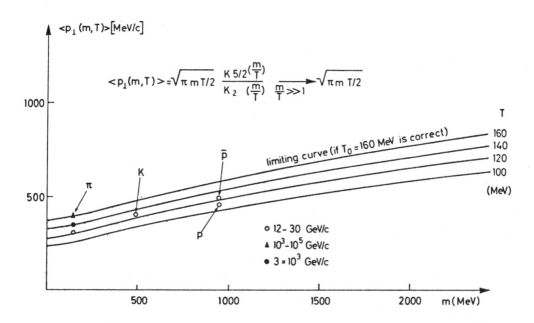

Figure 8  The mean transverse momentum $\langle p_\perp \rangle$ of particles produced in high energy collisions. In this theory $\langle p_\perp \rangle$ depends on the mass and the temperature of the emitting fireball according to the formula given above the curves. The points drawn in are experimental values.

$$\langle p_\perp(m,T)\rangle = (\pi m T/2)^{1/2}\frac{K_{5/2}(m/T)}{K_2(m/T)} \xrightarrow[m/T\gg 1]{} (\pi m T/2)^{1/2} \quad (2.13)$$

In Fig. 9, we have the comparison in particle production rate for $\bar{K}$, $\bar{p}$, $\bar{d}$ and $\overline{He^3}$. There is a difference between those particles which themselves have an excitation spectrum and those which do not. $\bar{d}$ does not because it will disintegrate if excited. In Serpukov, anti-He$^3$ was discovered. It also fits beautifully on the theoretical curve (solid). Note that for each further nuclear mass added, the predicted rate drops by $\sim 4$ orders of magnitude. To give an idea of the seriousness of the situation, we note that to produce an anti-He$^4$, we need to run on the Serpukov accelerator day and night for $\sim 300$ years for one event!

From these comparisons, we see that the theory is not just a speculation. No experiment has produced a single evidence that

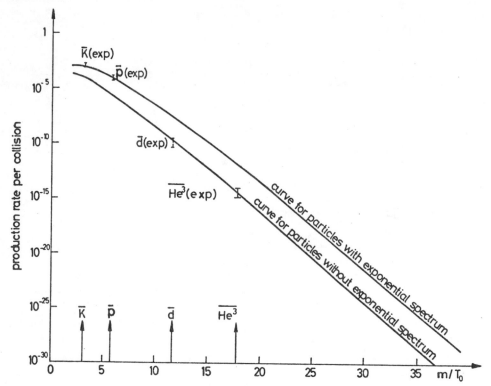

Figure 9  Production rate for heavy pairs.  It drops roughly $10^5$ for every increment by $m_p$ of the produced particle.  With some phase space correction applicable to anti-nuclei production, the recently produced $\overline{He^3}$ fits in very well.

disagrees with this theory.  Of course, this theory is not meant to predict the details of this reaction or that reaction, but rather, offers a "bird's eye view" of strong interactions.

FURTHER RESULTS

We have a few further results which we will just state here rather than going into the details (see Frautschi, 1971; Carlitz, 1972; Matthiae, 1968; Hagedorn, 1971 a, b; 1972 a):
i) If we take the solution proposed by Frautschi and Nahm, i.e. (2.12), then in each $V_o$ there is one heavy particle of mass M and a few (mostly one) lighter ones with a mass of the order of $(Mm_\pi)^{1/2}$ and with kinetic energies of the order of $\frac{3}{2}$ T.  So,

# Statistical Thermodynamics of Strong Interactions

if a heavy particle decays, it will prefer to emit a light one and remain a heavy one, which in turn decays in a similar way:

ii) The lifetimes of decays $M \to M' + m$ are independent of M and of the order of $10^{-23}$ sec. The probability of this mass m appearing is given by:

$$P(m) \sim m^{-3/2}.$$

iii) The pressure inside any given volume is constant because the number of particles divided by $V_o$ approaches $\sim \ln 2$. The number tends to be such that there is one heavy particle and a few light ones. Nearly all the energy we put into $V_o$ will appear as a heavy particle.

iv) For $T \to T_o$ the energy density and its fluctuations grow like

$$\varepsilon \to T_o^2/(T_o - T)^{1/2}$$

$$\Delta\varepsilon/\varepsilon \to (\varepsilon/V)^{1/2}$$

The last result is particularly interesting in cosmology, because, if V is fixed, increase in $\varepsilon$ means increase in fluctuations proportional to $(\varepsilon)^{1/2}$.

Lecture 3

APPLICATIONS TO COSMOLOGY (BIG BANG)

Here I will be discussing the recent Carlitz, Frautschi, Nahm (CFN) (1972) approach, which I have only seen in a draft manuscript form so far. Their approach will not invalidate the previous results of Kundt (1971), but the mechanism leading to the final state of clustered matter (galaxies) is entirely different. In the case which Kundt considered there was thermal equilibrium all the way from $10^{-23}$ sec to $10^{-4}$ sec, and at that time, at the end of the hadronic era, there remained enough fluctuations to have galaxies. In the CFN approach there is no thermal equilibrium, and one has to start with something else. CFN start instead from the <u>smoothest possible</u> initial state. Working down to $10^{-4}$ sec, they find the same size density fluctuations as Kundt found with the other method. Thus, either method gives at $t = 10^{-4}$ sec the same mass fluctuations, sufficient to lead to the formation of galaxies.

The Carlitz, Frautschi, Nahm paper starts by considering the case of zero initial baryon number. The hadronic era starts at $t = \tau \equiv 10^{-23}$ sec, when the horizon $ct$ is just equal to the size of the elementary volume, $m_\pi^{-1} = 10^{-13}$ cm. Before this time particle physics is incompetent to discuss what is going on. At this time $\tau$ the energy density is $\varepsilon \sim 10^{38} M_\pi/V_0$, where $V_0 = \frac{4\pi}{3} m_\pi^{-3}$, the size of a nucleon. Notice that the magic number $10^{38}$ appears here; later on we will see the other magic number, $10^{78}$.

Calculating in a straight forward way from the partition function, we find

$$\varepsilon(T) = \frac{CT_0^2}{4\pi}\left[\frac{2T_0}{T_0-T}\right]^{1/2}, \qquad (3.1)$$

with energy fluctuations

$$\frac{\Delta\varepsilon}{\varepsilon}(T) = \frac{2\pi}{CT_o^{3/2}} (\varepsilon/V)^{1/2} \tag{3.2}$$

From Frautschi's statistical bootstrap calculation (1971) we have

$$C = \frac{(2\pi)^{3/2} \ln 2}{2V_o} (m_\pi^3/T_o^3)^{1/2} \simeq 1.5 \, m_\pi^2 \tag{3.3}$$

This same constant as determined from the low-energy mass spectrum is about $8 \, m_\pi^2$, in reasonable agreement considering that some approximations were made in Frautschi's paper, and that in dealing with the universe we don't need to worry about a factor of 5 anyway.

Putting in the numbers for $t = \tau$, the elementary fluctuations inside an elementary volume $V_o$ are $\frac{\Delta\varepsilon}{\varepsilon}(V_o) \simeq 10^{13}$; that is, most of the little boxes of volume $V_o$ will be empty, but when you find a particle it will frequently be bigger than the average mass $10^{38} m_\pi$ by a factor of $\gg 10^{13}$! Does it make sense to start from such an equilibrium state? Obviously not, since the horizon is $10^{-13}$ cm, the size of one elementary volume, and outside of $V_o$ you don't know what is happening; the volume $V_o$ is isolated from the outside world. This means that you cannot apply <u>canonical</u> thermal equilibrium to it. At best, you can suppose that there is some given energy in the volume, and use a microcanonical description, and this is what CFN do. They in fact assume the smoothest possible initial conditions, and still get large enough fluctuations at $t = 10^{-4}$ sec.

To illustrate the problems with initial thermal equilibrium, CFN define a "fluctuation cell" $V^*$, such that the average energy fluctuation is 1, i.e.,

$$\frac{\Delta\varepsilon}{\varepsilon} = \frac{2\pi}{CT_o^{3/2}} (\varepsilon/V^*)^{1/2} = 1,$$

giving

$$V^* \simeq \frac{\varepsilon V_o^2}{m_\pi} \quad \text{and} \quad M^* \simeq \frac{\varepsilon^2 V_o^2}{m_\pi}.$$

At $t = \tau$ $(= 10^{-23}$ sec),

$$V^*(\tau) = 10^{38} V_o, \quad M^*(\tau) = 10^{76} m_\pi.$$

This means that in thermal equilibrium a large part of the mass $M^* = 10^{76} m_\pi$ (here is the other magic number!) would often be concentrated in <u>one</u> volume $V_o$ (as <u>one</u> particle!), due to the huge energy fluctuation for $V_o$, with in addition about $\bar{N} \simeq 10^{38}$ lighter masses, $\bar{m} \simeq (M^* m_\pi)^{1/2} = 10^{38} m_\pi$, in the other elementary volumes $V_o$ making up $V^*$. ($\bar{N}\bar{m} = 10^{76} m_\pi \simeq \tfrac{1}{2} M^*$) That is, in <u>one</u> volume $V_o$, one huge particle, and in the other $10^{38} - 1$ volumes $V_o$, next to nothing, or at least relatively small particles. Now the linear size of $V^*$ at $t = \tau$ is $D^*(\tau) = (10^{38} V_o)^{1/3} \simeq 10^{13} m_\pi^{-1} = 1$cm. This is $10^{13}$ times larger than the horizon, and one certainly can't speak of thermal equilibrium in such a large volume.

The approach of Carlitz, Frautschi, and Nahm is thus to <u>abandon</u> statistical equilibrium at $t = \tau$. They start instead with a homogeneous density distribution, putting in each volume $V_o$ one very big hadron of mass $10^{38} m_\pi$. This is the "smoothest possible" initial condition, and will clearly give smaller density fluctuations at a later time t than any other initial conditions. Starting with this initial energy density, they use a microcanonical description for each single isolated cell. They assume that even for these big masses the exponential mass spectrum still holds: $\rho(m) \sim Cm^{-3} e^{m/T_o}$. This mass spectrum then completely determines the decay of the original big single particle: it decays in steps of single emission of lighter particles, with each step taking about $10^{-23}$ sec, and it takes a long time for that big initial mass to disappear.

As the particle decays the cell $V_o$ is expanding, at the ex-

# Statistical Thermodynamics of Strong Interactions 283

pansion rate of the universe. The horizon is also growing, so that neighboring cells can start to interact, and the question is, what sort of density fluctuations will this interaction and expansion produce? The expansion is proportional to $t^2$:

$$V(t,V_o) = V_o (t/\tau)^2 \qquad (3.5)$$

and so

$$\varepsilon(t) = \frac{10^{38} m_\pi}{V_o} (\tau/t)^2 \qquad (3.6)$$

Now $V^*$, the volume with unity density fluctuations in thermal equilibrium, goes as

$$V^*(t) = 10^{38} V_o (\tau/t)^2 \qquad (3.7)$$

and at $t = 10^{-4}$ sec, $V = V_o$. This suggests already that one cannot hope to have thermal equilibrium before $10^{-4}$ sec, as we will see more clearly a little later. Furthermore, the distance between two particles which were in the same cell at $t = 10^{-23}$ sec will grow, due to the expansion:

$D(t) = $ dimension of cell $V(t,V_o)$

$$= v_{separation} \times t = (\tau/t)^{1/3} t; \qquad (3.8)$$

$D(\tau) = \tau = 1/m_\pi$.

As the volume $V_o$ expands, the original particle decays and more and more particles populate the volume $V(t,V_o)$. The mass of the original "grain" is $M_G(\tau) = 10^{38} m_\pi$. It decays, if $\rho(m)$ is exponential, in steps of 2-body decay, $M \to M' + m$, with m small. The distribution in m is $P(m) \simeq \frac{1}{2} m_\pi^{1/2} m^{-3/2}$, with a mean value $\bar{m} = (M m_\pi)^{1/2}$, strongly favouring the light masses. m carries away negligible kinetic energy, about $\frac{3}{2} T_o$. The differential equation for $M_G(t)$ is thus:

$$\frac{dM_G}{dt} = -\frac{\bar{m}}{\tau}, \qquad (3.9)$$

giving

$$M_G(t) \simeq M_G(\tau)\left[1 - 4\frac{t}{\tau}(m_\pi/M_G)^{1/2}\right] \qquad (3.10)$$

Here $(m_\pi/M_G)^{1/2} \simeq 10^{-19}$, and this solution is valid over the range $1 \ll t/\tau \ll 10^{18}$. So, we see that the grain mass $M_G(t)$ decreases rather <u>slowly</u>, remaining of the order of $M_G$ until $t \simeq 10^{19}\,\tau \simeq 10^{-4}$ sec. So, provided only that the exponential form for the mass spectrum holds and using elementary decay kinematics, the grains take $10^{-4}$ seconds to disappear. Compare this with the decrease of the (fictitious) mass $M^*$ in the cell $V^*$ of fluctuation unity:

$$\frac{dM^*}{dt} \simeq -10^{55}\left[\frac{10^{-14}\,\text{sec}}{t}\right]^5 m_o\,\text{sec}^{-1} \qquad (3.11)$$

This is a decrease which goes rather fast, by comparison with $M_G$ (the reason is that $M^*$ is <u>defined</u> to be proportional to $\varepsilon^2$ and it decreases by expansion).

Other processes which might contribute to thermalization, such as drift of the grains or recoil from the decay, have been showed by CFN to be negligible; $\gamma$, $\nu$, e, etc. have no effect because there are so few of them ($T < T_o$!).

In Fig. 10 the functions $M_G(t)$ and $M^*(t)$ are plotted over the time interval $10^{-23} < t < 10^{-4}$ sec. The final point at $t = 10^{-4}$ sec, where thermal equilibrium can be established, is where the grain has completely decayed and the mass for energy fluctuation unity is of the order of the pion mass. The first intersection of the two curves corresponds to an apparent equilibrium-like configuration but is not permanent. So we see that equilibrium does not hold until $t = 10^{-4}$ sec, when the big grain masses have completely decayed.

# Statistical Thermodynamics of Strong Interactions

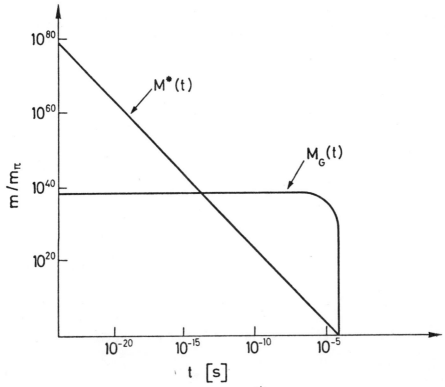

Figure 10  The time dependence of $M^*$ and $M_G$. The decrease of $M^*$ is purely formal and due to expansion, that of $M_G$ is the real decrease of a heavy grain by decay into lighter particles. Figure drawn according to Carlitz, Frautschi and Nahm.

As a final remark on the grain structure, let us imagine another consequence of thermal equilibrium at $t = 10^{-23}$ sec. As we said earlier, there would be in about one out of $10^{38}$ $V_0$'s a particle of mass $10^{78}$ $m_\pi$. The corresponding Schwarzschild radius is $2G \times 10^{78}$ $m_\pi \sim 10^{38}$ $m_\pi^{-1}$, and so such particles, if they did form, would probably just disappear into black holes. If for some reason they did not, they would require a much larger decay time, finishing their decay process at about $t = 10^{15}$ sec, and something would be seen in the sky corresponding to the remains of these massive particles. Nothing of this sort is seen (or is it?)

Now let us finally consider the crucial point, the growth of fluctuations. At $t = 2\tau$ each heavy particle has made one decay on the average, and the volume $V_0$ has expanded by a factor of 4: $V(2\tau) = V_0(t/\tau)^2 = 4V_0$. The volume $4V_0$ thus contains two particles, one heavy, $\simeq M_G$, and one light, $\simeq (M_G m_\pi)^{1/2}$. If we cut $V(2\tau)$ into two halves, of volume $2V_0$ each, the heavy mass is either in one or in the other. Thus the mass fluctuations at $t \simeq 2\tau$ are $\delta m/m \simeq 1$, while at $t = \tau$ there were <u>no</u> mass fluctuations at all. Now consider a larger region, with a volume at $t = 2\tau$ of $V = 4nV_0$. It contains n heavy grains and total mass $m = nM_G$. V can be divided into 2n cells of volume $2V_0$, for which the absolute mass fluctuations are about $\pm M_G/2$. Let us calculate the surface fluctuations of this volume due to moving in and out of the grains. The surface of V comprises $6(2n)^{2/3}$ cells of volume $2V_0$, each with a 50% chance of being nearly empty. The average fluctuation across the boundary surface is thus

$$\delta m \sim (3/2)^{1/2} (2n)^{1/3} M_G \simeq 1.5 \, n^{1/3} M_G, \tag{3.12}$$

and

$$\left(\frac{\delta m}{m}\right)_{V = 4nV_0} \simeq n^{-2/3}. \tag{3.13}$$

Now suppose that V is chosen to contain the mass of a protogalaxy of $10^{52}$ grams. It must then contain $n \simeq 10^{37}$ grains. At $t = 2\tau = 2 \times 10^{-23}$ sec it has therefore

$$\frac{\delta m}{m} \simeq 10^{-25}.$$

This provides an initial density contrast, and, in the subsequent expansion of the universe, denser regions expand less rapidly than more diffuse regions, which causes the density contrast to grow like (Kundt, 1971):

$$\frac{\delta m}{m}(t) = \frac{\delta m}{m}(\tau) (t/\tau)^{2/3} \tag{3.14}$$

Thus for the protogalaxy, at the end of the hadronic era,

$$\frac{\delta m}{m} (10^{-4} \text{ sec}) \simeq 10^{-13}$$

and

$$m (10^{-4} \text{ sec}) \simeq 10^{43} \text{ g},$$

as required for formation of galaxies by standard cosmologies. So, starting with the most homogeneous initial conditions at $t = 10^{-23}$ sec (not thermal equilibrium), density fluctuations large enough to form galaxies develop.

We have not explained at all, however, how we wind up with matter and antimatter separated. If this did not happen, there would be $10^9$ times less matter in the universe than we actually have, and there would probably be no galaxies or stars. A matter-antimatter mechanism has been proposed by Omnes (1953), consisting of two parts. The first is the initial separation of matter and antimatter, which Omnes describes as a phase transition setting in deep in the hadronic era, at a temperature of about 250 MeV. This is larger than our maximum limiting temperature of 160 MeV, but this difference between the two theories could perhaps be reconciled. The second part of Omnes's theory is the maintaining of the separation by the Leidenfrost phenomenon and other mechanisms at the interface between matter and antimatter. This part and the subsequent evolution into galaxies could go through just the same even if the separation mechanism doesn't work exactly as proposed. It is the first part of the process that is the most critical. So, my personal feeling is that the Omnes mechanism is not yet an established theory, and that much work is still necessary to bring it into agreement with the facts of high-energy physics, in particular with the exponential mass distribution and a limiting temperature.

The Carlitz, Frautschi, Nahm results for $B = 0$ are thus only partially satisfactory; they get the desired density fluctuations, but do not separate matter and antimatter. So, they considered also the possibility that the universe started out with the baryon number of the presently observed universe. There are two possibilities:
1) there exist superbaryons, single particles of arbitrarily large baryon number (not nuclei, more strongly bound), which can be fitted into one elementary volume $V_0$ at $t = \tau$ ; or

2) superbaryons do not exist, and one must squeeze into the same volume $V_0$ the necessary number of baryons, that is, a large number of particles and not just one. Most present experimental knowledge goes against the existence of such superbaryons. First, they have not been seen; this is an inconclusive argument, though, since the production could be too small, or the mass too large, say mass > 5 GeV. Second, the experimental mass distribution would rise more slowly if there existed superbaryons; the maximum temperature would be about 110 MeV, rather than the 150 MeV observed. (Frautschi, Hamer, 1971; Nahm, 1972)

Let us look all the same at the results we would get if superbaryons did exist. We get the same fluctuations as before; the initial condition of one particle in $V_0$ and the decay scheme go through in the same way. At $10^{-4}$ sec we would have the necessary density fluctuations to form galaxies, and they would be pure matter galaxies. The existence of superbaryons would evidently be most desirable.

If superbaryons do not exist, we must put into a volume $V_0$ at $t = \tau$ a very large number of baryons. The statistical mechanics can be worked out, using a chemical potential to take care of conservation of baryon number. One finds

$$\frac{\Delta \varepsilon_B}{\varepsilon_B} = \frac{\Delta \varepsilon_0}{\varepsilon_0} \frac{1}{(B_0)^{1/2}} \simeq 10^{-12} \text{ at } t = \tau . \qquad (3.15)$$

Now in the B = o case we started with density fluctuations $\Delta \varepsilon_0/\varepsilon_0$ of about 1 and got what we needed. Here we start with density fluctuations of $10^{-12}$, and one can show that the universe starts out in thermal equilibrium and stays in thermal equilibrium throughout the hadronic era. There is therefore no chance of having fluctuations big enough to make galaxies. Thus this possibility for the theory is excluded.

In the paper by Kundt (1971) the solution was considered where the universe is always in equilibrium, but it so happened that it gave the same density contrast at $10^{-4}$ sec as the CFN treatment. So, the major conclusions that Kundt finds are right, although the mechanism is completely different from that discussed here.

Now for a last word about the validity of the theory at $t = 10^{-23}$

sec. The only prediction that can be experimentally confirmed is that the mass spectrum grows exponentially, which can be checked up to about 15 proton masses. At $t = \tau$ we are talking about particles of about $10^{38} m_p$, and from 15 to $10^{38}$ is a rather long way. However, by comparison with all other proposed thermodynamical descriptions of the early universe, this one has <u>all</u> the presently known properties of the strong interactions in it and is the best candidate to be extrapolated to very high densities. Free-particle thermodynamics is clearly out, as it doesn't hold even at present accelerator energies.

## THE STATISTICAL BOOTSTRAP MODEL (THERMODYNAMICAL MODEL) OF STRONG INTERACTIONS; A GUIDE TO THE LITERATURE (status of June 1972)

This guide contains two tables and a list of references. It is neither complete nor unbiased.

Table I gives the recommended reading sequence. After having read the introductory lectures (vastly overlapping) the reader should be able to enter the lower boxes at any place, though it might be advantageous to follow the given sequence. The reader will notice overlaps and inconsistencies, because the reading sequence does not coincide with the historical development.

Table II tries to picture the logical (and roughly the historical) connections. The isolated box with the name Koppe shall indicate that he was the first to contemplate on statistical-thermodynamical interpretations of pion production. Unfortunately his two papers came too early and went unnoticed. When two years later Fermi elaborated the same idea in great detail, he obviously had no knowledge of Koppe's work.

In both tables the sequence goes along falling lines of connection, unless indicated otherwise by an arrow.

## Table I
### Reading Sequence

**INTRODUCTIONS**

Cargese Lectures                                      34
(with introduction to particle phys.)

CERN Lectures                                         33
(supposes knowledge of particle phys.)

---

**STRUCTURE/FOUNDATION**

Bootstrap model:
9, 15, 19, 20, 21, 22, 28, 32, 33, 35, 37, 50, 52

Phase space for $E \to \infty$:
2, 33, 60, 65

Resonances = partic.:
5, 6, 16, 33, 50

Velocity distrib.:
12, 38, 48, 56

Relation to dual models:
24, 33, 42, 49, 61

---

**ASTROPHYS. APPL.**

Big Bang Cosmology:
10, 23, 31, 40, 42, 47, 51, 53, 62

Dense Stars:
45, 49, 59, 69

---

**MODIFICATIONS**

Coalescense at $T > T_0$: 43

Superfluidity at $T > T_0$: 68

Distinguishable particles: 1

---

**PARTICLE PHYSICS**

Inclusive one particle disk:
25, 30, 36, 37, 38, 48

Non-symmetric collisions:
8, 41, 57

Two particle correlations:
37, 58

Annihilation:
37, 39

Heavy pair production:
25, 29, 32

General remarks:
32, 37

Ericson fluctuations:
20

Computer programs:
25, 55

Table II
Logical Flow

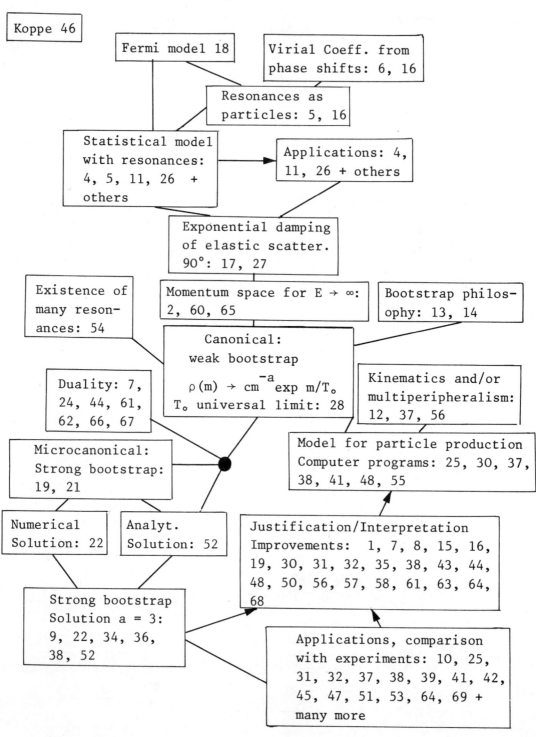

## REFERENCES

1. Alexanian, M., 1972, Phys. Rev., $\underline{D5}$, 922

2. Auberson, G., and Escoubès, B., 1965, Nuovo Cimento, $\underline{36}$, 628

3. Bahcall, J. N., and Frautschi, S. C., 1971, Astroph. Journ., $\underline{170}$, L81

4. Barashenkov, V. S., Barbashev, B. M., Bubelov, E. G. and Maksimenko, V. M., 1958, Nucl. Phys., $\underline{5}$, 17

5. Belenky, S. Z., 1956, Nucl. Phys., $\underline{2}$, 259

6. Beth, E. and Uhlenbeck, G. E., 1937, Physica, $\underline{4}$, 915

7. Brout, R., letter to R. Hagedorn, March 1969 (unpublished)

8. Buschbeck, B. and Hodl, J., 1972, Oesterr. Acad. d. Wiss., HEP-II

9. Carlitz, R. D., "Hadronic Matter at High Density", 1972, preprint Inst. Adv. Stud. Princeton

10. Carlitz, R. D., Frautschi, S. C. and Nahm, W., 1972, CERN preprint

11. Cerulus, F., 1959, Nuovo Cimento, $\underline{14}$, 827

12. Cerulus, F., 1970, Proceedings Coll. on High Multiplicity Hadron Interactions, Ecole Polytech. Paris

13. Chew, G. F., 1970, Physics Today, $\underline{23}$, 23

14. Chew, G. F. and Mandelstam, S., 1961, Nuovo Cimento, $\underline{19}$ 752

15. Chiu, C. B. and Heimann, R. L., 1971, Phys. Rev., $\underline{D4}$, 3184

16. Dashen, R., Ma, S. and Bernstein, H. J., 1969, Phys. Rev., $\underline{187}$, 345

17. Fast, G and Hagedorn, R., 1963, Nuovo Cimento, $\underline{27}$, 208
    Fast, G., Hagedorn, R. and Jones, L. W., 1963, Nuovo Cimento, $\underline{27}$, 856

18. Fermi, E., 1950, Prog. Theor. Phys., $\underline{5}$, 570

19. Frautschi, S.C., 1971, Phys. Rev., $\underline{D3}$, 2821

20. Frautschi, S. C., 1972, preprint CERN/TH, 1463

21. Frautschi, S. C., 1972, Lectures given at the Torino Colloqium on thermodynamics of strong interactions; forthcoming report of the Institute of Theor. Phys. Univ. Torino, Italy

22. Frautschi, S. C. and Hamer, C. J., 1971, Phys. Rev., $\underline{D4}$, 2125

23. Frautschi, S. C., Steigmann, A. and Bahcall, J., 1972, Ap. J., $\underline{175}$, 307

24. Fubini, S., Gordon, D. and Venewiano, G., 1969, Phys. Lett., $\underline{29B}$, 679

Fubini, S. and Veneziano, G., 1969, Nuovo Cimento, $\underline{64A}$, 811

25. Grote, H., Hagedorn, R., and Rahft, J., CERN-1970, "Areas of Particle Spectra"

26. Hagedorn, R., 1960, Nuovo Cimento, $\underline{15}$, 434

27. Hagedorn, R., 1965, Nuovo Cimento, $\underline{35}$, 216

28. Hagedorn, R., 1965, Suppl. Nuovo Cim., $\underline{3}$, 147

29. Hagedorn, R., 1968, Suppl. Nuovo Cim., $\underline{6}$, 311

30. Hagedorn, R., 1968, Nuovo Cimento, $\underline{56A}$, 1027

31. Hagedorn, R., 1970, Astron. and Astrophys., $\underline{5}$ 184

32. Hagedorn, R., 1970, Nucl. Phys., $\underline{B24}$, 93

33. Hagedorn, R., CERN 71-12 (yellow report)

34. Hagedorn, R., 1971, Lectures published in Proceedings of the Cargèse Summer Institute on Cosmology

35. Hagedorn, R., Torino lectures (see ref. 21)

36. Hagedorn, R., 1972, Lecture published in Proceedings of the Colloquium on high multiplicity hadron interactions, Zakopane, Poland

37. Hagedorn, R. and Ranft, J., 1968, Suppl. Nuovo Cim., $\underline{6}$, 169

38. Hagedorn, R. and Ranft, J., 1972, Nucl. Phys., $\underline{B48}$, 157

39. Hamer, C. J., 1972, preprint Calif. Inst. Techn., CALT-68-346

40. Harrison, E. R., 1970, Phys. Rev., D1, 2726

41. Htun Than and Ranft, J., 1972, preprint Leipzig Univ., KMU-HEP-7207

42. Huang, K. and Weinberg, S., 1970, Phys. Rev. Letters, 25, 895

43. Imaeda, K., 1971, Nuovo Cimento Letters, 1, 290

44. Krzywicki, A., 1969, Phys. Rev., 187, 1964

45. Koebke, K., Hilf, E. and Ebert, R., 1970, Nature, 226, 625

46. Koppe, H., 1948, Z. f. Naturforschung, 3A, 251
    Koppe, H., 1949, Phys. Rev., 76, 688

47. Kundt, W., 1971, Springer Tracts of Mod. Phys., 58, 1

48. Letessier, J, and Tounsi, A., 1972, Inst. Phys. Nucl. preprint IPNO/TH 72-6, Paris

49. Leung, Y. C. and Wang, C. G., 1971, Astrophys. Journ., 170, 499
    Lee, H., Leung, Y. C. and Wang, C. G., 1971, Astrophys. Journ., 166, 387

50. Matthiae, G., 1968, Nucl. Phys., B7, 142

51. Mollenhoff, C., 1970, Astron. and Astrophys., 7, 488

52. Nahm, W., 1972, Nucl. Phys., B45, 525

53. Omnes, R., 1972, Phys. Reports, 3C, 1

54. Particle Data Group, 1970, Phys. Letters, 33B, Number 1

55. Ranft, J., 1970, Leipzig Univ. Report TUL 37, reprinted as "Reprint 5" in ref. 25

56. Ranft, Gisela and Ranft, J., 1970, Phys. Lett., 32B 207

57. Ranft, J. and Mathaus, Leipzig Univ. Report KMU-HEP 7203

58. Ranft, Gisela and Ranft, J., Leipzig Univ. preprint KMU-HEP-7205

59. Rhvades, C. E. and Ruffini, R., 1971, Astrophys. Journ. (Lett.), 163, L 83

60. Satz, H., 1965, Nuovo Cimento, 37, 1407

61. Satz, H., Lecture given at Zakopane Colloquium (see ref. 36)
62. Schmid, C., 1968, Phys. Rev. Lett., $\underline{20}$, 628
63. Sertorio, L. and Toller, M., 1971, CERN-TH 1421
64. Stauffer, D., to be publ. in Phys. Rev. D, 1972
65. Vandermeulen, J., July, 1964, Preprint Univ. Liege (Belgium)
66. Veneziano, G., 1968, Nuovo Cimento, $\underline{57A}$, 190
67. Veneziano, G., 1969, Physics Today, $\underline{22}$, 31
68. Mann, A. and Weiner, R., 1971, Nuovo Cimento Letters, $\underline{2}$, 248
69. Wheeler, J. C., 1971, Calif. Inst. Techn. Orange Report OAP-239

Microphysics, Cosmology, and High Energy Astrophysics

Lectures by

F. Hoyle

Institute of Theoretical Astronomy, Cambridge

Notes by
S. Fulling, C. Dyer

Lecture 1

### 1. Units: $c=1$, $\hbar=1$

Consider at first a flat space-time where an observer can describe a general point by coordinates t and $\vec{x} = (x^1, x^2, x^3)$. The square of the 4-dimensional distance between this point and the origin is $t^2 - \vec{x}^2$. A second observer in motion relative to the first can arrange to have the same origin of coordinates, but he will assign different coordinates to the point. He will, however, get the same value for the 4-dimensional distance. This is the essence of special relativity.

To choose the velocity of light other than unity would be like measuring the spatial coordinates in different directions in different units (e.g., $x^1$ in meters, $x^2$ in yards, $x^3$ in cubits). The reason that the velocity of light is often not taken as 1 is that typically we compare two events 1 and 2 for which $|t_2 - t_1| \gg |\vec{x}_2 - \vec{x}_1|$; in other words, we live in a low-energy world. A convenient way to put in a scaling factor would be to write

t (scaled to convenient unit) $= 10^{-n}$ t

where n is a convenient integer. Instead, for historical reasons we write:

t (seconds) $= \dfrac{1}{299782500}$ t (meters),

and the large number in the denominator is called the velocity of light. But in fact it has nothing to do with light; it is a purely conventional scaling factor.

This eccentricity is sometimes pushed to absurdity in connection with Maxwell's equations. If $c \neq 1$ we must write, for instance,

$$\nabla \times \vec{B} = \dfrac{4\pi}{c}\vec{j} + \dfrac{1}{c}\dfrac{\partial \vec{E}}{\partial t}.$$

In the last term the $c^{-1}$ has arisen because of the redefinition of t; if we used the true special-relativistic time coordinates, the c would be absorbed into it. A similar remark applies to the $\vec{j}$ term, since a current density implies a sum of particle velocities and thus involves time derivatives. Unfortunately, some people try to get rid of the c by absorbing it into the $\vec{E}$, so that $\vec{E}$ and $\vec{B}$ have different dimensions. This is undesirable, since, from a relativistic point of view, $\vec{B}$ and $\vec{E}$ are components of the same tensor, $F_{ij}$.

Coordinates are actually measured by means of electromagnetic waves. Consider the plane wave:

$$\exp i\left[(kt - \vec{k}\cdot\vec{x})\right] \qquad (k = |\vec{k}|)$$

characterized by the vector $\vec{k}$. The quantity in brackets, like $t^2 - \vec{x}^2$, is a Lorentz invariant, although the individual components of $\vec{k}$ will of course appear different to different observers. The changes in $|\vec{k}|$ and in the direction of $\vec{k}$ are called, respectively, the Doppler effect and the aberration. If $\vec{x}$ is held fixed, the plane wave shows a phase oscillation $\exp(ikt)$ which has a period $\frac{2\pi}{k}$. Similarly, there is a behavior periodic in x when t is held fixed. Hence, some standard type of plane wave can be used to define a unit of time or length. The present metric standards of time and length are determined in this way from the light emitted by certain transitions of $Cs^{133}$ and $Kr^{86}$ respectively:

1 650 763.73   Kr fiducial marks = 1 meter
9 192 631 700  Cs fiducial marks = 1 second.

Thus the length and time scales are determined by the dynamics of atoms, and hence ultimately by the magnitudes of particle masses and charges. This brings us to the subject of Planck's constant $\hbar$ and the relation between the scales for mass and length.

In particle mechanics we define the action, S, for a given path as an integral, along the path, of the Lagrangian, L, so that

$$S = \int_{t_1}^{t_2} L dt .$$

Changing the path slightly from one with action S will give an action $S + \delta S$. When all such small displacements from the original path give $\delta S = 0$ to first order in small quantities, one has the classical path for the particle.

Consider a particle of mass m moving in an electromagnetic field with potential $A_i$. Then the non-relativistic Lagrangian is:

$$L = \frac{1}{2} m \dot{\vec{x}}^2 + e A_i \frac{dx^i}{dt} .$$

In the relativistic case we write for the 4-dimensional distance along the path:

$$da^2 = dt^2 - d\vec{x}^2 .$$

To lowest order we then have:

$$da = dt(1 - \frac{1}{2} \dot{\vec{x}}^2 + \ldots)$$

We postulate that the relativistic action is

$$S = -\int_{x_1}^{x_2} m\, da + e \int_{x_1}^{x_2} A_i dx^i .$$

One can easily see that, except for a term, $m(t_2 - t_1)$, which is independent of the path anyway, this is the same in the non-relativistic limit as the earlier prescription.

The condition $\delta S = 0$ determines the classical equations of motion. In the present case we derive in this way the Lorentz force,

$$F_i = e F_{ik} \frac{dx^k}{da} ,$$

where

$$F_{ik} = \frac{\partial A_k}{\partial x^i} - \frac{\partial A_i}{\partial x^k} .$$

We now turn to nonrelativistic quantum mechanics, in the formulation first proposed by Dirac and developed by Feynman. If a particle is at $x_1$, we want to know the probability that it will go to $x_2$. There is no unique path, and a wave function is formed by summing over all possible paths:

$$\psi = \sum_{\text{all paths}} \exp(iS/\hbar) \quad .$$

If S is large, a slight change in the path will change the phase angle greatly; then the contributions from neighboring paths will cancel each other except near where S has an extremum ($\delta S = 0$). Thus $S \gg \hbar$ is the condition yielding the classical limit. (More precisely a system is nonclassical if many paths differ in action by less than $\hbar$.)

In the phase function the h is needed to make $S/\hbar$ dimensionless. We could write

$$L = \frac{1}{2} \frac{m}{\hbar} \dot{x}^2 + \frac{e}{h} A_i \frac{dx^i}{dt} ,$$

making S already dimensionless. We can find the dimension of S by looking at the second term. The $A_i$ already involves the basic charge. The field due to a point charge in b can be written;

$$A^i(x) = e \int \delta(q^2(x, b)) db^i = e L^{-1}$$

where L has the dimension of length. (Here $q^2(x, b)$ is the distance between the field point x and the point b on the path and has dimensions $L^2$.) It follows that S has the dimension of $e^2$ (prior to division by $\hbar$). Hence $e^2/\hbar$ must be dimensionless, and since all particle charges are multiples of the fundamental charge unit, $e^2/\hbar$ is a fundamental number, the "fine structure constant":

$$e^2/\hbar = 7.297351 \times 10^{-3} .$$

Henceforth we shall absorb $\hbar$ into $e^2$.

Why not also absorb $\hbar$ into m in the first term of the action? From then on $\hbar$ will never be seen again. This can also be done in relativistic physics, for instance in the Dirac equation:

$$\gamma^k \frac{\partial \psi}{\partial x^k} + i \frac{m}{\hbar} \psi = 0.$$

This convention gives mass the dimension of $(\text{length})^{-1}$, and our physical constants are now the masses of particles and some dimensionless coupling constants. Take the unit of length to be the electron Compton wavelength:

$$m_e^{-1} = C = \frac{1}{2\pi} \, 2.4263096 \times 10^{-12} \text{ meters}.$$

(This equality is a statement about meters, rather than about electrons.) Then any physical quantity has the form:

(dimensionless number) $\times \, C^n$.

For instance, a typical distance between particles in a macroscopic body is $e^2 C$. The Rydberg constant, the basic frequency unit in atomic transitions, is $\frac{e^4}{4\pi} C^{-1}$.

## 2. Redshifts

Consider an atom at $x_1$ which emits light (monochromatic) and an observer at $x_2$ who receives the light. The frequencies involved are related to the number $C$, which cannot be determined absolutely. Nevertheless the frequencies (or, equivalently, the wavelengths) can be compared. If $\lambda_{x_1}$ and $\lambda_{x_2}$ are the emitted and received wavelengths, we write

$$\lambda_{x_2}/\lambda_{x_1} = 1 + z,$$

where, if $z > 0$, $z$ is termed the redshift.

In flat space there are only two explanations for a redshift:
1) $x_1$ and $x_2$ are in relative motion, so the $\vec{k}$ vector is changed, giving the Doppler effect.
2) The scale factors $C_1$ and $C_2$ (as determined by electron masses) are different at the two points.
On a Riemannian curved space-time, the geometry provides a third

possibility which, as we shall see, turns out to be related to 2).

Lecture 2

CONFORMAL TRANSFORMATIONS

As was pointed out earlier, one cannot find the local value of C, from physical measurements. One might think, though, that one can compare the values of C at different points. If we receive light at $x_2$ from matter at $x_1$, we can compare the wavelengths of spectral lines with those of light generated locally at $x_2$.

This would allow one to determine whether C is the same at both points, if one were working in a flat space-time (and one could be sure that the sources at $x_1$ and $x_2$ are at relative rest).

This procedure fails, however, if one admits Riemannian geometry. We shall see that in Riemannian geometry the wavelength at $x_1$ can be adjusted to anything we please without altering the observed wavelength ratio at $x_2$. This freedom is permitted by a conformal transformation of the Riemannian geometry.

Consider a Riemannian geometry with metric

$$ds^2 = g_{ik} \, dx^i dx^k$$

and a different geometry defined on the same coordinates:

$$ds^{*2} = g^*_{ik} \, dx^i dx^k .$$

If $g^*_{ik} = \Omega^2 g_{ik}$ where $\Omega(x) > 0$ and is non-singular, we say that the geometry has undergone a conformal transformation. There is a change in the geometrical length scale given by $ds^* = \Omega ds$.

Maxwell's equations are invariant under conformal transformations in the following sense. Let $F_{ik}$ be an electromagnetic field satisfying $F^{ik}_{;k} = 4\pi j^i$ for some current $j^i$. Let $F^*_{ik} = F_{ik}$ and $j^{*i} = \Omega^{-4} j^i$. Then the Maxwell equation continues to hold in the conformally transformed geometry:

$$F^{*ik}_{;k} = 4\pi j^{*i} .$$

The $\Omega^{-4}$ transformation law for the current is inevitable if all charged particles have the same coordinate paths in the two geometries. We have

$$j^i(x) = \sum e \frac{dx^i}{ds} \quad \text{and} \quad j^{*i}(x) = \sum e \frac{dx^i}{ds^*}$$

where each sum is over all particles in a box at x of unit <u>proper</u> 3-dimensional volume. Under the conformal transformation $dx^i$ remains the same, since both the particles and the coordinates have the same names as before. We pick up a factor $\Omega^{-1}$ from the way ds transforms and another factor $\Omega^{-3}$ from the transformation of the proper volume at x, so that we have $j^{*i}(x) = \Omega^{-4} j^i(x)$. (Note, incidentally, that $F^{*ik} = \Omega^{-4} F^{ik}$.)

If a system of charged particles following certain coordinate paths produces an electromagnetic wave $\exp(ik_i x^i)$ at x in a locally flat space-time in the unstarred geometry, then it produces the same wave in the starred geometry. This follows from the invariance of Maxwell's equations. However, the physical frequency will be different. The unstarred frequency is $k_4$, but the starred frequency is $\Omega^{-1} k_4$, since proper distance is $\Omega$ x coordinate distance in the starred geometry. A frequency transforms as the reciprocal of a length.

Suppose an observer at $x_2$ assigns coordinates $x^i$ to all points from which he receives radiation (such as a point $x_1$ on the world line of a charged particle). We ask whether he can determine uniquely 1) the geometry of the world, and 2) the value of C at each point (relative to C at some given point). The answer is <u>no</u>, as will now be shown.

Assume to the contrary that our observer claims to have determined the geometry and the scaling factor C uniquely. Introduce a new (starred) geometry by a conformal transformation with a function $\Omega$ such that $\Omega(x_1) \neq \Omega(x_2)$. Multiply the physical unit $C_1$ by $\Omega(x_1)$ and $C_2$ by $\Omega(x_2)$. Then nothing has changed as far as physical experiments, such as wavelength determinations, are concerned. Light leaves $x_1$ and arrives at $x_2$ with the same frequency relative to the physical unit as in the unstarred situation. Thus we have introduced a new geometry and new physical units but no change in observational physics. Hence <u>if the physical unit C is allowed to be a function of position, there is no unique geometry</u>.

The situation would be different if we knew that the physical unit was the same at all points. But since experiments can only determine dimensionless numbers, how could one know that? Any abstract theory that fixes C would seem to be based on prejudice.

In the early days of general relativity it was realized that physical laws are invariant under general coordinate transformations. This, which might seem to be an embarrassment, came to be regarded as a virtue. The present situation is similar. We ought to require also that physics be invariant under all conformal transformations of the metric. The choice of a particular metric should have no deeper significance that a particular choice of coordinate system.

Traditional physics is not conformally invariant because of the constant particle masses. The action for a particle of matter is $S = -m \int da$. A conformal transformation takes $da$ into $da^* = \Omega da$. Hence the action will be invariant only if $m^* = \Omega^{-1} m$ and $m^*$ is taken inside the integral (since $\Omega(x)$ may be a function of position). Then

$$S = - \int da\, m(A)$$

where $A$ parametrizes the point along the world line.

Next consider the Dirac equation:

$$\gamma^k \frac{\partial \psi}{\partial x^k} + im\psi = 0.$$

The first term is conformally invariant when $\psi$ transforms

Microphysics, Cosmology and High Energy Astrophysics        308

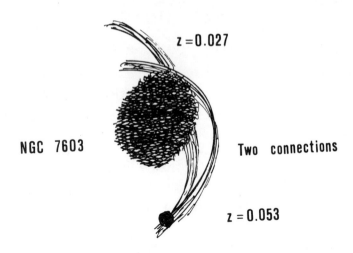

Figure 2.1

according to $\psi^* = \Omega^{-3/2} \psi$. The second term is conformally invariant if and only if m transforms just as described above.

It is known that by a suitable coordinate transformation any metric can be made flat in a neighborhood of a given point. The popular view is that:
1) Space can be regarded as flat for all practical purposes in a region as large as a cluster of galaxies.
2) The C values are the same throughout such a region of space.
   I want to assault this idea.
   Prof. Arp has loaned me his prize slide, which is of NGC 7603 (Fig. 2.1). The large object is a Seyfert galaxy with a redshift $z = 0.027$ while the redshift for the small object is $z = 0.053$. Interpreted as a Doppler shift, this indicates a relative speed of 8000 km/sec. Since the small object lies at the intersection of two filaments from the larger object, they seem to be clearly physically associated. It is true that if the small object is whizzing by the larger, the larger should be tidally disturbed; but this effect should not show up until the object has passed and can hardly account for the observed structure. If the galaxy shot the small object out of itself, one would expect it to move radially and would not expect these curved filaments. Why, in such associations, are the discrepant objects always red shifted relative to the larger object? Statistically half of them should be blue shifted relative to the larger object. Why is the central object a peculiar Seyfert galaxy when these are only about 1% of normal galaxies? The idea that we are seeing

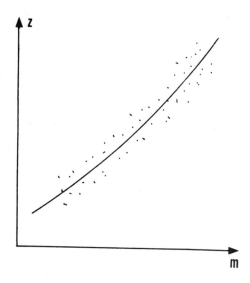

Figure 2.2

a chance coincidence with a more distant background object is very implausible, except to people with a very strong prejudice against a space-varying C.

Another Arp example (VV 172) is less convincing by itself

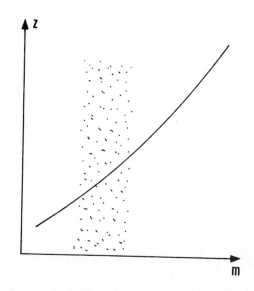

Figure 2.3

than the preceding example, since the discrepant object <u>could</u> be in the background, but it adds to the cumulative evidence.

Finally, consider the Hubble relation plotted so that the z scale is linear. Fig. 2.2 is a best fit for galaxies while Fig. 2.3 shows the relation for QSR's ($q_o$ = +1). In the second figure, the fit is atrocious.

It seems hard to explain such phenomena while maintaining the normal prejudice that C is constant. I am hopeful that the theory I am presenting to you can be used to explain them, although so far I do not know how to do it.

Lecture 3

## CONFORMALLY INVARIANT THEORY OF GRAVITATION

The simple classical action for a particle of mass m is

$$S = -\int m_a(A)\,da$$

where $m_a(A)$ is the mass of the particle a at the point A on its world line. Considering the conformal transformation where $g^*_{ik} = \Omega^2(x) g_{ik}$, it is obvious that conformal invariance requires $m^*_a(A) = \Omega^{-1}(A)\, m_a(A)$. Thus the masses of particles appear as fields of some kind.

Consider another particle b. This particle produces a field at each point x, which at A makes a contribution to the mass of a. Hence, summing over all such particles b, we have for the mass of a

$$m_a(A) = \sum_{b \neq a} m^{(b)}(A)$$

where self-interaction has been excluded. This is analogous to action-at-a-distance electromagnetic theory. The total action of matter can be written as

$$S_{mat} = -\frac{1}{2} \sum_a \int m_a(A)\,da$$

so that we have

$$S_{mat} = -\frac{1}{2} \sum_a \sum_{b \neq a} \int m^{(b)}(A)\,da$$

If a and b are similar, one should expect symmetry so that $m^{(b)}(A)$ is of the form:

$$\tilde{m}^{(b)}(A) = \int P(A, B) db$$

with $P(A, B) = P(B, A)$. We then have for $S_{mat}$

$$S_{mat} = - \sum\sum_{a<b} \int\int P(A, B) da\, db$$

If mass is to be a scalar, then $P(A, B)$ should be a scalar function, and in view of the transformation of da and db, $P(A, B)$ must transform under the conformal transformation as

$$P^*(A, B) = \Omega^{-1}(A)\, \Omega^{-1}(B)\, P(A, B).$$

The only freedom left, if $P(A, B)$ is to satisfy a second order wave equation, is a constant factor to be determined by experiment. That is

$$P(A, B) = \lambda^2\, \tilde{G}(A, B),$$

where $\tilde{G}(A, B)$ satisfies the equation

$$\square_x \tilde{G}(X, B) + \frac{1}{6} R(x)\, \tilde{G}(X, B) = [-g(B)]^{-1/2} \delta_4(X, B) \qquad (3.1)$$

It is also required that $\tilde{G}(X, B) = \tilde{G}(B, X)$, which implies an equal mixture of advanced and retarded propagation. If the constant $\lambda^2$ is positive, the masses will be positive. Conformal invariance forces the appearance of the contracted Riemann tensor term $R(X)$.

In flat space the equation for $G(X, B)$ reduces to $\square_x \tilde{G}(X, B) = \delta_4(X, B)$ so that $\tilde{G}(X, B) = (1/4\pi)\, \delta(q^2(X, B))$, where $q^2(X, B)$ is the squared distance from X to B.

The total action is $S_{tot} = S_{mat} + S_{em}$ + terms from other possible interactions. The electromagnetic action, $S_{em}$, can be obtained solely from particles and does not require additional degrees of freedom for the fields involved. In the usual Einstein theory the action has the form

$$S = -\sum_a m_a \int da + \frac{1}{16\pi G} \int R(-g)^{1/2} d^4x$$

which is a mixture of a particle term (the first) and a field term (the last).

If we now require $\delta S = 0$ for $g_{ik} \to g_{ik} + \delta g_{ik}$, in the purely gravitational case (i.e., no electromagnetic interaction), we have

$$\delta \left\{ -\lambda^2 \sum_{a<b} \sum \iint \tilde{G}(A, B) \, da\, db \right\} = 0.$$

We can then write

$$\delta \left\{ \sum_a \int m_a(A) da \right\} = 0.$$

We do not have a mixture of field and particle terms to consider as we would have in the usual theory. In the corresponding quantum theory, one must consider the sum $\sum \exp(i\, S_{mat})$ as before, only now summing over all metrics as well as all particle paths. The condition $\delta S = 0$ yields the gravitational field equations (Hoyle and Narlikar, 1964)

$$F(R_{ik} - \frac{1}{2} R\, g_{ik}) = -3(T_{ik} + \Phi_{ik}) + (g_{ik} \Box F - F_{;ik}) \qquad (3.2)$$

where

$$F = \frac{1}{\lambda^2} \sum_{a<b} \sum m^{(a)} m^{(b)},$$

with $m^{(a)}(x) = \lambda^2 \int \tilde{G}(X, A) da$,

$$\Phi_{ik} = -\sum_{a<b} \sum \left\{ m_i^{(a)} m_k^{(b)} + m_k^{(a)} m_i^{(b)} - g_{ik} m_l^{(a)} m^{(b)l} \right\}$$

with $m^{(a)}_i = \partial m^{(a)}/\partial x^i$, $m^{(a)1} = g^{il} m^{(a)}_i$, etc., and $T_{ik}$ is the usual energy-momentum tensor, which is most easily written in the contravariant form

$$T^{ik} = \lambda^2 \sum_a \int \delta_4(X, A) \left[-g(A)\right]^{-1/2} m_a(A) \frac{da^i}{da} \frac{da^k}{da} da.$$

It is striking to remark that the action prescription is simpler than in general relativity, but the field equation is considerably more complex. The problem is whether it is also more useful. In weak field problems, or problems involving a very large number of particles (cosmological problems) we can write

$$F \simeq \frac{1}{2\lambda^2} \left(\sum_a m^{(a)}\right)^2 \quad \text{and}$$

$$\Phi_{ik} \simeq -\left(\sum_a m^{(a)}\right)_i \left(\sum_b m^{(b)}\right)_k + \frac{1}{2} g_{ik} g^{lp} \left(\sum_a m^{(a)}\right)_l \left(\sum_b m^{(b)}\right)_p.$$

Since $F^* = \Omega^{-2} F$ under a conformal transformation, we can choose a conformal transformation such that $\Omega(x) \propto F^{1/2}$ and hence have $F^* = $ constant independent of X. Dropping the $*$, assume that we have F = constant. We then have $F = \frac{1}{2} m^2 = $ constant $> 0$, and

$$m(x) = \sum_b m^{(b)}(x) = \text{const.} = m_o.$$

Then all the derivatives in $\Phi_{ik}$ are zero, so that $\Phi_{ik} = 0$. The field equations then reduce to:

$$R_{ik} - \frac{1}{2} R g_{ik} = -\frac{3\lambda^2}{F} T_{ik}.$$

Define G by the relation

$$8\pi G = \frac{3\lambda^2}{F} = 6\lambda^2/m_o^2$$

so that

$$R_{ik} - \frac{1}{2} R g_{ik} = -8\pi G T_{ik}$$

We have arrived at the Einstein field equations. Note that there is no $\Lambda$ term nor can there be one in these field equations. Note also that we can always, by conformal transformation, make F constant, but we cannot always show that m(x) is constant. The constancy of m(x) required that the double sum in F reduce to the square of a single sum, which required in turn that we work in the weak field approximation or be considering a large number of particles.

Since F was shown to be a positive constant, it follows immediately that G is also a positive constant. Previously this was only a result of observation.

It is well known that Robertson-Walker spaces are conformally flat, but this is of little use if the gravitational theory is not conformal. One advantage of this theory is that we can now transform everything to flat space and obtain some simplification.

## Lecture 4

In this theory particle orbits can be found in two ways: One is to take the divergence of the field equations above. The other is to vary the action

$$S = -\int m_a(A)\,da + e_a \int A_i\,da^i$$

with respect to the particle paths and require that $\delta S = 0$. (The existence of two methods of deriving the particle equations of motion suggests that there is some redundancy in the theory, and in fact a closer analysis shows that besides the particles there are only 6 degrees of freedom in the gravitational field, not 10). The equations of motion which result are

$$\frac{d}{da}\left(m_a \frac{da^i}{da}\right) + m_a \Gamma^i_{kl} \frac{da^k}{da}\frac{da^l}{da} - g^{ik}\frac{\partial m_a}{\partial a^k} =$$

$$= e_a \sum_{b \neq a} F^{(b)i}_{\ \ k} \frac{da^k}{da},$$

where $F^{(b)}_{ik}$ is the electromagnetic field due to particle b. In the conformal frame where $m_o = \sum_b m^{(b)} =$ constant, we have, provided that

$$m^{(a)} \ll \sum_{b \neq a} m^{(b)},$$

constant particle masses:

$$m_a = \sum_{b \neq a} m^{(b)} \simeq \sum_b m^{(b)} = m_o.$$

Then the mass derivative term vanishes and we have the standard geodesic equation plus the electromagnetic interaction.

The approximation we have been making ignores the singularity

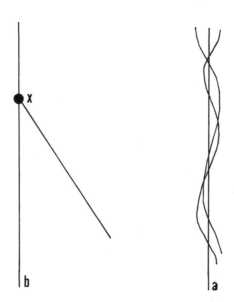

Figure 4.1

of $m^{(a)}(A)$ at points on the worldline of particle a itself. This approach should be all right if there are so many particles that they can be regarded as a smooth fluid. One might worry, however, about the effect on $m^{(a)}(A)$ of the metric singularity caused by the point particle a. Narlikar and I have some interesting unpublished results on this problem in the classical context, but the full resolution lies in quantum mechanics.

In the quantum treatment we must consider all possible paths. Now all the paths of particle a which have significant probability lie roughly in the timelike direction, with rather little spatial spread (Fig. 4.1). Therefore the classical approximation should make very little difference in the mass field $m^{(a)}(X)$ at a point X which is far away from the particle. Near the particle, however, one must take account of the spatial spread, which is of the order of the Compton wavelength, C. One expects, then, that on the worldline of particle a

$$m^{(a)}(A) \simeq \frac{\text{const.}}{C} \; .$$

On the other hand, most other particles in the universe will be at cosmological distances from a, so that typically

$$m^{(b)}(A) \simeq \frac{\text{const.}}{H^{-1}},$$

where H is the Hubble constant. If there are N such particles,

$$\sum_{b \neq a} m^{(b)}(A) \simeq N \frac{\text{const.}}{H^{-1}}$$

Now we have

$$m_a(A) = \sum_{b \neq a} m^{(b)}(A) + m^{(a)}(A),$$

and thus for the self-action term to be negligible we need NCH >> 1. In nature we have $N \simeq 10^{80}$, $CH \simeq 10^{-40}$, so the approximation is very good.

## PROBLEM OF A SPHERICALLY SYMMETRIC BODY

Consider a rather compact spherically symmetric collection of particles in a universe which is also spherically symmetric. Choose a conformal frame in which

$$m(x) = \sum_b m^{(b)}(x) = m_o \quad (\text{constant}).$$

Outside the body where the density of particles is very low, we can use the vacuum Schwarzschild solution for the metric:

$$ds^2 = (1 - \frac{S}{r}) dt^2 - \frac{dr^2}{1 - \frac{S}{r}} - r^2(d\theta^2 + \sin^2\theta d\phi^2).$$

S is the Schwarzschild radius, which is a constant of integration which must be matched to the interior solution. We can write $S \simeq 2Gnm_o$, where n is the number of particles in the body. Now $8\pi G = 6\lambda^2/m_o^2$ and $m_o \simeq \lambda^2 N/H^{-1}$,

where N is the total number of particles in the universe. Hence,

$$S \simeq H^{-1} n/N$$

The mass field of each particle obeys the equation

$$\Box_x m^{(b)}(x) + \frac{1}{6} R(x) m^{(b)}(x) = \lambda^2 \int [-g(B)]^{-1/2} \delta_4(x, B) \, db \tag{4.1}$$

Summing over b, we find

$$\frac{1}{6} R(x) m_o = \lambda^2 \sum_b \int [-g(B)]^{-1/2} \delta_4(x, B) \, db \tag{4.2}$$

The sum can be divided into a sum over the body particles and a sum over the rest of the universe. Let us study the relative importance of these two terms. Let

$$m_o = m_{ex}(r) + m_{in}(r) \,,$$

where

$$\Box_x m_{ex} + \frac{1}{6} R(x) m_{ex} = \lambda^2 \sum_{universe} \int [-g(B)]^{-1/2} \delta_4(x, B) \, db \tag{4.3}$$

$$\Box_x m_{in} + \frac{1}{6} R(x) m_{in} = \lambda^2 \sum_{body} \int [-g(B)]^{-1/2} \delta_4(x, B) \, db \tag{4.4}$$

With the help of the equation (4.2), $R(x)$ can be eliminated from equation (4.3) and (4.4). One obtains

$$\Box_x m_{ex} = -\lambda^2 \frac{m_{ex}}{m_o} \sum_{body} \int [-g(B)]^{-1/2} \delta_4(x, B) \, db + \text{universe terms} \tag{4.5}$$

$$\Box_x m_{in} = \lambda^2 (1 - \frac{m_{in}}{m_o}) \sum_{body} \int [-g(B)]^{-1/2} \delta_4(x, B) \, db +$$

$$+ \text{universe terms} \tag{4.6}$$

As in the case of the equations for the metric, we can treat the space outside the body as vacuum and drop the universe terms. Now outside the body, the body terms and the R dependent terms are also zero. For spherical symmetry

$$\Box_x \equiv \frac{1}{r^2} \frac{d}{dr}\left[r^2 (1 - \frac{S}{r}) \frac{d}{dr}\right],$$

so the equations can easily be solved:

$$m_{ex} = m_o + \frac{A}{S} \ln(1 - \frac{S}{r}),$$

$$m_{in} = -\frac{A}{S} \ln(1 - \frac{S}{r}) \qquad (\sim \frac{A}{r} \text{ at large } r)$$

where A is a constant of integration (which depends on the radius of the body). The other constant of integration, $m_o$, takes into account the effect of the mass in the outer universe which we neglected above.

Now, if $r_o$, the radius of the body, is much larger than S, it is easy to show that

$$A = \lambda^2 \frac{n}{4\pi},$$

and in this case $m_{ex} \gg m_{in}$. However, as $r_o \to S$ from above, $A \to 0$. (If this were not so, e.g., if $A > \varepsilon > 0$, we see from the solution with $r \to S$ that $m_{ex}$ becomes very large and positive, so that across the boundary the source term in

$$\Box_x m_{ex} = -\lambda^2 \frac{m_e}{m_o} \sum_{body} \int [-g(B)]^{-1/2} \delta_4(x, B) \, db$$

will be large and negative. In analogy with electrostatics, a negative source is inconsistent with a positive solution for $m_{ex}$. The same sort of argument excludes $A < 0$.) It follows that the body sources in the differentail equations for $m_{ex}$ and $m_{in}$ must tend to zero. Thus

$$\frac{m_{ex}}{m_o} \to 0, \quad 1 - \frac{m_{in}}{m_o} \to 0,$$

or

$$m_{ex} \to 0, \quad m_{in} \to m_o.$$

To summarize, when $r_o \gg S$, the theory is nearly classical: the masses of the body particles are determined by the external universe and hence fixed as regards the internal dynamics of the body. But when $r_o \to S$, the opposite is true. The influence of the universe is cut off, and the particles determine their own destiny. It is quite possible, therefore, that black hole physics is significantly different in our theory and in general relativity.

Besides this, however, one must check two things in the black hole problem. First, has our Einsteinian approximation broken down, that is, can we assume inside the body that

$$\sum_{a<b} m^{(a)} m^{(b)} \gg \frac{1}{2} \sum_a (m^{(a)})^2 ?$$

It is possible that in the (unknown) curved metric inside the body, some terms contribute to the sum on the left with a negative sign. Also, must this problem be treated quantum-mechanically?

Lecture 5

FRIEDMANN MODEL WITH k = 0

We are now ready to consider the consequences of the theory for the standard Friedmann cosmological models, in which space-time has the Robertson-Walker metric

$$ds^2 = dt^2 - Q^2(t)\left[\frac{dr^2}{1-kr^2} + r^2(d\theta^2 + \sin^2\theta d\phi^2)\right]$$

with k = 0 or ± 1. These spaces are conformal to flat Minkowski space. Since our gravitation theory is now conformally invariant, we can make more use of this fact than is usually done. At first let us consider just the case k = 0 (the Einstein-de Sitter model).

We assume, in keeping with the usual theory, that with respect to the (unstarred) Robertson-Walker metric $m_o$ is a constant and Einstein's equations hold. We shall introduce a starred conformal frame in which $m^*$ is variable and the metric is that of flat space. First we make a <u>coordinate</u> transformation

$$\tau = \int_0^t \frac{dt}{Q},$$

so that

$$ds^2 = Q^2\left[d\tau^2 - dr^2 - r^2(d\theta^2 + \sin^2\theta d\phi^2)\right].$$

Then it is obvious that one should choose $\Omega = Q^{-1}$, and thus pass to a starred conformal frame with

$$ds^{*2} = d\tau^2 - dr^2 - r^2(d\theta^2 + \sin^2\theta d\phi^2),$$

$$m^* = Qm_o.$$

We must determine the dependence of Q on $\tau$. Let n be the particle density in the unstarred metric. Then $n^* = \Omega^{-3}n = Q^3 n$. Now for the pressureless Einstein-deSitter model we have $Q^3 n =$ constant $\equiv L^{-3}$. Also, one of Einstein's equations is

$$\frac{\dot{Q}^2}{Q^2} = \frac{8\pi G}{3} \sigma,$$

where $\sigma = nm_o$ and $\dot{Q} = \frac{dQ}{dt}$. Since $8\pi G = 6\lambda^2/m_o^2$, we have

$$\dot{Q}^2 = \frac{2\lambda^2}{m_o L^3} \frac{1}{Q}$$

If $Q(0) = 0$, the solution is

$$Q(t) = \left[\frac{2\lambda^2}{m_o L^3}\right]^{1/3} (\frac{3}{2}t)^{2/3}.$$

Hence we find

$$\frac{1}{2}\tau = \left[\frac{2\lambda^2}{m_o L^3}\right]^{-1/3} (\frac{3}{2}t)^{1/3},$$

$$Q(\tau) = \frac{1}{2} \frac{\lambda^2}{m_o L^3} \tau^2$$

and, finally,

$$m^* = \frac{1}{2} \lambda^2 L^{-3} \tau^2.$$

All observational statements should be the same when made in either starred or unstarred terms. Let us verify this for the case of the redshift vs. magnitude relation. In the starred frame we can treat the behavior of light using flat-space ideas. Let the observer have the coordinate $r = 0$ at time $\tau$ (Fig. 5.1). Then he can receive light emitted at time $\tau - r$ from a galaxy at distance r. (Recall that the cosmological particles have fixed values of r, $\Theta$, and $\phi$ in these models.) If the luminosity of the galaxy is $L^*(\tau - r)$, the observed flux is

$$\frac{L^*(\tau - r)}{4\pi r^2}$$

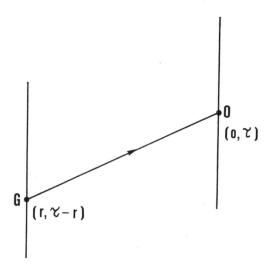

Figure 5.1

Suppose there are some standard galaxies which all have the same number and type of stars, and neglect stellar evolution effects. Then L, the intrinsic luminosity, is independent of r. Now dimensionally (luminosity) = (energy)/(time) = $c^{-1}/C = c^{-2}$ = (mass)$^2$, and therefore $L^* = Q^2 L$. Hence the observed flux is proportional to $\frac{(\tau - r)^4}{r^2}$. In the starred frame the observer and the galaxy are stationary, so the redshift is due entirely to the difference in the physical unit:

$$1 + z = \frac{m^*(\tau)}{m^*(\tau - r)} = \frac{\tau^2}{(\tau - r)^2}$$

Eliminating r from these two equations, one finds

observed flux $\propto (1 + z)^{-1} \left[ (1 + z)^{1/2} - 1 \right]^{-2}$,

which is the standard Hubble relation.

We have assumed that the masses are constant in the original Robertson-Walker frame, so that

$$m^* = \frac{1}{2} \lambda^2 L^{-3} \tau^2.$$

Can we check that this is consistent with the theory? In flat space the propagator is

$$\tilde{G}(X, B) = \frac{1}{4\pi} \delta(s^{*2}),$$

where $s^*$ is the four-dimensional distance from X to B. So the problem is equivalent to that of calculating the Coulomb potential. It is made even easier by the fact that we have a constant density of particles.

Let us compute the <u>retarded</u> contribution to the mass field at a point from all points on the past light cone back to $\tau = 0$, the origin of the universe. So we must include all particles out to a distance $r = \tau$. The Figure 5.2 shows that the other particles, at c, do not matter. From a single particle at distance r we have a contribution $\frac{\lambda^2}{8\pi r}$. (A factor of $\frac{1}{2}$ has come in because we are considering only the retarded half of the delta function.) Since the particle density is $L^{-3}$, the effect of all the particles is

$$\frac{\lambda^2}{8\pi} L^{-3} \int_0^\pi \sin\theta d\theta \int_0^{2\pi} d\phi \int_0^\tau \frac{r^2 dr}{r} = \frac{1}{4} \lambda^2 L^{-3} \tau^2,$$

which is exactly one half of what we want. We have yet to con-

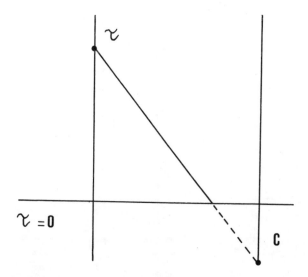

Figure 5.2

sider the advanced potential; by symmetry we expect an equal contribution from it, which will give us the desired result for the mass.

But there is a snag in this argument. Although the Green function G is symmetric between past and future, the sources apparently are not. In the retarded calculation there was a natural cutoff at $\tau = 0$, the beginning of the universe. In the advanced case there is no such cutoff, and a direct calculation of the advanced potential yields an infinite result. Yet I am convinced that the symmetrical result above is correct, as it is in electrodynamics, and that we must look around for some new conceptual idea which will lead to it.

Note that in the starred frame there is nothing strange about the geometry at $\tau = 0$. Why should we not go back beyond that point? It is true that Q is zero there, which violates one of the rules in the definition of a conformal transformation. But why must we regard the Robertson-Walker picture, rather than the flat one, as the correct one? The situation is like the case of the de Sitter model in the static form in which de Sitter first proposed it. There the universe seems to end at a horizon. But de Sitter's space can be embedded in a larger Robertson-Walker space, and it is seen that only the coordinate transformation functions are singular on the horizon. Let us take the similar

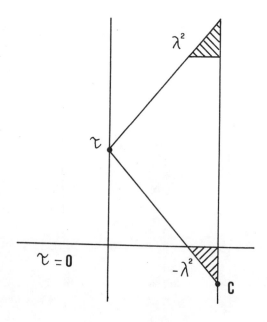

Figure 5.3

position in the present case that the singularity at $\tau = 0$ in the Friedmann model is fictitious.

Nevertheless, this singularity has some relevance to physics, since the masses become 0 there. We are able to assume that across the surface $\tau = 0$ the coupling constant $\lambda^2$ acts negatively. Then the contributions to the retarded potential from particles with $\tau < 0$ will cancel the embarrassing part of the advanced potential, as shown in Figure 5.3.

Lecture 6

## FRIEDMANN MODELS WITH k = ± 1

We now consider the cases $k = \pm 1$ in the Robertson-Walker metric:

$$ds^2 = dt^2 - Q^2(t)\left[\frac{dr^2}{1-kr^2} + r^2(d\theta^2 + \sin^2\theta d\phi^2)\right].$$

This is the frame in which particle masses are constant. Define

$$\alpha = \frac{8\pi G}{3} m_o (nQ)^3 = \frac{3\lambda^2}{m_o} nQ^3$$

where $m_o$ is the particle mass, $(nQ)^3$ the particle number density, and the second form of $\alpha$ comes from the earlier definition of G. There exists a conformal transformation to the Minkowski frame with metric:

$$ds^{*2} = d\tau^2 - d\rho^2 - \rho^2(d\theta^2 + \sin^2\theta d\phi^2),$$

where the angular coordinates are unchanged and $\rho = \rho(r, t)$ and $\tau = \tau(r, t)$.

1.) <u>k = +1</u>: In this case one finds that the mass transforms as:

$$m^* = \alpha\, C^{-1} m_o \frac{\{1 + (\tau+\rho)^2\}^{1/2} \{1 + (\tau-\rho)^2\}^{1/2} + \tau^2 - \rho^2 - 1}{\{1 + (\tau+\rho)^2\} \{1 + (\tau-\rho)^2\}}$$

where C is the basic physical unit discussed earlier. Since in the classical theory we are considering only one class of particles, we may take $m_o = C^{-1}$. Using the definition of $\alpha$, we then obtain:

$$m^* = 3\lambda^2 nQ^3 m_o \frac{\{1 + (\tau+\rho)^2\}^{1/2} \{1 + (\tau-\rho)^2\}^{1/2} + \tau^2 - \rho^2 - 1}{\{1 + (\tau+\rho)^2\} \{1 + (\tau-\rho)^2\}}$$

While in R-W the orbits of particles are given by r=const, in the Minkowski frame they are given by $\rho = (A^2 + \tau^2 - 1)^{1/2} - A$, where A is a constant. Hence a particle starting from $\tau = 0$ at some $\rho$ (which determines A) has $d\rho/d\tau = 0$ at $\tau = 0$ and $d\rho/d\tau = 1$

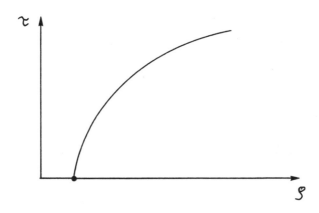

Figure 6.1

as $\tau \to \infty$ (Fig. 6.1).

If we return to the equations of motion we obtained earlier, we find that they only reduce to the familiar form in that conformal frame where masses are constant. Hence the odd particle trajectories obtained above are due to the non-zero 4-dimensional gradient of $m^*$. The density can be obtained for any $\tau$, and in particular, at $\tau = 0$ the density is proportional to $(1 + \rho^2)^{-3}$. In this coordinate system we then have a cloud of particles (concentrated mainly within $\rho \leq 1$ for $\tau = 0$) which disperses with time. We saw earlier that homogeneity and isotropy of the Einstein-deSitter universe was not destroyed by the conformal transformation to the Minkowski frame. One has to have very real doubts about the validity of the $k = +1$ model for it is a local cloud and one could start with any number of local clouds. The case $k = +1$ seems to be a pathological case which happens to give homogeneity and isotropy in the particular R-W frame. $m^*$ is zero at $\tau = 0$ and $m^* \to 0$ as $\tau \to \infty$. One finds that in taking $\tau$ from 0 to $\infty$, only the expanding half of the cycloid giving $Q(t)$ is traced out (Fig. 6.2). If we want the collapsing half to follow, the cloud must, for some reason, start falling back in on itself. In the Minkowski frame $d\rho/d\tau \to 1$ as $\tau \to \infty$ so the particles "arrive" at $\infty$ with non-zero velocity and hence it is hard to see why they should start falling back again.

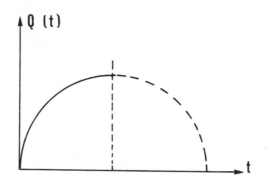

Figure 6.2

2.) <u>k = -1</u>: This case is similar to the previous one, except that trigonometric functions are replaced by hyperbolic functions. This gives rise to some changes of sign:

$$m^* = -\alpha m_o^2 \frac{\{1 - (\tau+\rho)^2\}^{1/2} \{1 - (\tau-\rho)^2\}^{1/2} + \rho^2 - \tau^2 - 1}{\{1 - (\tau+\rho)^2\} \{1 - (\tau-\rho)^2\}}$$

$$\rho = -(A^2 + \tau^2 - 1)^{1/2} + A$$

Density at $\tau = 0$ is proportional to $(1 - \rho^2)^{-3}$. We thus see that every particle orbit collapses to $\rho = 0$ at $\tau = 1$ (Fig. 6.3). The density distribution lies completely between $\rho = 0$ and 1 for $\tau = 0$ and is singular at $\rho = 1$. This model becomes quite singular in the Minkowski frame and seems pathological

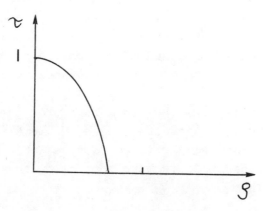

Figure 6.3

in the sense that in the R-W frame it becomes homogeneous. Considering that there could be a number of such clouds, even an infinity of them, in the Minkowski frame, one must have grave doubts about the applicability of this model to the whole universe.

The Einstein-deSitter model thus seems to be the only Friedmann model demanding serious attention unless we are willing to admit that what we investigate with telescopes, etc., is only a piece of our universe and thus may be described by an expanding cloud model in a hierarchial structure. (The $\Lambda \neq 0$ models are excluded because they are not conformally invariant.)

## Physical Constants

We have introduced the dimensionless coupling constant $\lambda^2$ and the mean particle spacing $L^{-3}$ over and above the physical unit C. In Minkowski frame we had $m^* = \frac{1}{2}\lambda^2 L^{-3} \tau^2$. Then $C^* = m^{*-1}$ shows the time dependence of $C^*$. In the R-W frame we have $C = m_o^{-1}$ which is a constant. Consider now the Einstein-deSitter model. We have

$$1 + z = \left(\frac{\tau}{\tau-r}\right)^2$$

for a galaxy at radial distance r. Expanding in powers of $r/\tau$, we have

$$1 + z = 1 + 2r/\tau + O(r/\tau)^2,$$

and hence

$$z \simeq 2r/\tau.$$

The Hubble constant is defined empirically by $z = H\tau$ and hence $H^{-1} = \tau/2$. In Minkowski space $\sigma = m^* L^{-3}$ is the density and since $8\pi G = 6\lambda^2/m^{*2}$, we have

$$8\pi G \sigma = \frac{6\lambda^2}{m^*} L^{-3} = 12\tau^{-2} = 3H^2.$$

Hence we have for the density

$$\sigma = 3H^2/8\pi G$$

This is the same expression one obtains in the Robertson-Walker frame.

So far nothing has been said about which particle we are considering, but for the sake of argument take it to be the proton. Once G has been determined from planetary orbits, etc., and we have determined $m_p$ empirically, we can determine the dimensionless coupling constant, $\lambda^2$. Using $G \simeq 10^{-66}$ cm$^2$ and $m_p \simeq 10^{14}$ cm$^{-1}$, we find from $\lambda^2 = 4\pi G/3 m_*^2$ that $\lambda^2 \simeq 5 \times 10^{-38}$. Collecting the numbers together, we have:

$$\lambda^2 \simeq 5 \times 10^{-38}, \quad L^{-3} \simeq 2 \times 10^{-6} \cdot cm^{-3}, \quad H^{-1} \simeq 2 \times 10^{28} \text{ cm},$$

$$\tau \simeq 4 \times 10^{28} \text{ cm}, \quad L^{-3}\tau^3 \simeq 10^{80}.$$

$L^{-3}\tau^3 \simeq$ the number of particles with which we are in communication at the present moment. Note the coincidence of $(L^{-3}\tau^3)^{1/2}$ and $\lambda^{-2}$ which arises also in other cosmologies. Since the coupling between particles was of the form $\lambda^2 \iint \tilde{G}(A, B) \, dadb$ and we should associate one power of $\lambda$ with each of the integrals, changing $\lambda^2$ from $\simeq 10^{-40}$ to $\lambda^2 \simeq 1$ should increase all masses by a factor of $10^{20}$. In such conditions the interaction of two particles placed at 1 Fermi distance would compete in intensity with strong interactions.

# Lecture 7

## HISTORY AND PRESENT STATUS OF THE STEADY-STATE COSMOLOGY

It was T. Gold who first suggested in 1946 that the universe be in a steady state. H. Bondi pointed out that creation of matter would be required to reconcile this idea with the observed recession of the galaxies. Bondi and Gold worked from the "perfect cosmological principle", a philosophical assumption that the universe must be homogeneous in both space and time. In 1948 I made a crude modification of Einstein's equations to allow creation of matter. A steady-state model was very attractive at that time, because the age of the universe in a Friedmann model seemed to be only $2 \times 10^9$ years, given the then accepted value of the Hubble constant, in contradiction to geological and astrophysical evidence.

In 1958 or thereabouts M. H. L. Pryce suggested a better modification of Einstein's equations, which Narlikar and I subsequently adopted. One writes

$$R_{ik} - \tfrac{1}{2} R g_{ik} = - 8\pi G \left[ T_{ik} + \lambda (C_i C_k - \tfrac{1}{2} g_{ik} C_l C^l) \right],$$

where $C_i = \partial C / \partial x^i$, etc., are derivatives of a scalar field, C. We became convinced, however, that the theory ought to be conformally invariant. This idea led ultimately to the equations presented above:

$$F(R_{ik} - \tfrac{1}{2} R g_{ik}) = - 3(T_{ik} + \Phi_{ik}) + (g_{ik} \Box F - F_{;ik}).$$

As before, whenever one can make the approximation

$$F = \sum\sum_{a<b} m^{(a)} m^{(b)} \simeq \tfrac{1}{2} \left[ \sum_a m^{(a)} \right]^2,$$

one can choose a conformal frame where $\sum_a m^{(a)} = m_0$ is constant. Before, when creation of matter was not considered, we saw that in this frame $\Phi_{ik} = 0$, since this quantity depends (in this approximation) only on the derivatives of the mass fields in the combination

$$\sum_a m_i^{(a)} = \left[\sum_a m^{(a)}\right]_i = \frac{\partial m_o}{\partial x^i} = 0.$$

If particles are being created, however, the leftmost equality of the string is not valid: $m_o(x+dx)$ may involve new terms which were not present in $m_o(x)$, so the derivative of $\sum_a m^{(a)}(x)$ is not just the sum of the derivatives of the individual terms of $m^{(a)}(x)$ present at x. When creation is allowed, there is an extra term in Einstein's equations, and thus one can obtain the steady-state model out of the conformally invariant theory.

The theory reached this state of development in 1965. By this time, however, the observational outlook for the steady-state model had grown rather bleak, primarily because of the discovery of microwave background radiation. Since the position is now somewhat better than it was in 1965 I want next to review the observational evidence.

<u>Age of stars and universe (1950)</u>. A strong early argument for the steady-state theory was that some stars seemed to be older than the apparent age of the universe. (The oldest stars are about 11 billion years old, according to a calculation of mine made about 1958.) In 1952, however, Baade began to reduce the observed value of the Hubble constant $H_0$, and the accepted value has now fallen to about 50 km/sec/Mpc from the original 500. Thus this argument in favour of the steady-state theory has disappeared.

<u>Origin of the elements</u> (1950; Helium problem, 1964). Gamow originally saw the possibility of building the elements in a hot big bang, but detailed calculations revealed a serious difficulty in passing mass number A = 5. It was then demonstrated that the heavier elements can be produced astrophysically in the course of stellar evolution. The abundances of helium and especially deuterium are perhaps harder to explain.

<u>Stebbins-Whitford effect</u> (1952). An excess reddening was thought to be observed in the light of distant galaxies; this was interpreted as evidence of galactic evolution. It was found, however, to be due to erroneous calibrations. In 1971

a new search for the effect yielded a null result, agreeing with the steady-state theory.

$q_0$ for galaxies (1956). Humason, Mayall and Sandage found that the most distant clusters fell above the linear Hubble curve. The data seemed to indicate $q_0 = +1$, not $q_0 = -1$ as required by the steady-state theory. However, later corrections have moved these objects back to the main line so that one can say $q_0 = 0 \pm 1$. Thus this kind of observation does not seem to rule out the steady-state theory.

$q_0$ for QSO's (1964). The earliest QSO's seemed to fit the curve for $q_0 = +1$. By now, as I pointed out before, the QSO's don't seem to fit a Hubble line at all, so the steady-state theory also survives the test.

log N - log S curve (1958). The slope of the curve for the distribution of radio sources was first stated as -3, but by the time of the 3CR catalogue it was down to - 1.8. This was on the basis of 50 sources per steradian. The 4C left the slope at - 1.8, with 500 sources/sterad. This could not be attributed to local irregularities, and was very troublesome to the steady-state theory. The Parkes survey at 408 MHz confirmed - 1.8. The steady state theory with a Euclidean space requires a value not greater than - 1.5.

Later Bolton, Shimmins, and Walls at 11 cm. found a slope of - 1.5 for about 1000 sources in a limited area of the sky. The Parkes flux scale was found to be somewhat in error. The 4C fluxes are probably also wrong, since they disagree with recent

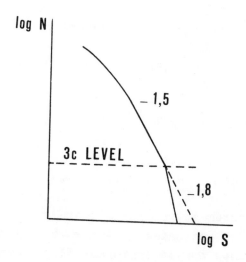

Figure 7.1

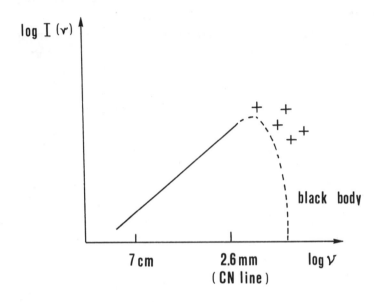

Figure 7.2

measurements at Arecibo and Molonglo. The present situation is thus that the curve has slope - 1.5 except at the bright end, where the 3C value of - 1.8 applies (Fig. 7.1). The difference of about 5 sources/sterad in this region is not statistically significant. (The curve flattens out at the top because of Riemannian curvature.) Preliminary results of the NRAO survey show that all <u>identified</u> sources (QSO's or radio galaxies) yield a slope of - 1.5. The unidentified sources account for the extra steepness. The latter are anisotropically distributed and they may be very local. The Euclidean character of the curve for the identified QSO's may indicate that they are not really at cosmological distances.

<u>Microwave background</u> (1965). This is the deadliest thrust against the steady-state cosmology, and the only observational item which is serious for the theory at the present time. Penzias and Wilson found isotropic microwave radiation which fits a Rayleigh-Jeans law for 3°K. Field and Hitchcock extended the curve by observing the 2.6 mm excitation of cyanogen (Fig. 7.2). Measurements at large $\nu$ are very difficult. Rocket measurements indicated a flux much higher than expected. Balloon observations and later rocket flights disagreed. The

high flux may possibly be from water outgassing from the rocket. The evidence seems to be tending toward a black body curve (dashes in figure). However, if the correct curve were to turn out to be like that indicated in the figure by stars, the situation would be equivocal.

In any case, a steady-state theory must explain the radiation in terms of discrete sources. To account for the observed isotropy one needs about $10^{15}$ sources. A ray of hope lies in the evidence that about 10% of the original hydrogen in the universe has been converted to helium. This implies an energy release of $6 \times 10^{-13}$ ergs/cm$^3$. If this energy could be thermalized, it would yield a black body spectrum of about 3°K.

Here is a sketch of a possible mechanism to accomplish this thermalization in the framework of conformal gravitation theory. We have seen that $m^* = \frac{1}{2}\lambda^2 L^{-3} \tau^2$, where $\lambda^2 \simeq 5 \times 10^{-38}$. The right-hand side can be written in the form

$$\lambda_a \sum_b \lambda_b \int \tilde{G}(A, B) \, db,$$

where the responsibility for the value of $\lambda^2$ is divided among the particles. Consider now the possibility that the $\lambda$ of a particle can be either positive or negative. (Particles with negative mass might be identifiable with antiparticles.) If the signs are distributed randomly, one has in effect the square root of the former number of terms. Then the dependence of m on $\tau$ becomes linear. (This means that the Einstein-deSitter universe becomes a perfect absorber (cf. Narlikar lectures). A major theoretical objection to the Friedmann cosmology is thereby eliminated.) In the steady-state model we now consider the possibility of local fluctuations in $m^*$ which bring it through, or at least close to, zero. The electron-photon scattering cross-section, proportional to $e^2/m_o^2 c^2$, then tends to infinity. Thus electrons become very powerful thermalizers. A quantitative model of this effect has not yet been developed.

## Lecture 8

### THE NATURE OF MASS

First I want to indicate how the Feynman path integral approach to quantum mechanics extends to relativistic particles with spin $\frac{1}{2}$ (see also Prof. Narlikar's lectures). We consider at first a standard theory where the masses are absolute constants, and also space is flat. The classical action for a path of particle a is

$$S_a = -\tfrac{1}{2} \sum_{r=0}^{n-1} m_a \, da_r$$

Then the amplitude for a particle to go from $A_o$ to $A_n$ (Fig. 8.1) is the sum (over all paths connecting $A_o$ and $A_n$)

$$\sum_{\text{paths}} \exp(iS_a) = \sum_{\text{paths}} \prod_{r=0}^{n-1} \exp(-i \tfrac{1}{2} m_a \, da_r),$$

which can be written in the general form

$$\sum_{\text{paths}} K(A_n, A_{n-1}) \ldots K(A_1, A_o)$$

In relativistic quantum mechanics each $K(A_{r+1}, A_r)$ becomes a 4 × 4 matrix. The classical action involves a hidden square root:

$$-\tfrac{1}{2} m_a \, da = -\tfrac{1}{2} m_a (da_i \, da^i)^{1/2}$$

The Dirac theory deals with this square root in an elegant way.

Figure 8.1

Let
$$\gamma_i = \begin{pmatrix} 0 & \sigma_{i\alpha\dot{\beta}} \\ \sigma_i^{\alpha\dot{\beta}} & 0 \end{pmatrix}$$

Each $\sigma_i$ is a 2 x 2 matrix (the Pauli matrices).

(Spinor indices are raised and lowered by the rule

$$u^\alpha = u_\delta \varepsilon^{\delta\alpha}, \quad \text{where } \varepsilon = \begin{pmatrix} 0 & 1 \\ -1 & 0 \end{pmatrix}$$

Note also that $u_{\dot{\alpha}} = u_\alpha^*$ (complex conjugate).)

The $\gamma$ matrices obey the anticommutation rules:

$$\gamma_i \gamma_k + \gamma_k \gamma_i = 2\eta_{ik},$$

where $\eta_{ik}$ is the four dimensional analogue of Kronecker's symbol. Thus $da^2 I = (\gamma_i da^i)^2$, as one can easily check.

By choosing the appropriate square root of this last equation one can replace the classical $-\tfrac{1}{2} m_a da$ by $+\tfrac{1}{2} m_a d\!\!\!/a$, where $d\!\!\!/a = \gamma_i da^i$. (For reasons beyond the scope of this lecture there are additional factors in $K(A_{r+1}, A_r)$.) The correct result is

$$K(A_{r+1}, A_r) = \frac{1}{2\pi} d\!\!\!/a_r\, \delta^1(da_r^2) \exp\left(\frac{i}{2} m_a d\!\!\!/a_r\right).$$

Ultimately after summing over all paths one gets a propagator which satisfies

$$(\nabla\!\!\!\!/_n + i m_a) K(A_n, A_o) = \delta_4(A_n, A_o),$$

where $\nabla\!\!\!\!/ = \eta^{ik} \gamma_k \dfrac{\partial}{\partial x^i}$

This is a time-symmetric propagator. To get the Feynman propagator one would use the positive-frequency part of the $\delta$ function, $\delta_+$, instead of the symmetric $\delta$. We want, however, in the spirit of Prof. Narlikar's lectures, to start from a time-symmetric formalism and to derive the retarded physical effects from the response of the universe.

The equation for the propagator can be solved by iteration, starting from the propagator for zero mass. First, note that $1 + (i/2)m\, d\slashed{a}$ is a sufficiently good approximation to the exponential factor when the limit of small $da$ is taken. Then one obtains the series solution.

$$K(A_n, A_o) = \sum_{r=0}^{\infty} K^{(2r)}(A_n, A_o) - im \sum_{r=0}^{\infty} K^{(2r+1)}(A_n, A_o).$$

Let $I(2, 1) = \frac{1}{4\pi} \delta\left[s^2(2, 1)\right]$ (or $\frac{1}{4\pi} \delta_+\left[s^2(2, 1)\right]$ for the Feynman propagator). Then

$$K^{(o)}(A_n, A_o) = -\slashed{\gamma}_o I(A_n, A_o),$$

$$K^{(1)}(A_n, A_o) = I(A_n, A_o),$$

$$K^{(2r)}(A_n, A_o) = -(-im)^{2r} \slashed{\gamma}_o \int \cdots \int I(A_n, r) I(r, r-1) \cdots$$
$$\cdots I(2, 1) I(1, A_o)\, d\tau_1 \cdots d\tau_r,$$

$$K^{(2r)}(A_n, A_o) = (-im)^{2r} \int \cdots \int I(A_n, r) I(r, r-1) \cdots$$
$$\cdots I(2, 1) I(1, A_o)\, d\tau_1 \cdots d\tau_r.$$

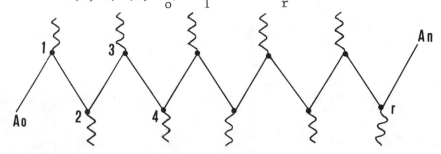

Figure 8.2

The idea here is that the particle propagates along infinitesimal null paths like a free massless particle, but the mass field occasionally causes it to change its null direction.

Since mass can be treated in this way as something external, we are ready to construct a conformally invariant quantum theory like the classical theory described earlier. How must the definition of the mass field be modified to fit the Dirac theory? In the classical theory we had the action $m_a(A) da/2$, where

$$m_a(A) = \sum_{b \neq a} m^{(b)}(A), \quad m^{(b)}(A) = \lambda^2 \int \tilde{G}(A, B) \frac{db}{2}$$

(where the definition of $\lambda^2$ has been altered by a factor of 2), and hence finally an interaction

$$\lambda^2 \frac{da}{2} \tilde{G}(A, B) \frac{db}{2}.$$

In quantum mechanics the action becomes $m_a(A) d\rlap{/}a/2$. Now b should enter symmetrically to a, so we postulate the interaction

$$\lambda^2 \frac{d\rlap{/}a}{2} \tilde{G}(A, B) \frac{d\rlap{/}b}{2} .$$

(The $\gamma$ matrices in $d\rlap{/}a$ and $d\rlap{/}b$ are not multiplied together; they are standing free, ready to contract onto those belonging to the next segment of the path.)

In other words,

$$m^b(A) = \lambda^2 \sum \tilde{G}(A, B) \frac{d\rlap{/}b}{2} ,$$

where we must sum over all paths involving $d\rlap{/}b$ and over all $d\rlap{/}b$. The result is

$$m^b(A) = 2\lambda^2 \int \tilde{G}(A, B) \left[ \bar{\psi}_b \psi_b \right]_B \left[ -g(B) \right]^{1/2} d\tau_B$$

where $\left[ \bar{\psi}_b \psi_b \right]_B = \bar{\psi}_b(B) \psi_b(B)$. In the spinor notation we can write

$$\psi_b = \begin{pmatrix} u_\alpha \\ v^{\dot\beta} \end{pmatrix} \qquad \bar\psi = (v^\alpha, \; u_{\dot\beta})$$

so that

$$\bar\psi_b \psi_b = u_\alpha v^\alpha + u_{\dot\beta} v^{\dot\beta}$$

For a neutrino one or the other of the spinors u and v is identically zero. In this case $\bar\psi \psi = 0$, so a neutrino doesn't produce a mass field. This is what one would expect for a particle without a mass of its own.

Finally, we present a model of the leptons (the muon, the electron, and the two associated neutrinos). Write the interaction

$$\lambda^2 \, L_{\Gamma\Lambda}^{(a)} \, \frac{d\tilde{\mathcal{A}}}{2} \, \tilde{G}(A, B) \, \frac{d\tilde{\mathcal{B}}}{2} \, L^{(b)\Gamma\Lambda}$$

where

$$L_{\Gamma\Lambda}^{(a)} = \begin{pmatrix} \nu_e & e \\ -\mu & \nu_\mu \end{pmatrix}$$

$$L^{(b)\Gamma\Lambda} = \begin{pmatrix} \nu_\mu & \mu \\ -e & \nu_e \end{pmatrix}$$

We have introduced an abstract spin space (analogous to isospin space). Now

$$L_{\Gamma\Lambda}^{(a)} L^{(b)\Gamma\Lambda} = \nu_e^{(a)} \nu_\mu^{(b)} + \nu_\mu^{(a)} \nu_e^{(b)} + e^{(a)} \mu^{(b)} + \mu^{(a)} e^{(b)}.$$

It follows that

$$m_e(A) = 2\lambda^2 \sum_\mu \int \tilde{G}(A, B) \left[\bar\psi_\mu \psi_\mu\right]_B \left[-g(B)\right]^{1/2} d\tau_B,$$

$$m_\mu(A) = 2\lambda^2 \sum_e \int \tilde{G}(A, B) \left[\bar\psi_e \psi_e\right]_B \left[-g(B)\right]^{1/2} d\tau_B,$$

$$m_{\nu_e}(A) = 2\lambda^2 \sum_{\nu_\mu} \int \tilde{G}(A, B) \left[\bar{\psi}_{\nu_\mu} \psi_{\nu_\mu}\right]_B \left[-g(B)\right]^{1/2} d\tau_B$$

$$m_{\nu_\mu}(A) = 2\lambda^2 \sum_{\nu_e} \int \tilde{G}(A, B) \left[\bar{\psi}_{\nu_e} \psi_{\nu_e}\right]_B \left[-g(B)\right]^{1/2} d\tau_B.$$

The ascription of zero mass to the neutrinos is self-consistent (although not required by anything put into the theory previously). We shall put in an electromagnetic interaction and regard the nonzero mass of the charged particles as electromagnetic in origin. The electromagnetic coupling for Minkowski space is

$$e_a e_b da^i L^{(a)\Gamma}_\Gamma \delta\left[s^2(A, B)\right] \eta_{ik} L^{(b)\Lambda}_\Lambda db^k$$

(The $\delta(s^2) \eta_{ik}$ is the standard electromagnetic propagator.) Since $L^\Gamma_{\ \Gamma} = e + \mu$, the interaction vanishes if either of the particles involved is a neutrino.

There is no asymmetry between the electron and the muon in the equations of the theory. However, if there are more electrons in nature than muons, the muons will be more massive. So the theory is in qualitative agreement with nature. The problem is to explain the observed mass ratio of 207. The weak interaction causes the muon to decay. Therefore, according to the theory the mass ratio should grow in time even if it begins by being nearby unity until eventually all the muons have disappeared and the electron mass falls to zero.

REFERENCES

Hoyle, F., and Narlikar, J. V., 1964, Proc. Roy. Soc. A., 282, 191.

Electrodynamics and Cosmology

Lectures by

J. V. Narlikar

Institute of Theoretical Astronomy, Cambridge

Notes by

J. Kiskis, H. Tesser

Electrodynamics and Cosmology

Lecture 1

SUMMARY OF BASIC IDEAS IN CLASSICAL ELECTRODYNAMICS

My first lecture will deal with the early development of electromagnetism. In later lectures I shall discuss the role of electromagnetism in cosmology.

When you look at the history of physics you find that the relative importance of astronomy has undergone great variation from time to time. For example, before Newton, astronomy was the most important branch of physics. However, with Newton came the important developments in the laws of motion and laboratory physics.

Still later, toward the end of the $19^{th}$ century, laboratory physics attained a dominant role despite important advances in astronomy. The relative importance of laboratory physics has lasted until recently. However, as you have heard in the other lectures here, recent developments in astrophysics have brought it once again to the forefront of physics.

After Newton's law of gravitation

$$F = \frac{Gmm'}{r^2} \tag{1.1}$$

received general acceptance, physicists tried to formulate a similar mathematical law for electromagnetism

$$F = \pm \frac{ee'}{r^2} \tag{1.2}$$

Coulomb's law also seemed to work very well for a while but not as successfully as Newton's law. The reason is easy to see. The law of gravitation, although it worked very well, could not be applied with the same degree of freedom as Coulomb's law. The smallness of the gravitational constant prevents us from using laboratory masses in a variety of experiments. Large masses are only available to us through astronomy where we may observe the masses but we cannot disturb them. The interaction of isolated charges is much more easily studied in the laboratory since the interaction is relatively large. For typical elementary particles (such as an electron and a proton) the ratio of the electrical to the gravitational force is $\sim 10^{40}$. When people began to investigate Coulomb's law in the laboratory they found that it does not work very well. In particular, the discrepancy from Coulomb's law arises when the interacting charged particles are in motion relative to each other. For rapidly moving or oscillating charges, Coulomb's law simply breaks down.

Originally, physicists tried to patch up Coulomb's law by adding terms depending on the particles' velocity and acceleration, but this does not lead to a neat and simple law. Gauss realized that the problem was in part due to the use of an instantaneous action between particles. His was the first suggestion that the velocity of propagation between charged particles might propagate with the velocity of light. Gauss did not pursue this work and progress was left to others. Maxwell finally solved the problem in an altogether different way by introducing the notion of the electromagnetic field. Thus particles influenced one another through disturbances in this field. Maxwell also derived the equation showing that disturbances in the field propagated with the velocity of light. Maxwell's theory not only led to predictions in agreement with experiment but also inspired other theoretical advances such as the Special Theory of Relativity. Einstein wondered whether the role of velocity of light could be only an "accident" in Maxwell's theory, or part of some more general scheme. In the Special Theory of Relativity it seems very natural that Maxwell equations are invariant under Lorentz transformations which preserve the basic nature of the velocity of light.

In the beginning of this century a number of people felt that Gauss idea of a direct interparticle interaction should be revived. This was done by Schwarzschild (1903), Tetrode (1922) and Fokker (1929 a, b; 1932). Basically they suggested that the interaction takes place via retarded, rather than instantaneous, interaction at a distance. Consider the world lines of two particles of charges e and e' (Figure 1.1). The old formulation of instantaneous action would occur along lines AC while the Special Theory demanded that the propagation proceeded along DA or AB. In order to retain the Newtonian concept of action and reaction we must allow advanced effects along the light cone as well as retarded effects. If the concept of the law of action and reaction is to be valid then one must allow advanced interactions to proceed from B to A and A to D. We might ask "Why not drop the concept of action and reaction?" It turns out that you cannot formulate a Lagrangian for action at a distance consistent with the ideas of classical physics without the notion of action and reaction. So, if you want to have retarded action at a distance you must also have advanced interactions. Now advanced interactions seem to violate our sense of causality since the future will influence the past. Consider the following typical paradoxical situation. In figure 1.2 A and B are separated by one light hour and operate according to the rules: 1) A sends a signal to B at 4 p.m. if and only if he does not hear from B at 4 p.m.; 2) B sends a signal to A at 5 p.m. if and only if he hears from A at 5 p.m.. What does A do at 4 p.m.? If A sends a signal at 4 p.m. it means he did not receive a signal from B. Then A's signal would reach B at 5 p.m.. B would in turn send a signal to A. However, the advanced effect of B's signal would reach A at 4 p.m. which contradicts the assumption that A

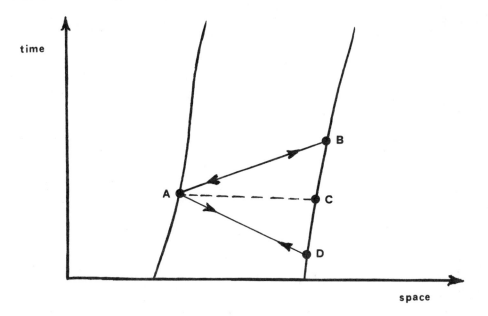

Figure 1.1

Figure 1.2

did not receive a signal at 4 p.m.. If conversely you assume that A did not send a signal at 4 p.m. you come to the conclusion that A in fact should have sent a signal at 4 p.m.. We find then that neither possibility is self consistent. In the 1940's Wheeler and Feynman addressed themselves to this problem. They suggested that you can make use of the Fokker, Tetrode and Schwartzschild formalism provided you take into account the rest of the universe. This is where cosmology first comes into picture. Suppose particle (a) and particle (b) described by world lines in figure 1.3 are separated by a distance r. Then if (a) sends a foreward signal to (b) it will reach (b) at a time r/c later. The advanced reaction from (b) will come back along the same path so that (a) will experience a reaction which is instantaneous with the emission of the signal regardless of the distance of separation, r. Therefore, we must include the interaction with all the matter in the universe

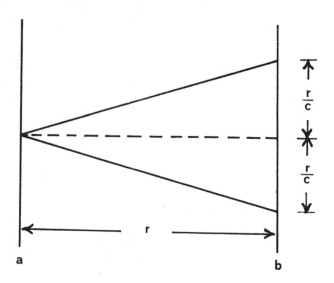

Figure 1.3

and calculate the response on particle (a).

First, I will describe what we expect from the response of the universe and in later lectures I will describe the detailed calculations. Normally, the particles radiate both forward and backward in time i.e. ½ (Retarded + Advanced) signals. We want the universe to yield a response which is a ½ (Retarded - Advanced) signal so that the sum of the two signals is just the usual retarded signal. We shall see later that the Wheeler-Feynman analysis gives just this result in certain cases.

We now present some mathematical relations leading up to the Wheeler-Feynman result. We start with a review of Hamilton's principle of least action.

## Classical Action Principle

A free particle of mass m, moving along a path $\underset{\sim}{a}$ between two points fixed at times $t_1$ and $t_2$ possesses an action J, where

$$J = \frac{1}{2} \int_{t_1}^{t_2} m \dot{\underset{\sim}{a}}^2(t) dt. \qquad \text{(non-relativistic)} \qquad (1.3)$$

Now Hamilton's principle requires that we choose that path, $\underset{\sim}{a}(t)$, which makes the action stationary

$$\delta J = 0. \qquad (1.4)$$

Using the usual Euler-Lagrange variational techniques we obtain

$$m \ddot{\underset{\sim}{a}} = 0 \qquad (1.5)$$

# Electrodynamics and Cosmology

Of course this is Newton's law restated for a free particle. We use the variational principle because it allows us to treat more complicated problems in a simpler manner than if treated by Newton's law. One cannot hope to be like Newton and guess the right answer!

We can extend the non-relativistic to a relativistic treatment. The simplest scalar quantity one can associate with a particle's motion is $m\,da$ and the action is given by

$$J = -m \int da, \tag{1.6}$$

where $da$ is an element of proper time along the path of the particle defined by $da^2 = \eta_{ik} da^i da^k$, where the $a^i$ are the coordinates of the particle and $\eta_{ik}$ is the Minkowski metric of signature $-2$. The speed of light $c = 1$.

If we again take the variation of the action equal to zero we obtain the relativistic equation of motion of a free particle,

$$m \frac{d^2 a^i}{da^2} = 0. \tag{1.7}$$

We can extend this idea to other branches of physics such as Maxwell's electrodynamics. Here the simplest action which describes the equations of motion of both fields and particles is

$$J = -\sum_a m_a \int da - \frac{1}{16\pi} \int F_{ik} F^{ik} d^4x - \sum_a e_a \int A_i da^i \tag{1.8}$$

where again the speed of light set equal to one, $A_i$ is the four potential $A_i = [-\vec{A}, \Phi]$, $F_{ik} = \partial_i A_k - \partial_k A_i$, and $e_a$ is the charge of particle (a).

Variation of the action with respect to the world-line of particle (a) yields the equation of motion

$$m_a \frac{d^2 a^i}{da^2} = e_a F^i{}_k \frac{da^k}{da} \tag{1.9}$$

Varying the action with respect to the fields leads to the equation

$$F^{ik}{}_{,i} = 4\pi j^k, \tag{1.10}$$

where

$$j^k(x) = \sum_a e_a \int \frac{da^k}{da} \delta_4(x, A(a))\, da \tag{1.11}$$

and $A(a)$ is a point on the world-line of particle (a). We can compare the action just treated with that given in an action at a distance formulation

$$J = -\sum_a m_a \int da - \sum_{a<b} \sum e_a e_b \iint \delta(s^2_{AB}) \eta_{ik} da^i da^k \tag{1.12}$$

Here $s^2_{AB}$ is the square of the four-distance between point A, on the world-line of particle (a), and B on the world line of particle (b). In Minkowski coordinates

$$s^2_{AB} = (t_B - t_A)^2 - (\vec{x}_B - \vec{x}_A)^2 \tag{1.13}$$

The symbol $\delta$ denotes the usual Dirac delta function. Finally the double sum does not permit self interactions.

The delta function implies that in order that there be a non-zero contribution to the integral, $s^2_{AB} = 0$. Therefore we get interactions between particles (a) and (b) whenever

$$t_B - t_A = \pm |\vec{x}_B - \vec{x}_A| \tag{1.14}$$

The two possibilities account for the retarded and advanced interactions. We define the four-potential produced by a particle (a) at a field point X

$$A_i^{(a)}(X) = e_a \int \delta(s^2_{AX}) \eta_{ik} da^k \tag{1.15}$$

$$= \frac{1}{2} A_i^{(a)}(ret) + A_i^{(a)}(adv)$$

and the current by

$$j_i^{(a)}(X) = e_a \int \delta_4(X,A) \eta_{ik} da^k \tag{1.16}$$

From these definitions we immediately get

$$\Box A_i^{(a)} = 4\pi j_i^{(a)}$$
$$A^{(a)i}{}_{,i} = 0 \tag{1.17}$$

That is, Maxwell's equations and the gauge condition are identically satisfied. When we vary the world line of particle (a) in the action, we obtain

$$m_a \frac{d^2 a_i}{da^2} = \sum_{b \neq a} F^{(b)i}{}_k \frac{da^k}{da} \qquad (1.18)$$

The particle then moves under the influence of all the other particles in the universe.

## Lecture 2

### THE WHEELER-FEYNMAN ABSORBER THEORY

I will now review the Wheeler-Feynman treatment of action at a distance electrodynamics. As mentioned in the previous lecture the particle produces a signal which is the sum of one-half the retarded and one-half the advanced *fields*[1]. We will show that the contribution of the universe is just one-half the retarded minus one-half the advanced *field* of the particle, leaving just the retarded field of the particle. Moreover the half retarded minus half advanced effect of the universe is just what is needed to account for radiative damping, an effect which one ordinarily does not expect from a theory devoid of self-interactions.

Because the expression describing the effect of the absorber is quite long I present it term by term together with explanation. Since we expect the universe to make a contribution equal to one-half the retarded minus one-half the advanced field, and further the source contributes one-half the sum of advanced and retarded fields, we expect to see the full retarded wave leaving the source. Therefore I shall use the full retarded wave for the calculation, and look for a self-consistent solution. For simplicity we can imagine the source to be located at the center of a cavity in the distribution of absorber particles. The absorber extends outward from a radius R from the source. The calculation proceeds as follows:

1) $\vec{U} = \vec{U}_0 e^{-i\omega t}$ - It represents the acceleration of the source. We assume a general acceleration can be Fourier analyzed, in which case this term represents a particular component.

2) $-\frac{e}{r} \sin \theta$ - Here e is the charge of the source particle. The angle $\theta$ is the angle between the vector $\vec{U}_0$ and the line joining the source and a point in the absorber.

The product of 1) and 2) describes the strength of the retarded field at sufficiently large distances from the source.

3) $e^{i\omega r}$ - This is the phase change of the wave at a radial distance r from the source.

4) $2(1 + n - ik)^{-1}$ - A factor by which the outgoing disturbance must be reduced by reflection at the boundary of the absorber cavity.

---

1. In the discussion of the action at a distance theory the word *field* will continue to be used. It is used for convenience and does not imply any independent existance of the electromagnetic disturbance.

5) $e^{i\omega(n-ik-1)(r-R)}$ - A factor allowing for the change of phase and amplitudes as the disturbance propagates through the absorber medium.

The product of the five terms above yields the magnitude of the electric field

$$\vec{E} = -\vec{U}_o (\frac{e}{r} \sin \theta)(\frac{2}{1+n-ik}) e^{-i\omega(t-r)} e^{i\omega(n-ik-1)(r-R)}. \quad (2.1)$$

This field acts upon the absorber particles setting them in motion. We turn now to evaluating the response of the absorber to this field

6) $\frac{e_k}{m_k} P(\omega)E$ - This term is the ecceleration of the absorber particle to the *electric field* E. $P(\omega)$ is the frequency dependence of the response of the particle. The complex index of refraction n-ik is related to $P(\omega)$ by the relation

$$(n-ik)^2 = 1 - \frac{4\pi e_k^2 N}{m_k \omega^2} P(\omega),$$

where N is the number of the absorber particles per unit volume.

7) $-\frac{e_k}{2r} \sin \theta\, e^{-i\omega r}$ - These terms correspond to terms 2) and 3) above, but are evaluated for the advanced field of the absorber at the source. The factor $\sin \theta$ arises because we are considering the resolved part of the absorber field parallel to the acceleration of the source.

So far we have calculated the field of an absorber particle back on the source particle. You may ask why I do not put the respective index into this part of the calculation. The reason is that we needed the net field that acted to accelerate the absorber particles, but now we want to calculate the direct elementary action of the absorber back on the source and not the net field back on the source. The product of terms 6) and 7) give the field of a single absorber particle on the source. Integrating the contribution of all the absorbers acting at the source and parallel to the acceleration $\vec{U}_o$ yields

$$\vec{R} = \frac{e\vec{U}_o}{1+n-ik} \frac{e_k^2}{m_k} \int_R^\infty \int_0^\pi \int_0^{2\pi} dr d\theta d\phi\, P(\omega) N \sin^3\theta\, e^{-i\omega t} e^{i\omega(n-ik-1)(r-R)}$$

$$= -\frac{2}{3} ei\omega \vec{U}_o e^{-i\omega t} \quad (2.2)$$

We see that all of the terms describing the characteristics of the universe have cancelled out. We are left with a contribution of the electric field which is parallel to the acceleration. If we now note that the Fourier component $(-i\omega)$ corresponds to differentiation, we have for the force acting on the source particle

$$e\vec{R} = \frac{2e^2}{3} \frac{d\vec{U}}{dt} \tag{2.3}$$

This is the usual form of the non-relativistic radiation reaction. There is no net response force in any other direction.

In the calculation just completed we calculated the effect of the universe at the position of the particle. One can also calculate the *field* of the absorber in the neighborhood of the source. Wheeler and Feynman have done this. When you calculate this field, you find that the effect of the absorber is a field which is half of the difference of the retarded and advanced *fields* of the particle.

## GENERAL METHOD

As we have seen the particular characteristics of the absorber have dropped out of the final result. This leads us to seek a more general way to arrive at the same result without going through all the intermediate details. This method is also found in Wheeler and Feynman (1945). We consider a completely absorbing universe, i.e. there is some boundary beyond which the *field* of the particles vanish:

$$\frac{1}{2}\sum_b (F_{ret}^{(b)} + F_{adv}^{(b)}) = 0 \qquad \text{(outside the absorber)} \tag{2.4}$$

Here $F_{ret}^{(b)}$ ($F_{adv}^{b}$) is the retarded (advanced) field of the $b^{th}$ particle. However, if the sum vanishes then

$$\sum_b F_{ret}^{(b)} = 0 \quad \text{and} \quad \sum_b F_{adv}^{(b)} = 0,$$

since there cannot be complete destructive interference between the outgoing retarded and incoming advanced waves. This implies that

$$\frac{1}{2}\sum_b (F_{ret}^{(b)} - F_{adv}^{(b)}) = 0 \qquad \text{(outside the absorber)} \tag{2.5}$$

The half retarded minus the half advanced field satisfies the homogeneous Maxwell equation and, if it is zero everywhere on a closed boundary, then it must also vanish everywhere inside the boundary. We can now calculate the force acting on the charge "a". According to Wheeler-Feynman theory the field acting on "a" is

$$\frac{1}{2}\sum_{b\neq a}(F_{ret}^{(b)} + F_{adv}^{(b)}). \qquad (2.6)$$

This can be broken up as follows:

$$\frac{1}{2}\sum_{b\neq a}(F_{ret}^{(b)} + F_{adv}^{(b)}) = \sum_{b\neq a}F_{ret}^{(b)} + \frac{1}{2}(F_{ret}^{(a)} - F_{adv}^{(a)}) - \frac{1}{2}\sum_{b}(F_{ret}^{(b)} - F_{adv}^{(b)}),$$

the last term of which vanishes. We are left with

$$\sum_{b\neq a}F_{ret}^{(b)} + \frac{1}{2}(F_{ret}^{(a)} - F_{adv}^{(a)}). \qquad (2.7)$$

This means that charge "a" is acted upon by the retarded field of all the other particles in the universe and its radiation reaction.

## TIME SYMMETRY

We started out with a theory which was explicitly time symmetric but which now, through the radiation reaction force, is no longer symmetric. It seems that we should be able to reverse the sign of the time and not affect the universe but we know that advanced and retarded fields would switch roles. Wheeler and Feynman (1945) recognized that the arrow of time is selected by the boundary conditions placed on the absorbing universe. These boundary conditions are thermodynamic in character, connected with the asymmetry between initial conditions and final conditions. In obtaining the usual radiation damping we assumed that the particles of the absorber were at rest before the source is set in motion, and move later as a result of absorbtion of energy from the source. In the time reversed picture the absorber particles were moving before the source is accelerated in just such a way as to come to rest and give up their energy to the source particle. Thermodynamics argues against such an unlikely situation. Hogarth (1962) examined the effect of cosmology on the Wheeler-Feynman theory. He noted that if the universe is expanding the invariance under $t \to -t$ is no longer valid for the expanding universe becomes a contracting one. This means we must check to see if the retarded or advanced solutions work in an expanding or contracting universe. In this way cosmology can lead us to a particular choice of models.

## SELF INTERACTION

I will now turn to a discussion of the solutions of the equation on the non-relativistic level. The equation of motion in one dimension reduces to

$$m\ddot{x} = F + \frac{2e^2}{3}\dddot{x}; \qquad (2.8)$$

where F is an external force. This is the same equation as one obtains from classical field theory. In the case F = 0 it reduces to

$$m\ddot{x} = \frac{2e^2}{3} \dddot{x}. \qquad (2.9)$$

This equation has, as one solution,

$$\dot{x} = e^{\lambda t}, \text{ where } \lambda = \frac{3m}{2e^2}, \qquad (2.10)$$

the classic runaway solution.
Dirac noted an equivalent problem in the relativistic treatment. Dirac suggested an ingenious way out of this difficulty. He considered what happens if the particle is acted upon by a delta function force

$$m\ddot{x} - \frac{2e^2}{3} \dddot{x} = \delta(t). \qquad (2.11)$$

By applying final boundary conditions, $\dot{x}$ = const. for t > 0, he found he must have a pre-acceleration before the pulse hit the particle as shown in figure 2.1. The time scale over which the particle suffers a pre-acceleration is very short, of the order $10^{-23}$ seconds. Dirac attempted to relate the premonitory effects to light propagation inside the electron. In the Wheeler-Feynman theory such effects arise from the advanced disturbance of the universe. One advantage which is present in the Wheeler-Feynman theory is the impossibility of the runaway solution. If you impose the solution (2.10) you find that the fields become infinite in extent, and you can no longer satisfy the condition of complete absorption.

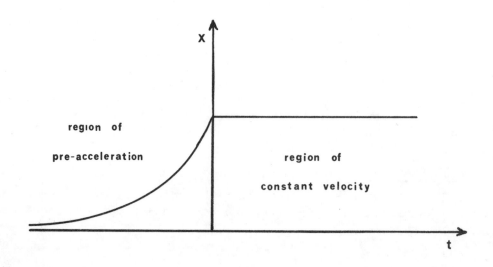

Figure 2.1

Electrodynamics and Cosmology

Lecture 3

THE RESPONSE OF THE UNIVERSE

This lecture will be concerned with the application of the Wheeler-Feynman theory to cosmological problems. However, before one can proceed, there are some technical problems which must be solved. The flat space action contained the expression $\delta(s^2)\eta_{ik}$ where $s^2$ and $\eta_{ik}$ refer to flat space. Also, the $\delta$ - function is the propagator in flat space. It will be necessary to make this expression generally covariant. I will outline the way this is done while leaving out the details. The expression $\delta(s^2)\eta_{ik}$ is replaced by $\vec{G}_{i_A i_B}$. Let me explain what this means. We recall that in the picture of the worldlines of particles (a) and (b), $\delta(s^2)\eta_{ik}$ spanned the space between the points A and B. It has a foot on each end and must transform as a vector at each end. In the flat space case, an object which transforms as a vector at one point transforms as a vector everywhere. But, this is not true in curved space. Thus, the $i_A$ subscript indicates that $\vec{G}_{i_A i_B}$ transforms as a vector at point A and the $i_B$ that it transforms as a vector at point B. We call such an object a bivector.

If we consider the point A fixed, then $\vec{G}_{i_A i_B}$ is a function of B. It behaves like a $\delta$ - function on the light cone, but it also has support inside the light cone. The part of $\vec{G}$ with support inside the light cone is a result of curved space and is sometimes called gravitational scattering. It turns out to be not very important in our calculations as I will show. But it has the basic property that in curved space light travels not only on the light cone but also inside of it. Now, what one must do is take a cosmological model with curved spacetime and work out $\vec{G}$. At first sight this appears to be a complicated problem. However, one may utilize the conformal invariance of this theory to transform to a flat space where $\vec{G}_{i_A i_B} = \delta(s^2)\eta_{ik}$. I will describe how this is done.

Consider the Robertson-Walker line element

$$ds^2 = dt^2 - S^2(t)\left[\frac{dr^2}{1-kr^2} + r^2(d\theta^2 + \sin^2\theta d\phi^2)\right]. \qquad (3.1)$$

S(t) is the expansion factor, and k is a parameter which takes the values 0, ± 1. The line element can be rewritten as

$$ds^2 = e^{2\zeta}\left[d\tau^2 - d\rho^2 - \rho^2 d\Omega^2\right], \qquad (3.2)$$

where $\rho = \rho(r,t), \tau = \tau(r,t)$ and $d\Omega^2 = d\theta^2 + \sin^2\theta d\phi^2$.

Expression (3.2) is just a factor $e^{2\zeta}$ multiplying a line element for flat space. Ignoring this factor you are in flat space where you can use $\vec{G} = \delta(s^2)\eta_{ik}$. Now, we can make a conformal transformation to flat space and use the simple flat space form of the Wheeler-Feynmann theory.

As an example, I will do the case k = 0. The cases k = ± 1 can be done, but they are more complicated. For k = 0, the $(t,r) \to (\tau,\rho)$ transformation is

$$\tau = \int^t \frac{dt}{S(t)}, \quad \rho = r, \text{ and } e^{2\zeta} = S^2(t).$$

In the Einstein-deSitter model

$$S(t) = \left(\frac{3Ht}{2}\right)^{2/3} = \frac{1}{4} H^2 \tau^2, \qquad (3.3)$$

where H is the present value of the Hubble constant. In this case, both t and $\tau$ range between 0 and $+\infty$. The present time corresponds to t = 2/3H, or $\tau = 2/H$.

In the steady-state model we have

$$S(t) = e^{Ht} = -\frac{1}{H\tau}. \qquad (3.4)$$

In this case t and $\tau$ range in the intervals $-\infty < t < \infty$ and $-\infty < \tau < 0$, while the present time corresponds to t = 0, or $\tau = -H^{-1}$.

Since physical processes are measured relative to the t coordinate, one does not need to be bothered by the finite range of the $\tau$ coordinate.

I now wish to consider the redshift phenomena. A retarded wave in the flat space which is conformal to the model universes we are considering will have the form $\exp\{-i\omega_o(\tau-r)\}$. Since the equations of electrodynamics are conformally invariant, it will have the same form in the geometries we are considering when measured relative to the $\tau$ coordinate. But $d\tau = dt/S(t)$, so that the wave will appear redshifted relative to the t coordinate. The observed frequency will be

$$\omega = \omega_o e^{-\zeta} \qquad (3.5)$$

I must now repeat the calculation of the response of the universe which I did in the previous lecture. I will not go through it all but will point out where the essential differences arise. In the flat space case the wave contained a phase factor: $\exp\{i\omega_o(n-ik-1)r\}$. In the curved space case this must be replaced by:

$$\exp\left[i\omega_o \int_o^r (n - ik - 1)dr\right].$$

The integral appears because the index of refraction is measured relative to the t coordinate and the frequency changes in the t coordinate system as the wave propagates. We have taken the cavity radius R which appeared in the earlier calculation to be zero since we are only interested in the large r part of the integral. The argument of the exponential contains a real part $\omega_o \int kdr$. If we are going to get a damping over and above the $1/r$ dependence, it must come from here. Thus, the condition for a complete absorption of retarded waves is

$$\int_o^r kdr \to -\infty \qquad \text{as r approaches } \infty,$$

or whatever is the maximum of the r variable in the geometry. A similar analysis for advanced wave absorption gives

$$\int_o^r kdr \to +\infty.$$

We now examine particular universes to see what type of refractive index we would actually have. First, I consider the case where the retarded wave is affected by radiative damping. If one is looking for self-consistent retarded solutions, then the radiative damping force and the equation of motion are given by (2.7) and (2.8). This leads to a refractive index

$$(n - ik)^2 = 1 - \frac{4\pi N e^2}{m\omega^2}\left[1 - \frac{2ie^2}{3m}\omega + O(\omega^2)\right] \tag{3.6}$$

(If we were looking for self-consistent advanced waves the $\dddot{x}$ term would have the opposite sign and things would be worked out in a similar way). In the steady-state model, N = constant, and if we recall that $\omega$ is getting red-shifted to small values, then we find that $k \propto 1/\omega$, the constant of proportionality being a negative quantity, and that

$$\int kdr \sim -\int \frac{dr}{\omega} \sim -\int \frac{d\tau}{\omega} \sim -\int \frac{d\omega}{\omega} \to -\infty.$$

This shows that the absorption is complete. For the Einstein-de Sitter model $N \propto S^{-3} \propto \omega^3$, and we get $k \sim \omega^2$. The result is that

the integral ∫kdr converges and the absorption is not complete.
When the same calculation is carried out for advanced waves, the
result is that the absorption is complete for the Einstein-de Sitter model but not for the steady-state universe. If we recall that
the advanced wave is indefinitely blue shifted in both cases, we
can understand this result by noticing that the particle density
rises in the Einstein-de Sitter model to absorb this energy, but
it remains constant in the steady-state universe. These results,
however, are not as reliable as for the retarded waves since we
are ignoring quantum effects for high frequency waves, but are
valid if quantum cross sections converge at high energies - as
they must do. Figure 3.1 shows the models we have considered. The
case $k = +1$ is also shown. In this case there are infinite density states in both the past and the future which could lead to
complete absorption and the consistency of both advanced and retarded solutions. The table below shows the results for the various models.

| Model | Future Absorber | Past Absorber | Nature of e.m. wave propagation |
|---|---|---|---|
| Minkowski | perfect | perfect | ambiguous |
| Einstein-de Sitter | imperfect | perfect | advanced |
| Closed Friedmann | perfect | perfect | ambiguous |
| Steady-state | perfect | imperfect | retarded |

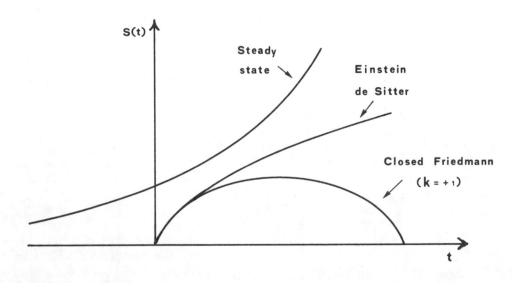

Figure 3.1

As another mechanism for absorption of the retarded wave we may consider collisional damping. In this case, for a particle in the absorber, we have

$$\ddot{x} + \nu \dot{x} = \frac{E}{m}$$

where $\nu$ is the collision frequency. The radiative damping $\dddot{x}$ term would also be here, but it is dominated by the $\nu\dot{x}$ term for the small red-shifted frequencies we are considering. If we wish to consider the absorption of advanced waves, it is not clear what to do with the sign of the $\nu\dot{x}$ term. Its origin is thermodynamical and it is difficult to say whether thermodynamics would go backward or not. When we work this out for retarded waves, we must be careful about the $\omega \to 0$ limit. If $\omega$ is very small, then the wave is damped by collisions before it can set up oscillations. But, if we take this into account correctly and work everything out, we get essentially the same result as for the radiative damping case.

Other types of cosmological models can be considered to see how far one can go and still get perfect absorption. The flat space with constant density was a perfect future absorber, but the Einstein-de Sitter model was not. Thus, there should be some cases in between where the absorption is still perfect. Or, if you want to bring in creation of matter, you can work out what sorts of expansion factors will give perfect absorption. If you insist on solutions to Einstein's equations, the choices are limited. But, if you do not want to limit yourself, you can consider more general models where the expansion factor goes like $S(t) \sim t^n$, and the density is not proportional to $S(t)^{-3}$, but has some dependence of the type $N \propto t^m$. Hogarth and Davies have looked at some models like this one. It turns out that the conditions for perfect absorption are sensitive to the details of the models. Thus we see that the large scale nature of the universe bears on the results of laboratory experiments which measure the electromagnetic waves from accelerated particles.

We found that the steady-state model satisfies the condition of perfect absorption of the future waves. There are other models which do this, and one which has been considered is shown in figure 3.2. It is a steady state in the asymptotic past and future, and involves matter creation. A random observer in this model would most likely be at $|t|$ very large. If he is toward $t = +\infty$ he is essentially in the steady-state universe which gives consistent retarded waves. His universe would be expanding. An observer toward $t = -\infty$, where the model is contracting, would have consistent advanced waves. But, since he uses electromagnetic waves to view his universe, he would see it as expanding too! To make this more certain, one should examine the way thermodynamics goes relative

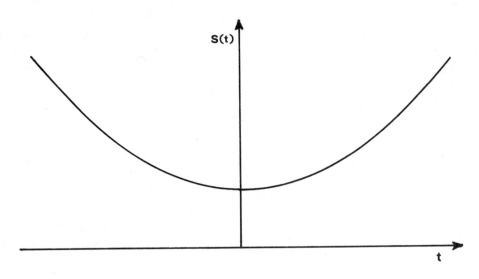

Figure 3.2

to the direction of electromagnetic wave propagation. Black body radiation shows a time asymmetry in the existence of spontaneous downward transitions. One should try to understand this asymmetry. To do so, we will need a quantum theory and this will be the topic of my next lecture. Let me briefly indicate which problems will arise. In the Maxwell theory, we have both the particle trajectories and the electromagnetic fields as dynamical variables. The quantization of the fields leads to the concept of photons and is responsible for the spontaneous downward transitions. However in the Wheeler-Feynman theory the fields are not separate dynamical variables to be quantized. Also, the non-local nature of the Wheeler-Feynman theory could be a source of difficulties. We will face these problems in the next lecture.

Electrodynamics and Cosmology

Lecture 4

QUANTUM FORMULATION

I would like to begin by discussing the ideas of path integration which were introduced by Feynman in 1948. We will use this method of quantization in order to avoid the problems which I discussed at the end of the last lecture.

This method is based on the following expression for the amplitude K (2, 1) that a particle at the space-time point 1 will be found at the space-time point 2:

$$K(2,1) = \sum_{\Gamma_{2,1}} \exp\left[\frac{iS}{\hbar}\right], \qquad (4.1)$$

where S is the classical action for the system and the sum is extended over all paths $\Gamma_{2,1}$ that connect the point 1 to 2. This is a non-relativistic theory, so we will restrict ourselves to paths which do not go backward in time. That is, paths such as (a) and (b) in figure 4.1 are allowed, but not (c). What we are saying is that there is an amplitude $\exp\left[iS(\Gamma_{2,1})/\hbar\right]$ for the particle to go from 1 to 2 by the path $\Gamma_{2,1}$. To get the total amplitude for the transition we must sum the contributions for all the paths. Now, we must recognize that there are not just a discrete number of paths but rather a continuum. The sum becomes an integral over the space of paths which I indicate by

$$K(2,1) = \int^P \exp \frac{iS(\Gamma)}{\hbar} D^3\Gamma \qquad (4.2)$$

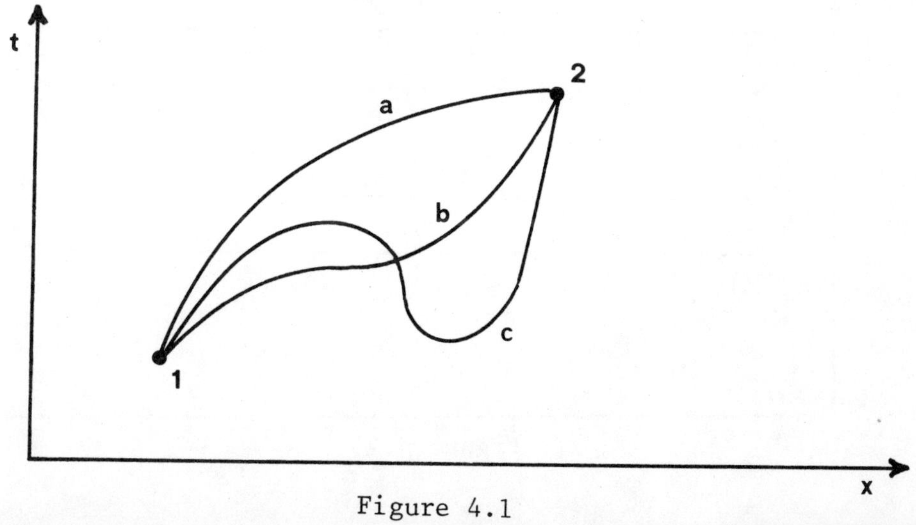

Figure 4.1

Rigorous mathematical treatments of this integral over paths are still in their early stages, but Feynman showed that the correct physical answers can be obtained if one is sufficiently clever! We can now see how the classical prescription $\delta S = 0$ arises. In a macroscopic system, $S \gg \hbar$, and a small change in the path will result in a huge change in the phase. For most paths there will be many others around it with very different phases, and these will tend to all add up to zero. This will be the case except for those paths very near the classical path $\delta S = 0$. In this region, the action is stationary for a small change of path, so these paths make a finite contribution. This idea is due to Dirac originally, and is described in great detail by Feynman and Hibbs (1965). Let us consider the example of a free particle with the action given by (1.3). We proceed by dividing the space up into small time steps of length $\varepsilon$, with $N\varepsilon = t_2 - t_1$. We will approximate the path between two neighbouring points $\underline{x}_{k+1}$ and $\underline{x}_k$ by a straight line. For a particular path, we add up the contributions from each little interval to get

$$S \simeq \frac{1}{2} m \sum_{k=0}^{N-1} \left[ \frac{\underline{x}_{k+1} - \underline{x}_k}{t_{k+1} - t_k} \right]^2 (t_{k+1} - t_k)$$

$$\simeq \frac{1}{2} m \sum_{k=0}^{N-1} \frac{(\underline{x}_{k+1} - \underline{x}_k)^2}{\varepsilon} \qquad (4.3)$$

We must sum this over all paths. To do this, we integrate over each of the points $\underline{x}_k$ ($k = 1, \ldots, N-1$) to generate all paths. Finally, we will pass to the limit $\varepsilon = 0$. Thus

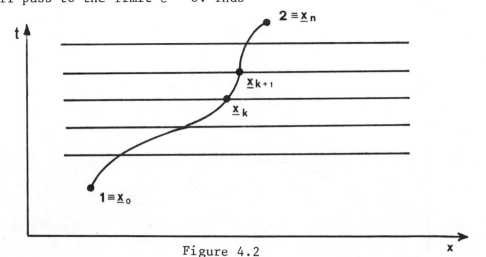

Figure 4.2

$$K(2,1) = \lim_{\varepsilon \to 0} \int \cdots \int A^{-N} \exp\left\{\frac{i}{\hbar} \frac{1}{2} m \sum_{k=0}^{N-1} \frac{(\underline{x}_{k+1} - \underline{x}_k)^2}{\varepsilon}\right\} d^3\underline{x}_1 \cdots d^3\underline{x}_{N-1}. \quad (4.4)$$

The factor $A^{-N}$ represents the difficulties with the integration over paths. Feynman showed how to find this factor for various special cases. For this case it is

$$A = \left(\frac{m}{2\pi\hbar i \varepsilon}\right)^{-3/2}.$$

If we carry out all the integrations and take the limit $\varepsilon \to 0$ ($N \to \infty$), we get

$$K(2,1) = \left[\frac{m}{2\pi\hbar i(t_2-t_1)}\right]^{3/2} \exp \frac{im(\underline{x}_2-\underline{x}_1)^2}{2\hbar(t_2-t_1)} \quad \text{for } t_2 > t_1$$

$$= 0 \text{ for } t_2 < t_1. \quad (4.5)$$

You will recognize this to be the *propagator* for the Schrödinger equation, and recall that it satisfies the equation

$$\left[-\frac{\hbar^2}{2m}\nabla_2^2 - i\hbar\frac{\partial}{\partial t_2}\right] K(2,1) = \delta^4(2,1). \quad (4.6)$$

The $\delta^4(2,1)$ comes from the discontinuities of $K(2,1)$ at $2 = 1$. Thus, we ended up with the Schrödinger equation rather than having begun with it, as in the conventional formulations of the quantum theory. Now, if I do not know where the particle came from, but I have given the amplitude $\psi(\underline{x}_1,t_1)$ that the particle is at a point $\underline{x}_1$ at time $t_1$, I can calculate the amplitude $\psi(\underline{x}_2,t_2)$ that it will be at a point $\underline{x}_2$ at time $t_2$:

$$\psi(\underline{x}_2,t_2) = \int K(2,1) \psi(\underline{x}_1,t_1) d^3\underline{x}_1. \quad (4.7)$$

A more general situation would be represented by a Lagrangian

$$L = \frac{1}{2}m\dot{\underline{x}}^2 - V(\underline{x}).$$

The propagator then will give a wave function which satisfies

$$-\frac{\hbar^2}{2m}\nabla_2^2 \psi + V(2)\psi = i\hbar\frac{\partial \psi}{\partial t_2}. \quad (4.8)$$

Now, if we use the principles of quantum mechanics, we can write down the expression for the transition amplitudes between states $\psi_i(1)$ and $\psi_f(2)$

$$\langle\psi_f|\psi_i\rangle = \iint \bar{\psi}_f(2) K(2,1)\psi_i(1) d^3\underline{x}_1 d^3\underline{x}_2$$

$$= \iiint^P \bar{\psi}_f(2) \exp\left[\frac{iS(\Gamma_{2,1})}{\hbar}\right]\psi_i(1) d^3\underline{x}_1 d^3\underline{x}_2 D^3\Gamma_{2,1}, \qquad (4.9)$$

and for the transition probability

$$|\langle\psi_f|\psi_i\rangle|^2 = \iiiint\int^P\int^{P'} \bar{\psi}_f(\underline{x}_2,t_2)\bar{\psi}_i(\underline{x}'_1,t_1)\psi_i(\underline{x}_1,t_1)\psi_f(\underline{x}'_2,t_2)$$

$$\times \exp\left[\frac{i}{\hbar}\{S(\Gamma)-S(\Gamma')\}\right]d^3\underline{x}_1 d^3\underline{x}'_1 d^3\underline{x}_2 d^3\underline{x}'_2 D^3\Gamma D^3\Gamma'. \qquad (4.10)$$

In general, we would like to be able to calculate the transition probability for a particle (a) under the influence of all the other particles (b) in the universe. However, we are not interested in the final states of the particles (b) and will sum over them. These effects are represented by the influence functional. For a simple case we have the action divided up in the following way:

$$S = S_o\left[q(t)\right] + S_E\left[Q(t)\right] + S_I\left[q(t),Q(t)\right]. \qquad (4.11)$$

$S_o$ is the action for the free particle with coordinates $q(t)$; $S_E(Q(t))$ is the free action for the particles of coordinates $Q(t)$ which cause the influence, $S_I\left[q(t),Q(t)\right]$ represents the interaction between them. We are now interested in calculating the transition probability between a state $\psi_i(q_i,t_i)$ and a state $\psi_f(q_f,t_f)$ regardless of what happens to the other particles. Substituting in our previous expressions we get

$$P(\psi_i\to\psi_f) = \iiiint\int^P\int^{P'} \bar{\psi}_f(q_f)\psi_f(q'_f)\psi_i(q_i)\bar{\psi}_i(q'_i) \qquad (4.12)$$

$$\times \exp\left[\frac{i}{\hbar}\{S_o(q)-S_o(q')\}\right] F(q,q') dq_i dq'_i dq_f dq'_f D^3q D^3q',$$

where $q$ and $q'$ represent paths from $q_i$ to $q_f$, and the influence functional $F(q,q')$ is defined as:

Electrodynamics and Cosmology

$$F(q,q') = \sum \int \int_{}^{P} \int_{}^{P} \exp\left[\frac{i}{\hbar}\{S_E(Q) - S_E(Q')\}\right]$$
$$\times \exp\left[\frac{i}{\hbar}\{S_I(q,Q) - S_I(q',Q')\}\right] D^3 Q D^3 Q', \qquad (4.13)$$

where the sum is over the final states of Q.
This is what we will need for our calculation of the quantum response of the universe in various cosmological models.

For the classical case we showed that the effect of all other particles (b) on the particle (a) is given by the expression (2.7), where the first term is the normal retarded field and the second term is the effect of all the advanced fields. I will show that in the quantum mechanical case the retarded effects of the rest of the universe give rise to the induced transitions, and the advanced effects give rise to spontaneous transitions. Classically, a particle in an energy level above the ground state might stay there forever since there is no field to induce a transition. However, in our formulation, there are paths to the lower state which must be taken into account. Since the particle changes its energy on these paths, it is accelerated and the advanced field of the universe responds with a field at the particle which can be viewed as causing the transition. One might then ask the question:"What about spontaneous upward transitions?". We will be able to see later why these cannot happen. Let us now write down the transition probability for a particle to go from a state m of higher energy to a state n of lower energy. We make use of our previous results to get

$$P(m \to n) = \iiiint \bar{\psi}_n(\underline{a}_f) \psi_n(\underline{a}'_f) K \psi_m(\underline{a}_i) \bar{\psi}_m(\underline{a}'_i) d^3\underline{a}_i d^3\underline{a}'_i d^3\underline{a}_f d^3\underline{a}'_f , \qquad (4.14)$$

where the a's are particle coordinates and

$$K = \int_{}^{P} \int_{}^{P} \exp\left\{\frac{i}{\hbar}\left[S_o\{\underline{a}(t)\} - S_o\{\underline{a}'(t)\}\right]\right\} F\{\underline{a}(t),\underline{a}'(t)\} D^3\underline{a} D^3\underline{a}',$$

with

$$F\{\underline{a}(t),\underline{a}'(t)\} = \prod_{b \neq a} F^{(b)}\{\underline{a}(t),\underline{a}'(t)\},$$

$$F^{(b)}\{\underline{a}(t),\underline{a}'(t)\} = \sum \iiiint \bar{\psi}_f(\underline{b}_f) \psi_f(\underline{b}'_f) J^{(b)} \psi_i(\underline{b}_i) \bar{\psi}_i(\underline{b}'_i) d^3\underline{b}_i d^3\underline{b}'_i d^3\underline{b}_f d^3\underline{b}'_f,$$

where the sum is over the final states of (b), and

$$J^{(b)} = \int_{}^{P} \int_{}^{P} \exp\left\{\frac{i}{\hbar}\left[S_o(\underline{b}) + S_I(\underline{a},\underline{b}) - S_I(\underline{a}',\underline{b}') - S_o(\underline{b}')\right]\right\} D^3\underline{b} D^3\underline{b}'.$$

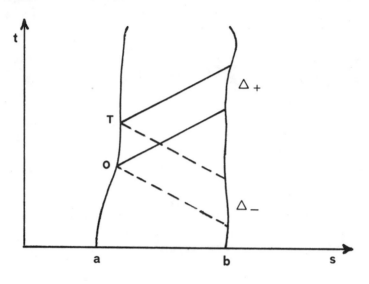

Figure 4.3

We are interested in the same sort of self-consistent calculation as before so we take the retarded interaction

$$S_I(\underline{a},\underline{b}) = -\int_{\Delta_+} \underline{A}^{(a)}_{ret}(\underline{b})\underline{\dot{b}}dt. \qquad (4.15)$$

Since we are interested in the behaviour of particle (a) between $t = o$ and $t = T$, the integral is over $\Delta_+$ which is the part of the world-line of particle (b) that is between the forward light cones from particle (a) at $t = o$ and $t = T$. It is at this point that we must consider the role played by the universe, and the distinction that it makes between upward and downward transitions. In order to evaluate $F\{\underline{a}(t),\underline{a}'(t)\}$ we consider a Fourier decomposition of the path just as we did in the classical case. As a wave of frequency $\omega$ from particle (a) travels into the future in the expanding universe, it is red-shifted. When it reaches an absorber particle it will cause it to make a transition. We shall assume that an expanding universe is cold in the future. That is, all of the absorber particles are in their ground states, and the wave from particle (a) will induce only upward transitions. With this in mind, the calculation of $F\{\underline{a}(t),\underline{a}'(t)\}$ is performed by resolving $a(t)$ and $a'(t)$ into Fourier frequencies. If we assume perfect absorption in the future, and combine the contributions from all absorber particles, the parameters of the absorber again drop out, and we get:

$$F\{\underline{a}(t), \underline{a}'(t)\} = \exp\left[\frac{e^2}{4\pi^2\hbar}\int d\Omega \int_0^\infty kdk \sum_{j=1,2}\left\{\int_0^T (\underline{\alpha}_{-k}^{(j)}\cdot \underline{\dot{a}})\times\right.\right.$$

$$\times \exp(-i\underline{k}\cdot\underline{a} + ikt)dt \int_0^T (\underline{\alpha}_{-k}^{(j)}\cdot \underline{\dot{a}}')\exp(i\underline{k}\cdot\underline{a}' - ikt')dt' -$$

$$-\int_0^T (\underline{\alpha}_{-k}^{(j)}\cdot \underline{\dot{a}})\exp(-i\underline{k}\cdot\underline{a} - ikt)dt \int_0^t (\underline{\alpha}_k^{(j)}\cdot\underline{\dot{a}})\exp(i\underline{k}\cdot\underline{a} + ik\tilde{t})d\tilde{t} -$$

$$-\int_0^T (\underline{\alpha}_k^{(j)}\cdot \underline{\dot{a}}')\exp(-i\underline{k}\cdot\underline{a}' + ikt)dt \int_0^t (\underline{\alpha}_{-k}^{(j)}\cdot\underline{\dot{a}}')\exp(i\underline{k}\cdot\underline{a}' -$$

$$\left.\left.- ik\tilde{t})d\tilde{t}\right\}\right], \tag{4.16}$$

where $\int d\Omega$ is the solid angle summation of the type I did in the classical calculation. The $\underline{\alpha}_{-k}^{(j)}$ are the polarization vectors of the k Fourier component of the field. Note that the expression is not symmetric in k and -k. When it is substituted back into the rest of the expression for $P(m\to n)$, it gives the correct downward spontaneous transitions and no spontaneous upward transitions.

We will now consider an actual cosmological model. As we have seen, the only requirements are that the universe should be perfectly absorbing in the future, and that it should have a cold environment in the future. We have seen classically that the steady-state model meets these requirements. The numbers which follow refer to this model. One should not worry too much exactly what Hubble constant or particle density I put in; it won't make much difference. Collisional damping has been taken as a typical absorbing process. The collisional frequency for ionized hydrogen is given by

$$\nu_{eff} \simeq 2\pi N v (e^2/mv^2)^2 \ln(mv^2/\hbar\omega) \tag{4.17}$$

If we take $\rho = 3H^2/4\pi G$ to be $\sim 4 \times 10^{-29}$ gm/cm$^3$, then $N \sim$ $\sim 2 \times 10^{-5}$ atoms/cm$^3$. The average electron velocity v/c should be $\sim 1/300$. $\omega$ is the frequency of the wave when it is absorbed. We will anticipate the result and take $\omega \gg \nu_{eff}$ in which case the attenuation factor is $\sim (2\pi Ne^2/m\omega^2)\nu_{eff}$. The attenuation over a range $0 \le r < R$ is

$$\exp\left[ -\frac{2\pi Ne^2 \nu_{eff}}{mk^2} \int_0^R \frac{dr}{(1-Hr)^3} \right] \qquad (4.18)$$

where we have ignored the dependence of $\nu_{eff}$ on $\omega$. The factor $(1-Hr)$ is a redshift factor similar to those that appeared in the classical calculation, and k is the frequency which appeared in our previous expression for the influence functional and arose from a transition $E_i \to E_f = E_i + \hbar k$. We are measuring this frequency with respect to the time of the conformal space, so the proper frequency is redshifted:

$$\omega = (1 - Hr)k \qquad (4.19)$$

Appreciable absorption from the integral requires that

$$(1 - Hr)^2 \simeq \frac{2\pi Ne^2}{mk^2 H} \nu_{eff} \qquad (4.20)$$

or, from (4.19)

$$\omega^2 \simeq \frac{2\pi Ne^2}{mH} \nu_{eff} \qquad (4.21)$$

If we solve this equation along with our other equation between $\omega$ and $\nu_{eff}$ we get $\omega \sim 10^6$ sec$^{-1}$ and $\nu_{eff} \sim 10^{-10}$ sec$^{-1}$, justifying the $\omega \gg \nu_{eff}$ assumption. These are just some numbers to show where the absorption is taking place. If the frequency emitted in the laboratory is $> 10^6$ sec$^{-1}$, the radiation is first redshifted until it is below that critical frequency, and is then absorbed over a distance of $10^{28}$ cm. If it is emitted at $< 10^6$

# Electrodynamics and Cosmology

$\sec^{-1}$ it is absorbed right away.

I will conclude this lecture with a discussion of what is meant by the concept of a photon in the Wheeler-Feynman theory. You will recall that the photon normally comes from quantizing the free Maxwell field. In the Wheeler-Feynman theory, there is no electromagnetic field to quantize, so one would expect there to be no such thing as a photon. One can introduce the concept of a photon in a straight forward way having obtained the transition probability formula. You are used to talking about the photon as a solution of the source-free equations. In practice, a photon is always emitted somewhere and absorbed somewhere else. The source of a photon is a particle in an upper energy state. As we have seen, the universe offers a cold environment and the particle tends to jump to a lower state and lose some energy. We can take that spontaneous downward transitions as a unit and speak in the following way. For $E_m > E_n$, we can always write

$$\frac{P(E_m \to E_n)}{P(E_n \to E_m)} = \frac{n+1}{n} \qquad (4.22)$$

and call n the number of photons of energy $E_m - E_n$ present. We are able to do this because of the asymmetry between upward and downward transitions which, as we have seen, is due to the asymmetry between future and past of the universe.

Lecture 5

## WHEELER-FEYNMAN WITHOUT WHEELER-FEYNMAN

I now wish to continue the discussion of the Quantum aspects of the Wheeler-Feynman theory. Whenever I talk to Professor Wheeler about the Wheeler-Feynman theory he says that his view is like that of a converted alcoholic who started out pro action at a distance but realized the errors of his ways and is now strongly pro local field theory. Professor Feynman's view is that it is something that ought to be done, but he is not going to do it. Hence Wheeler-Feynman without Wheeler-Feynman.

In the last lecture I discussed the non-relativistic treatment of Wheeler-Feynman theory and showed how the universe played a role in atomic transitions. In this lecture I will discuss the treatment of relativistic electrons. The first difficulty we run into is writing down the path integral formulation of spin $\frac{1}{2}$ particles. There is presently no path integral formulation for spin $\frac{1}{2}$ particles which fits into the Wheeler-Feynman description. As Professor Hoyle will discuss this topic in his lectures, I will just treat it briefly.

Consider a relativistic, spin $\frac{1}{2}$ particle going from point 1 to point 2. Feynman has considered this problem non-relativistically. He has shown that the probability amplitude of a particle going from 1 to 2 can be understood as a product of a chain of propagators between the point 1 and 2. If $K(i, i+1)$ represents the propagator between two neighboring points $i$, $i+1$, then the probability amplitude for the particle going from 1 to 2 is

$$\text{Probability Amplitude } (1 \to 2) = \prod_{i=1}^{N-1} K(i, i+1) \, A. \qquad (5.1)$$

Here A is a factor of dimensionality $(\text{length})^3$ put in to reconcile the dimensionless probability amplitude and the propagator whose dimensionality is $(\text{length})^{-3}$. We let $N \to \infty$ and the probability amplitude is the equivalent to that defined by the sum over paths. Feynman applied this procedure to relativistic particles by using the propagator, K, appropriate to the Dirac equation. He used a propagator which was zero for motion into the past and non-zero for motion into the future. The difficulty with this approach is the same one which occurs for the Dirac equation, namely the existence of negative energy states. In the non-relativistic approach one need only consider positive energy solutions propagating forward in time, while in the Dirac theory both positive and negative solutions propagate forward in time. To get around this difficulty it is necessary to postulate a sea of electrons which fill the negative energy states.

Relativistic Electrons

I will now discuss a somewhat different approach to the problem. Feynman's treatment, with the filled sea of negative energy states is intrinsically a many body problem. To avoid this problem we use another idea of Feynman, treating the negative energy solutions as going backwards in time, or positrons.

From the Dirac equation

$$(\not{\partial} + im)\psi = 0 \tag{5.2}$$

one gets an equation for the propagator

$$(\not{\partial}_2 + im)K_o^{\pm}(2, 1) = \delta_4(2, 1). \tag{5.3}$$

The solutions for the propagators are

$$K_o^+ = \Theta(t_2 - t_1) \sum_n U_n(2) \bar{U}_n(1) \tag{5.4a}$$

$$K_o^- = -\Theta(t_1 - t_2) \sum_n U_n(2) \bar{U}_n(1) \tag{5.4b}$$

Here $\Theta(x)$ is the Heaviside function $\Theta(x) = 1$ for $x > 0$, $\Theta(x) = 0$ for $x < 0$; $U_n$ are just the free particle solutions of the Dirac equation. We see that $K_o^-$ is just the time reversed solution of $K_o^+$. Thus $K_o^+$ describes propagation forward in time and $K_o^-$ is used to describe those paths which propagate backwards in time. Thus we can define the probability amplitude for a path going forward in time as

Electrodynamics and Cosmology 376

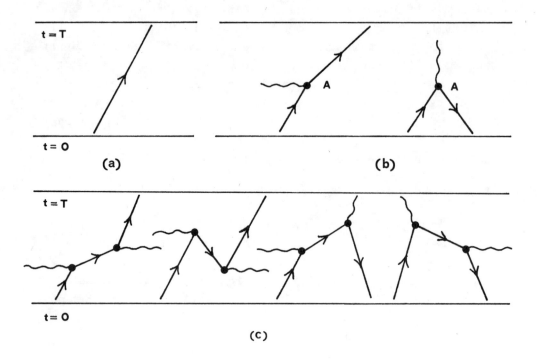

Figure 5.1

$$P(\Gamma^+_{21}) = \prod_{i=1}^{N-1} A_i \, K_o^+(i, i+1) \tag{5.5}$$

where $\Gamma^{\pm}$ represents paths going forward (backward) in time. Thus, electron wavefunctions are built up of $\Gamma^+$ paths and of positron wavefunctions of $\Gamma^-$ paths.

We now come to the more complicated case where an external potential can cause reversals in the paths of the particle. Consider the motion in a time slab between t = 0 and t = T. Looking at Fig. 5.1 you can see the various probabilities that must be computed for zero, one and two interactions of the electron field. In order to calculate the wavefunction at time T from that at time t = 0 we must take into account the various possibilities in calculating the propagator. If the propagator from t = 0 to t = T is $K^A(2, 1)$ we have

# Electrodynamics and Cosmology

$$K^A(2, 1) = K_+(2, 1) - ie \int K_+(2, 3) K_+(3, 1) d^3x_3 + \ldots \quad (5.6)$$

Here the first term describes a path which goes through without interacting with the electromagnetic field, and the second term describes a single interaction. As seen from Fig. 5.1(b) we have two possibilities: the particle can propagate forward, or can scatter off the external field and travel backward in time. So $K_+$ must include the two possibilities

$$K_+(2, 1) = \sum_{E_n > 0} U_n(2) \bar{U}_n(1) \quad t_2 > t_1$$

$$= -\sum_{E_n < 0} U_n(2) \bar{U}_n(1) \quad t_2 < t_1 \quad (5.7)$$

which you recognize as the Feynman propagator for electrons and positrons.

These possibilities describe whether the electron has scattered off the external field and remains an electron or if the situation is one of pair annihilation. Of course, the description can be carried out to higher order of scattering

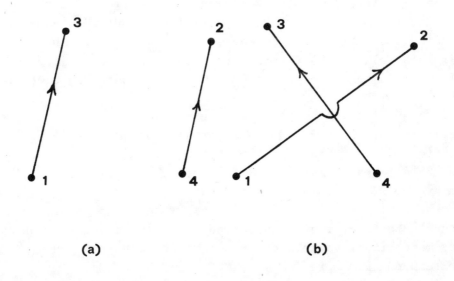

(a)            (b)

Figure 5.2

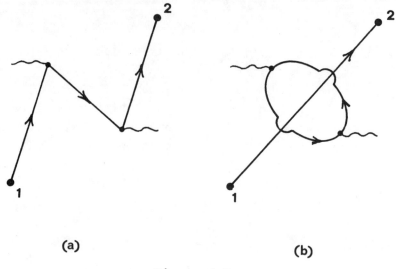

Figure 5.3

with the external field. You should note that $K_+$ contains very neatly the positive energy states propagated by $K_o^+$ and the negative states propagated by $K_o^-$. It must be used each time there is a possibility of scattering of the electron.

## Closed Loops

Consider the case where an electron goes from point 1 to point 3, and another electron goes from point 4 to point 2 as in Fig. 5.2(a). Antisymmetrization requires that the amplitude represented by 5.2(b) must also be included and subtracted from the amplitude represented by 5.2(a). The two diagrams (a) and (b) together represent, of course, the case of two incoming electrons at t = 0 and two outgoing electrons at t = T with no interactions.

Now consider the case of a particle which comes in at point 1, leaves at point 2 and the external field scatters twice. In Fig. 5.3(a) we represent the particle being scattered twice by the external potential. If we apply the same principle as in Fig. 5.2(a) and (b) to Fig. 5.3 we are lead to the closed loops represented in 5.3(b). These closed loops must occur in all orders of the perturbation expansion since you can allow for more time reversals of the electron paths and you get correspondingly more loops.

I would now like to consider the question of self-interaction.

Electrodynamics and Cosmology

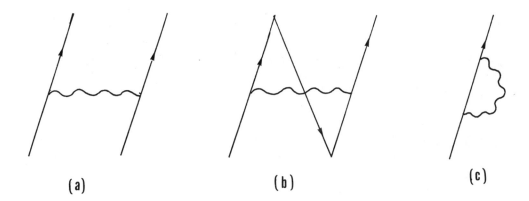

Figure 5.4

We saw that there was no self-action in the classical Wheeler-Feynman theory and we would like to see to what extent this is true in the relativistic quantum theory. An interaction between two worldlines is drawn in Fig. 5.4(a). The relativistic theory allows worldlines which go backwards so we might expect that the picture in Fig. 5.4(b) would also be possible. However, this represents self-interaction and we might want to eliminate it. Experimentally we cannot determine if such an interaction has taken place, so we have no reason to eliminate the possibility. If we are going to include graphs of that type, then ones of the type in Fig. 5.4(c) should be possible also.

In the classical theory all paths are time-like and we would not get a contribution because the $s^2$ in the $\delta$-function would never be zero. An exception to this would arise if the points 1 and 2 coincide. This problem is removed in the quantum theory by putting in a requirement which does not allow the two points to coincide.

I turn now to the relativistic quantized Wheeler-Feynman theory. The interaction is now written

$$R = -\tfrac{1}{2} \sum_a \sum_b e_a e_b \iint \delta(q_{AB}^2) da_i db^i \qquad (5.8)$$

R is what we have previously called $S_I$ and $q^2_{AB}$ was $s^2_{AB}$. The restriction $a \neq b$ has been dropped. I will look at these terms later; they are important only in discussing self-interaction. We will consider a time slab $t = 0$ to $t = T$ which we would like to detach from the rest of the universe $t < 0$ and $t > T$. I will now write the interaction as

$$-\sum_a e_a \int B_i \, da^i - \tfrac{1}{2} \sum_a \sum_b e_a e_b \iint \delta(q^2_{AB}) \, da^i \, db_i \qquad (5.9a)$$

where the first term represents contributions from outside the slab and the second term is the contribution from inside the slab. The potential $B_i$ will have two parts, $B_i(t < 0)$ and $B_i(t > T)$ arising from contributions before and after the slab respectively.

My aim is to determine what the universe must do in order to give results consistent with local experiments.

First, look at the classical case. The potential produced by particle b is

$$A^{i(b)}(X) = e_b \int \delta(q^2_{XB}) \, db^i \qquad (5.10)$$

The total potential which acts on the particle a is

$$A^i_{(a)}(X) = \sum_{b \neq a} A^{i(b)}(X)$$

$$= \tfrac{1}{2} \left[ A^i_{(a)\,\text{ret}}(X) + A^i_{(a)\,\text{adv}}(X) \right] \qquad (5.11)$$

The classical action can then be rewritten as

$$-\sum_a e_a \int B^i_{t>T} \, da_i - \tfrac{1}{2} \sum_a e_a \int \left[ A^i_{(a)\,\text{ret}} + A^i_{(a)\,\text{adv}} \right] da_i \qquad (5.9b)$$

# Electrodynamics and Cosmology

The $B^i_{t<0}$ part has not been included because I am only interested in considering the contributions which might lead to spontaneous transitions.

If there is complete absorption in the future, then the action would become

$$-\sum_a e_a \int \left\{ A^i_{(a)\,ret} + \tfrac{1}{2}\left[ A^{i(a)}_{ret} - A^{i(a)}_{adv} \right] \right\} da_i$$

What we can do is subtract this from our action and require that the difference be zero. The result is that we must have

$$B_i(X)_{t>T} = \tfrac{1}{2} \sum_a \left[ A^{(a)}_{i\,ret} - A^{(a)}_{i\,adv} \right] \tag{5.12}$$

For the quantum mechanical case we must multiply the action by $i/\hbar$ and exponentiate it, and, if you recall the form of a transition probability, then you will see that the expression

$$-\sum_a e_a \left[ \int \left\{ \tfrac{1}{2}\left[ A^i_{(a)\,ret} + A^i_{(a)\,adv} \right] + B^i(t>T) \right\} da_i \right.$$

$$\left. - \int \left\{ \tfrac{1}{2}\left[ A'^i_{(a)\,ret} + A'^i_{(a)\,adv} \right] + B'^i(t>T) \right\} da'_i \right]$$

will appear in the exponent. The primes indicate conjugate paths and the sum is over all a. We must now lay down a condition corresponding to the perfect absorption condition that will give us the right kind of QED and that will reduce to the classical result when $\hbar \to 0$. The result is that we must have

$$B_i(t>T) = B'_i(t>T) = \tfrac{1}{2} \sum_b \left[ A^{(b)}_{i\,ret+} - A^{(b)}_{i\,adv+} \right] + \tfrac{1}{2} \sum_b \left[ A'^{(b)}_{i\,ret-} - A'^{(b)}_{i\,adv-} \right] \tag{5.13}$$

where the + and − indicate the positive and negative frequency parts respectively. This expression is relativistically invariant. In the classical theory, there is a unique path and the conjugate path is thus equal to the path. In this case, the above expression for $B_i(t>T)$ reduces to the classical condition. This expression for $B_i(t>T)$ leads to the influence functional

$$F(\underline{a}, \underline{a}') = \exp\left\{ie_a^2\left[\iint_{t_{A'}>t_A}\delta_+(q_{A'A}^2)da_i'da^i - \iint_{t_A>t_{A'}}\delta_-(q_{AA'}^2)da_i'da^i \right.\right.$$

$$\left.\left. - \tfrac{1}{2}\iint \delta_+(q_{A\tilde{A}}^2)da^id\tilde{a}_i + \tfrac{1}{2}\iint \delta_-(q_{A'\tilde{A}'}^2)da'^id\tilde{a}'_i\right]\right\} =$$

$$= \exp\left[\frac{e_a^2}{4\pi^2}\int d\Omega \int_0^\infty kdk \left\{-\iint \exp\left[ik(t_A - t_{A'}) + \right.\right.\right.$$

$$\left.\left. + i\underline{k}\cdot(\underline{x}_{A'} - \underline{x}_A)\right]da^ida_i' + \iint_{t_A>t_{\tilde{A}}} \exp\left[ik(t_{\tilde{A}} - t_A) + \right.\right.$$

$$\left.\left. + i\underline{k}\cdot(\underline{x}_{\tilde{A}} - \underline{x}_A)\right]da^id\tilde{a}_i + \iint_{t_{A'}>t_{\tilde{A}}} \exp\left[ik(t_{A'} - t_{\tilde{A}'}) + \right.\right.$$

$$\left.\left.\left. + i\underline{k}\cdot(\underline{x}_{\tilde{A}'} - \underline{x}_{A'})\right]da'^id\tilde{a}'_i\right\}\right] \quad (5.14)$$

which is equal to the corresponding expression for the non-relativistic case. This gives the interaction of a single worldline with the rest of the universe. It would be applicable for the calculation of processes such as pair creation or annihilation. For the interaction of two particles we require the influence functional

# Electrodynamics and Cosmology

$$F(\underline{a}, \underline{b}; \underline{a}', \underline{b}') = \exp\left\{ie_a e_b \left[\iint_{t_{A'}>t_B} \delta_+(q^2_{A'B}) da'_i db^i - \iint_{t_B>t_{A'}} \delta_-(q^2_{BA'}) da'_i db^i + \iint_{t_{B'}>t_A} \delta_+(q^2_{B'A}) da^i db'_i \right.\right.$$

$$\left.\left. - \iint_{t_A \geq t_{B'}} \delta_-(q^2_{AB'}) da^i db'_i - \iint \delta_+(q^2_{AB}) da^i db_i + \iint \delta_-(q^2_{A'B'}) da'_i db'^i \right]\right\}$$

$$= \exp\left[\frac{e_a e_b}{4\pi^2} \int d\Omega \int k\, dk \left\{\iint \exp\left[-ik|t_A - t_B| + i\underline{k}\cdot(\underline{x}_B - \underline{x}_A)\right] da_i db^i + \iint \exp\left[ik|t_{A'} - t_{B'}| + i\underline{k}\cdot(\underline{x}_{A'} - \underline{x}_{B'})\right] da'_i db'^i \right.\right.$$

$$\left.\left. - \iint \exp\left[ik(t_A - t_{B'}) + i\underline{k}\cdot(\underline{x}_{B'} - \underline{x}_A)\right] da_i db'^i - \iint \exp\left[ik(t_B - t_{A'}) + i\underline{k}\cdot(\underline{x}_{A'} - \underline{x}_B)\right] da'_i db^i \right.\right] \quad (5.15)$$

This expression contains the information necessary for calculating many processes in electrodynamics including scattering and energy level shifts. One should note that it is given as an exponential in closed form. In principle, one could work out the answers for various processes in closed form. In practice, we expand the exponential and this leads to the standard perturbation expansion. We can now consider any model of the universe and check in detail whether it satisfies the condition on $B^i(t>T)$ that we have derived. The necessary condition for this to work is first of all that the classical theory should work and second that the model should provide a cold environment in the future.

I would now like to return to the discussion of the self-interaction problem. We saw before that in a self energy graph we did not want the points 1 and 2 to get too close together. A close look at this condition reveals that we must know what shape the worldline takes in the quantum sense. It turns out that the line is not smooth but is made up of small null steps. This comes from the fact that we wrote K(2, 1) as a product of propagators for very short steps, and that for short steps the main contribution to such a propagator comes from null directions. So, we must decide how small we will allow the steps to get. We can get a finite answer for $\delta m/m$ if we do not allow the step size to go to zero. If we take the size equal to the gravitational radius of the electron we get a result like Prof. Salam showed. Similar ideas apply to the calculation of closed loops.

In their book, Feynman and Hibbs (1965) showed in a problem that for an electron in one space and one time dimension, the Dirac equation comes out of the following consideration. The electron is assumed to do a random walk with velocity plus or minus c. The amplitude for a particular path is $(im\varepsilon)^R$, where R is the number of velocity reversals or corners in the path. If N(R) is the number of paths with R corners, the propagator is proportional to

$$\sum (im\varepsilon)^R N(R)$$

and leads to the Dirac equation when $\varepsilon \to 0$. One wonders what sort of magic leads to this result. Dr. Hoyle will talk more about the role of the mass as a scatterer.

This result, along with the considerations that arose in the self energy problem, suggests those problems should be looked into more carefully. We saw that in the interaction part of the action a $\delta(q^2)$ was converted into a $\delta_{\pm}(q^2)$ by the response of the universe. One wonders whether the universe plays some role in converting the $K_o^{\pm}$ propagators that we began with to the $K_+$ propagator.

## REFERENCES

Feynman, R. P., 1948, Rev. Mod. Phys., 20, 367.

Feynman, R. P., Hibbs, A. R., 1965, "Quantum Mechanics and Path Integrals", McGraw-Hill, New York.

Fokker, A. D., 1929a, z.f. Physik, 58, 386.

Fokker, A. D., 1929b, Physica, 9, 83.

Fokker, A. D., 1932, Physica, 12, 145.

Hogarth, J. E., 1962, Proc. Roy. Soc., A267, 365.

Hoyle, F., Narlikar, J., 1963, Proc. Roy. Soc., A277, 1.

Hoyle, F., Narlikar, J., 1969, Ann. Phys., 54, 207.

Hoyle, F., Narlikar, J., 1971, Ann. Phys., 62, 44.

Schwartzschild, K., 1903, Göttinger Nachrichten, 128, 132.

Tetrode, H., 1922, Z.f. Physik, 10, 317.

Wheeler, J. A., Feynman, R. P., 1945, Rev. Mod. Phys., 17, 157.

Wheeler, J. A., Feynman, R. P., 1949, Rev. Mod. Phys., 21, 425.

Weak Interactions and Cosmology

Lectures by

**Y. Ne'eman**

Tel Aviv University

(Owing to the loss of magnetic tapes of Professor Ne'eman's lectures, he has kindly given us permission to reprint here his articles covering the substance of each of these lectures. Ed.)

Lecture 1

## TIME-REVERSAL SYMMETRY VIOLATION AND THE OSCILLATING UNIVERSE[1]

### 1. Introduction

In a previous article (Ne'eman, 1970), the experimentally established violation of CP symmetry in the decay of the long-lived K meson (Christenson et al., 1964), and the further possibility of CPT violation, were studied in the context of time-symmetric oscillating models of the universe. It was shown that the current assumption, according to which the contracting phase of the oscillation is reinterpreted as a time-inverted expansion, cannot be retained at all if CPT is violated; if only CP is violated, the assumption is allowed and involves inverting the definitions of matter and antimatter.

To describe the evolution of a $K^°-\bar{K}^°$ complex, the Lee-Oehme-Yang formula (Lee et al., 1957) was used. This formula predicts the number of neutral K mesons remaining in the beam at any time t. In the present article we refine the argument and apply it to a different set of formulae which emphasize the observables involved.

In Section 2, the formulae for the fractional number of $K^°$'s and $\bar{K}^°$'s in a neutral kaon beam are discussed, in an ordinary and in a time-inverted coordinate schemes. In Section 3 these formulae are used for defining a relation between the cosmological arrow of time and the behaviour of the microscopic $K^°-\bar{K}^°$ system.

### 2. Time Reversal Asymmetry in $K^°-\bar{K}^°$ Distinguishing Formulae

We first consider at $t = 0$ a beam of pure $K^°$'s (strangeness $= +1$). For $t > 0$, these particles will decay via the weak Hamiltonian eigen-states $K_S$ and $K_L$, thus forming at time t, aside from the decay products, a number of $\bar{K}^°$'s. Using the parametrization of Lee and Wu (Lee et al., 1967), with $\delta$ the CPT non-invariance parameter ($\delta = 0$ if CPT is conserved) and $\varepsilon$ the T non-invariance parameter, a direct calculation gives

---

1. Reprint of the article by A. Aharony and Y. Ne'eman which appeared in the Int. Journ. Theor. Phys., Vol. 3, No. 6 (1970), pp. 437-441.

for the above beam (Aharony, 1970)

$$R^{K^\circ}(K^\circ, t) = \frac{1}{4}\left\{(1-4\text{Re}\delta)\exp(-\gamma_L t) + (1 + 4\text{Re}\delta)\exp(-\gamma_S t) + \right.$$
$$\left. + \left[(1 + 4i\text{Im}\delta)\exp(i\Delta mt) + c.c.\right]\exp\left[-\frac{1}{2}(\gamma_L + \gamma_S)t\right]\right\} \quad (1.1)$$

$$R^{K^\circ}(\bar{K}^\circ, t) = \frac{1}{4}(1-4\text{Re}\varepsilon)\left\{\exp(-\gamma_S t) + \exp(-\gamma_L t) - 2\cos\Delta mt \times \right.$$
$$\left. \times \exp\left[-\frac{1}{2}(\gamma_L + \gamma_S)t\right]\right\} \quad (1.2)$$

$R^{K^\circ}(K^\circ, t)$ and $R^{K^\circ}(\bar{K}^\circ, t)$ are, respectively, the fractions of $K^\circ$ and of $\bar{K}^\circ$ particles in the beam at the kaon's proper time t. $\gamma_L$ and $\gamma_S$ are the inverse lifetimes of $K_L$ and of $K_S$, and $\Delta m$ is their mass difference. For a beam initially made of pure $\bar{K}^\circ$'s, $R^{\bar{K}^\circ}(\bar{K}^\circ,t)$ and $R^{\bar{K}^\circ}(K^\circ, t)$ will be given by the same formulae, except for a change in the signs of $\varepsilon$ and $\delta$. (All expressions are to first order in $\varepsilon$ and $\delta$.)

We now wish to consider the same beam in a time reversed coordinate system. The expressions for reversed time, -t, are obtained from those for t by applying the time-reversal operation T. As shown in the previous article (Ne'eman, 1970) and by Zweig (1967), the equation of motion

$$i\frac{d\psi}{dt} = (M - i\Gamma)\psi \quad (1.3)$$

for the two-dimensional state-vector $\psi$ describing the $K^\circ - \bar{K}^\circ$ complex (M and $\Gamma$ are the 2 x 2 mass and decay matrices) is transformed under T to

$$i\frac{d}{dt'}\psi_T^* = (M^* - i\Gamma^*)\psi_T^* \quad (1.4)$$

where t' = -t. Therefore, $\psi_T^*(t')$ will exhibit a time evolution similar to that of $\psi(t)$, except for the transformation

$$\varepsilon \to -\varepsilon, \quad \delta \to \delta \quad (1.5)$$

Since equation (1.1) describes the fractional number of $K^\circ$'s in a beam beginning at t = 0 with pure $K^\circ$, we can deduce that the fractional number of $K^\circ$'s in the time-reversed coordinate system, described by $P^{K^\circ}(K^\circ, t')$, will have the same time dependence (note that equation (1.1) involves only $\delta$, which does not change under T):

$$P^{K^\circ}(K^\circ, -t) = R^{K^\circ}(K^\circ, t) \qquad (1.6)$$

Similarly, we obtain from equations (1.2) and (1.5)

$$P^{K^\circ}(\bar{K}^\circ, -t) = (1 + 8\mathrm{Re}\,\varepsilon)R^{K^\circ}(\bar{K}^\circ, t) \qquad (1.7)$$

3. A Relation Between Microscopic and Cosmological Arrows of Time

In a time-symmetric oscillatory model, we would expect every physical situation to repeat itself after a time $\tau$, $\tau$ being the oscillation period of the universe. Thus, the fractional number of $K^\circ$ and of $\bar{K}^\circ$ particles in the universe at the points A and B of maximum contraction (Fig. 1.1) should coincide. We can in fact restrict ourselves to a given volume element and discuss a subsector of the universe containing initially (at A) a beam of pure $K^\circ$ particles.

At times $t > 0$, the beam will decay, leading to a decrease in the number of $K^\circ$'s and an increase in the number of $\bar{K}^\circ$'s, described by equations (1.1)-(1.2). These formulae give the time-evolution only for short times, since they are based on the Wigner-Weisskopf approximation (Lee et al., 1957). Still, if the beam exists at times larger than $\tau/2$, then we must assume that the decay products are contracted to reproduce the initial $K^\circ$ beam, since we demand complete identity of the physical

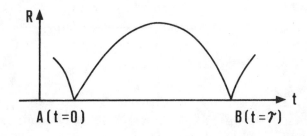

Figure 1.1

states at A and at B.

By the time symmetry of the oscillating model, we may assume that the behaviour of the beam at the time $(\tau-t)$ approaching the point B is the same as at the time $-t$, given by eqs. (1.6)-(1.7). The left-hand sides of these equations were derived as the results of a decay process in the time-inverted contracting universe, but they may be reinterpreted as the fractional numbers of $K^°$'s and $\bar{K}^°$'s needed at the time $(\tau-t)$ in order that the beam will end at the time $\tau$ as pure $K^°$.

Equation (1.6) thus represents a complete symmetry of the fractional number of $K^°$'s in the beam with respect to the time $\tau/2$. Note that this symmetry is independent of CPT invariance, since it holds for any value of $\delta$. There is no such symmetry in equation (1.7); the fractional number of $\bar{K}^°$'s in a beam of $K^°$'s will reveal a symmetry with respect to $\tau/2$ only if $Re\epsilon = 0$, hence if T is conserved!

If T is not conserved for the $K^°-\bar{K}^°$ system, as it now seems to be established (Casella, 1968, 1969; Achiman, 1969), we may use (1.7) to define the direction of the arrow of time. There are several possible ways to determine $Re\epsilon$ experimentally, e.g. by measuring the charge asymmetry in $K_L$ leptonic decay (Schwartz et al., 1967) or by other experiments measuring the overlap of the states $K_L$ and $K_S$. In most of these experiments one has expressions with combinations of $\epsilon$ and of $\delta$, but $Re\epsilon$ can be deduced from them. A direct measurement of $Re\epsilon$ is presented by equation (1.2) (Aharony, 1970): One has to measure the number of $\bar{K}^°$'s in a beam initiated by $K^°$ and determine the coefficient of equation (1.2). Such experiments have been discussed by Crawford (Crawford, 1965). In a time inverted contracting world this experiment will give a different sign for $Re\epsilon$. Note, that we have yet no theoretical way to relate the sign of $Re\epsilon$ with the oscillation phase of the universe (Zweig, 1967). Still, if the definitions of matter and antimatter are agreed, the sign of $Re\epsilon$ is fixed by them since it is related to the difference between the fractions of $K^°$ and of $\bar{K}^°$ in $K_L$ and $K_S$, and is thus determined by the experiments mentioned above. (These differences involve $Re(\epsilon + \delta)$ and $Re(\epsilon - \delta)$, and thus fix both $Re\epsilon$ and $Re\delta$.)

Because of this ambiguity, it is interesting to consider a beam ending (or starting, in the time-inverted contracting world) at time t = 0 as pure $\bar{K}^\circ$. Using the rule following equation (1.2) and equation (1.5) we find

$$P^{\bar{K}^\circ}(K^\circ, -t) = R^{K^\circ}(\bar{K}^\circ, t) \tag{1.8}$$

One might thus think that there is no distinction between the two oscillation phases if we invert the definitions of matter-antimatter. But this is not so, since in this case we shall have no symmetry between $P^{\bar{K}^\circ}(\bar{K}^\circ, t')$ and $P^{K^\circ}(K^\circ, t)$ unless $\delta = 0$. Thus, only if CPT is conserved, one regains a complete symmetry if one defines $\bar{K}^\circ$ as $K^\circ$, and vice versa.

One might try to avoid the difference between the decay formulae in the expanding and in the time-inverted contracting phases, by assuming that in the time-inverted universe one deals with different kinds of basis states $|K^\circ{}'\rangle$ and $|\bar{K}^\circ{}'\rangle$, instead of $|K^\circ\rangle$ and $|\bar{K}^\circ\rangle$, for which the physical behaviour is similar to equations (1.1)-(1.2). But it is easy to check that no combination of $|K^\circ\rangle$ and $|\bar{K}^\circ\rangle$ will give tha same behaviour unless $\text{Re}\,\varepsilon = 0$.

## 4. Conclusion

If both CPT and T are not conserved, an experiment counting the fractions of $K^\circ$'s and of $\bar{K}^\circ$'s in a $K^\circ$ beam will distinguish between the expanding and contracting phases of the universe oscillation, thus forming a relation between microscopic and cosmological arrows of time.

If CPT is conserved, this distinction may be resolved by inverting the definitions of matter-antimatter.

Even if CPT is conserved, there remains the question of the behaviour of a beam of $K^\circ$'s beginning at $t = 0$ and remaining, through the maximum expansion phase at $\tau/2$, until the end of a period, $\tau$. Had it been possible to conserve such a beam, including information concerning its behaviour at short times, an asymmetry is due to appear at the time $\tau - t$.

Lecture 2

## THERMODYNAMICS AND STATISTICAL MECHANICS OF THE CP VIOLATION[1]

1. Introduction

Two of the main results in the theory of statistical mechanics of irreversible processes are the H-theorem, or the Second law of Thermodynamics, and the Onsager relations between such processes. In both cases, the proofs of these results are usually based on the assumptions of microscopic reversibility, namely that the Hamiltonian of the system is invariant under time reversal, T.

The discovery of CP violation in the decay of the neutral kaon (Christenson et al., 1964), with the assumption of CPT conservation, implied a violation of time reversal symmetry in this decay. Moreover, it has recently been established, that the experimental results are indeed consistent with a violation of T, and that they cannot be explained assuming T conservation (Dass, 1971). There is thus evidence for a microscopic "arrow of time", defined by the behaviour of the neutral kaon system (Ne'eman, 1968 c; Zweig, 1957; Aharony and Ne'eman, 1970 b). Although no other experimental proof of time reversal violation has yet been found (Cannata et. al., 1970), it is still of interest to investigate the necessity of the assumption of time reversal symmetry in the proofs of both the H-theorem and the Onsager relations. Since the system of the neutral kaons is the only known time reversal non-invariant system, we shall discuss these questions in relation with it.

The basic idea of the present discussion is to describe the $K^\circ$-$\bar{K}^\circ$ system as interacting with one or more thermal baths, which contain the kaons' decay products. The derivation of a Master Equation for the reduced density matrix of the system is reviewed in Sec. 2. In Sec. 3, this reduced density matrix is used to define an H-function, for the description of the irreversible approach of the system to thermal equilibrium; the validity of the H-theorem is then discussed. Sec. 4 includes a generalization to the case of several thermal baths

---

1. Reprint of the article by A. Aharony and Y. Ne'eman presented at the 1972 Coral Gables Conference on Fundamental Interactions.

The Onsager coefficients for the energy currents between the system and the baths in a stationary state are defined, and the Onsager relations are checked. In Sec. 5, our results are compared with those for a time inverted coordinate scheme, and the relations between the microscopic, thermodynamical and cosmological arrows of time are discussed.

## 2. The Master Equation Description of the Neutral Kaon System

The derivation of the Master Equation for the reduced density matrix of the neutral kaons' system has been described in detail elsewhere (Aharony, 1971 b, c), so that we shall give here only a short review.

We describe the kaon's decay as being due to the interaction between the kaonic system, which has the three basis states $|0\rangle$, $a_1^\dagger|0\rangle$ and $a_2^\dagger|0\rangle$ ($|0\rangle$ is the vacuum, $a_i^\dagger$ creates $K_i^\circ$, $K_1^\circ$ and $K_2^\circ$ being the CP eigenstates with eigenvalues +1 and -1), with a thermal bath which contains all the possible final decay states, e.g. $2\pi$, $\pi\ell\nu$, etc., in thermal equilibrium.

The Hamiltonians of the kaonic system, the bath and the interaction are respectively

$$H_0^a = \hbar\omega_K (a_1^\dagger a_1 + a_2^\dagger a_2) \tag{2.1}$$

$$H_0^B = \sum_{rk} \hbar\omega_{rk} b_{rk}^\dagger b_{rk} \tag{2.2}$$

and

$$H_I = -i\hbar \sum_i \sum_{rk} g_{rk}^i a_i^\dagger b_{rk} + h.c. \tag{2.3}$$

($b_{rk}^\dagger$ creates an eigenstate of the strong Hamiltonian in the decay channel r with quantum numbers k, $-i\hbar g_{rk}^i$ is the appropriate matrix element of the weak Hamiltonian).

The reduced density matrix of the kaon system is now defined as

$$\rho_a = Tr_B \rho \tag{2.4}$$

where $\rho$ is the total density matrix, and the trace is taken over the states of the bath. The time evolution of $\rho_a$ is given by the Wangness-Bloch master equation (Wargness and Bloch, 1953; Aharony, 1971 a). If we start at $t = 0$ with

$$\rho_a(0) = \begin{pmatrix} 0 & 0 & 0 \\ 0 & & \\ 0 & \bar{\rho}(0) & \end{pmatrix} \tag{2.5}$$

(the first row and column stand for $|0\rangle$), then at time $t$ we have

$$\rho_a(t) = \begin{pmatrix} 1-\text{Tr}(t) & 0 & 0 \\ 0 & & \\ 0 & \bar{\rho}(t) & \end{pmatrix} \tag{2.6}$$

with

$$i \frac{\partial \bar{\rho}}{\partial t} = \Lambda \bar{\rho} - \bar{\rho} \Lambda^\dagger + i\Omega(1-\text{Tr}\bar{\rho}), \tag{2.7}$$

$$\Lambda_{ij} = \omega_K \delta_{ij} - \int \frac{d\omega}{\omega - \omega_K + i\epsilon} \sum_r N_r(\omega) \left( \overline{g_{rk}^i g_{rk}^{j*}} \right)_{\omega=\omega_k} (e^{-\beta\hbar\omega}+1) \tag{2.8}$$

$$\Omega_{ij} = 2\pi \sum_r N_r(\omega_K) e^{-\beta\hbar\omega_K} \text{Re}\left( \overline{g_{rk}^i g_{rk}^{i*}} \right)_{\omega_K=\omega_k} +$$
$$+ 2P \int \frac{d\omega}{\omega - \omega_K} \sum_r N_r(\omega) e^{-\beta\hbar\omega} \text{Im}\left( \overline{g_{rk}^i g_{rk}^i} \right)_{\omega=\omega_k} \tag{2.9}$$

($N_r(\omega)$ is the density of states in the channel $r$; $\beta = 1/k_B T$).

Clearly, for zero temperature Eq. (2.7) reduces to the usual Wigner-Weisskopf result (Lee, Oehme, and Yang, 1967; Michel, 1968):

$$i \frac{\partial \bar{\rho}}{\partial t} = \Lambda^\circ \bar{\rho} - \bar{\rho} \Lambda^{\circ\dagger} \tag{2.10}$$

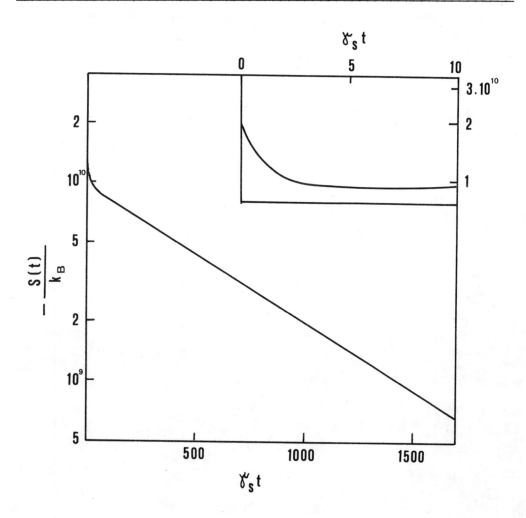

Figure 2.1 The inverse entropy $H_a = -\frac{S}{k_B}$ for $K^\circ$-decay with T and CPT symmetry.

Since $\hbar\omega_K \simeq 500$ MeV, $e^{-\beta\hbar\omega_K}$ will be very small compared to unity, unless the temperature approaches $10^{12}$ degrees. In this case, we can express $\Lambda_{12}$ and $\Lambda_{21}$ by the usual (Lee and Wu, 1967) T- and CPT- violation parameters $\varepsilon$ and $\delta$,

$$\Lambda_{12} = \left[ \Delta m - \frac{i}{2}(\gamma_L - \gamma_S) \right] (\varepsilon - \delta) \qquad (2.11)$$

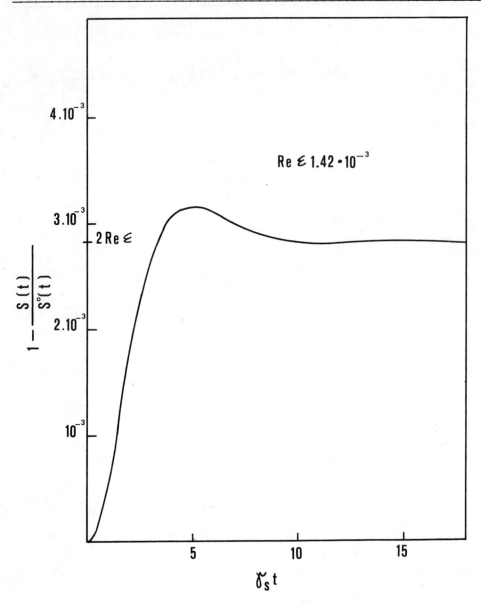

Figure 2.2 The oscillatory part of the entropy function due to T-violation.

$$\Lambda_{21} = \left[ \Delta m - \frac{i}{2} (\gamma_L - \gamma_S) \right] (\varepsilon + \delta) \qquad (2.12)$$

where $\gamma_S$, $\gamma_L$ are the inverse lifetimes of $K_S^0$, $K_L^0$, $\Delta m = m_L - m_S$ their mass difference ($K_L^0$ and $K_S^0$ are the eigenstates of $\Lambda$). Since $\varepsilon$ and $\delta$ are of the order $10^{-3}$, $\varepsilon$, $\delta$ and $\Omega_{ij}$ may be regarded as small perturbations, and Eq. (2.7) may be solved, to first order in them.

For any initial condition the solution approaches the same equilibrium

$$\rho_a^{eq} = \begin{pmatrix} 1-2e^{-\beta\hbar\omega_K} & 0 & 0 \\ 0 & e^{-\beta\hbar\omega_K} & 0 \\ 0 & 0 & e^{-\beta\hbar\omega_K} \end{pmatrix} \qquad (2.13)$$

Explicit expressions for the time dependence of $\rho_{ij}$ may be found in Aharony (1971 b, c).

## 3. The H-Theorem

We define the H-function of the total system as

$$H_a(t) = \text{Tr}\left[\rho_a(\ln\rho_a - \ln\rho_a^{eq})\right]. \qquad (2.14)$$

Inserting the solutions of (2.7) for a pure $K°$ at $t = 0$ we find

$$H_a(t) = K_a°(t)\left[1 + \text{Re}\varepsilon \cdot K_a^1(t) + \text{Re}\delta \cdot K_a^2(t) + \text{Im}\delta \cdot K_a^3(t)\right] \qquad (2.15)$$

$K_a°(t)$ and $-\text{Re}\varepsilon \cdot K_a(t)$ are given in Figs. 2.1 and 2.2 for $T = 300°K$ and for experimental value $\text{Re}\varepsilon = 1.42 \times 10^{-3}$ (Steinberger, 1969), $\delta = 0$ (detailed expressions may be found in Aharony, 1971 b, c). As seen from the figures, the violation of time reversal symmetry, $\varepsilon \neq 0$, results in an oscillation in the entropy, superimposed on the otherwise regularly monotonic increase. Obviously, $H_a(t)$ described in Figs. 2.1 and 2.2 obeys the H-theorem namely

$$\frac{dH_a}{dt} \leq 0 \qquad (2.16)$$

In order to check the possibility of a violation of the H-theorem for higher values of $\text{Re}\varepsilon$, Fig. 2.3 gives $dH_a/dt$ for various values of $\text{Re}\varepsilon$. It is seen that $dH_a/dt$ becomes negative only for $\text{Re}\varepsilon$ higher than $3.4 \times 10^{-2}$. This value is higher than the limit given by the Unitarity Sum Rule (Steinberger, 1969):

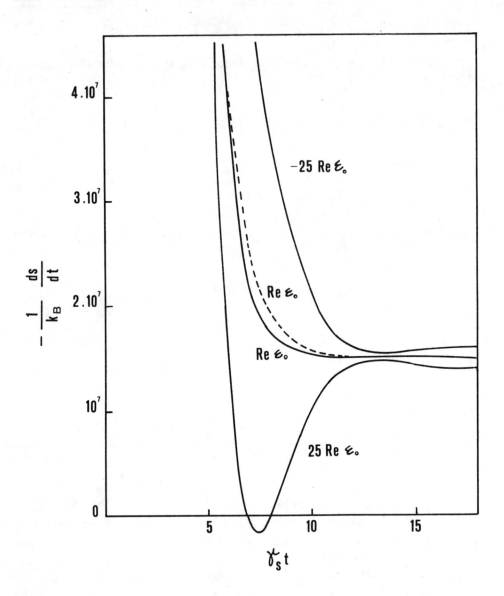

Figure 2.3 The time derivative of the entropy function for $K^\circ$-decay, with several values of the time reversal violation parameter $Re\varepsilon$. The numbers indicate the value of this parameter, with $Re\varepsilon_o = 1.42 \times 10^{-3}$.

$$\gamma_S = \sum_f | <f|T|K_S> |^2 \tag{2.17}$$

$$\gamma_L = \sum_f |<f|T|K_L>|^2 \tag{2.18}$$

$$\left[(\gamma_S - \gamma_L) - i2\Delta m\right](\text{Re}\varepsilon - i\text{Im}\delta) = \sum_f <f|T|K_L>^* <f|T|K_S> \tag{2.19}$$

The Schwartz inequality gives, for $\delta = 0$:

$$|\text{Re}\varepsilon| \leq \frac{(\gamma_L \gamma_S)^{1/2}}{|(\gamma_S - \gamma_L) - 2i\Delta m|} \simeq 0.03 \tag{2.20}$$

where the experimental values $\gamma_S = (1.17 \pm 0.1) \times 10^{10}$ sec$^{-1}$, $\gamma_L = (1.89 \pm .05) \times 10^7$ sec$^{-1}$ and $\Delta m/\gamma_S = .46 \pm .02$ (Bell and Steinberger, 1965) have been used.

Thus, unitarity is sufficient to ensure the validity of the H-theorem for this case. In fact, it may be shown that the proof of the H-theorem may be based on the unitarity of the S-matrix, and is thus independent of time reversal symmetry.

## 4. The Stationary State and the Onsager Relations

In order to discuss the Onsager relations, we now extend the above formalism to the case where the bosons may form several heat baths, with unequal temperatures.

We assume that the particles created in the channel r enter a bath with temperature $T_r$. All the expressions in Eqs. (2.7)-(2.9) remain unchanged, except that $\beta$ is replaced everywhere by $\beta_r = 1/k_B T_r$. The stationary state, given above by Eq. (2.13), will now change, and we shall have

$$\bar{\rho}_{ii}^{st} = \sum_r a_i^r \exp(-\beta_r \hbar \omega_K) \tag{2.21}$$

with

$$a_i^r = \frac{\pi N_r(\omega_K) \overline{|g_{rk}^i|^2}\Big|_{\omega_k = \omega_K}}{\sum_{r'} \pi N_{r'}(\omega_K) \overline{|g_{r'k}^i|^2}\Big|_{\omega_k = \omega_K}} = \frac{A_i^r}{\sum_{r'} A_i^{r'}} \tag{2.22}$$

Clearly, if all $\beta_r$'s are equal this reduces to (2.13).

As the thermodynamical currents we now define the energy flows from the system to the r'th bath,

$$J_r = \hbar\omega_K \left(\frac{\partial}{\partial t} \mathrm{Tr}\bar{\rho}\right)_r = \hbar\omega_K \sum_i \left[ A_i^r \exp(-\beta_r \hbar\omega_K) - A_i^{r-st} \rho_{ii} \right] \quad (2.23)$$

(we have neglected here terms of the order of $\exp(-2\beta_r \hbar\omega_K)$).

Since the sum of all the $J_r$'s is zero, we can express one particular current $J_o$ in terms of the others. By the first law of Thermodynamics, the entropy production in the system due to the energy flow is

$$\dot{S} = -\sum_r k_B \beta_r J_r = \sum_{r \neq 0} k_B (\beta_o - \beta_r) J_r \quad (2.24)$$

Thus, the $J_r$'s and the $k_B(\beta_o - \beta_r)$'s are the currents and the forces respectively, which are used in the Onsager theory (Onsager, 1931 a, b). The Onsager coefficients are now defined, by

$$L_{rr'} = -\frac{1}{k_B} \left(\frac{\partial J_r}{\partial \beta_{r'}}\right)_{\beta_{r'} = \beta_o} \quad (2.25)$$

By Eqs. (2.22), (2.23) we now find

$$L_{rr'} = \left[ -(\hbar\beta_K)^2 / k_B \right] \sum_i \left[ A_i^r A_i^{r'} / \sum_{r''} A_i^{r''} \right] \exp(-\beta_o \hbar\omega_K) \quad (2.26)$$

and clearly, to the first order in $\epsilon$, $\delta$ and $\exp(-\beta_o \hbar\omega_K)$ the Onsager relations hold,

$$L_{rr'} = L_{r'r}. \quad (2.27)$$

To obtain a higher approximation, we retain terms to order $\epsilon \exp(-\beta_r \hbar\omega_K)$ and $\delta \exp(-\beta_r \hbar\omega_K)$, and find that we must keep $\rho_{12}^{-st}$,

$$\bar{\rho}_{12}^{st} = \left[ \Omega_{12} + i\,(\Lambda_{21}^{*}\,\bar{\rho}_{11}^{st} - \Lambda_{12}\,\bar{\rho}_{22}^{st}) \right] / \left[ i\Delta m - \tfrac{1}{2}(\gamma_L + \gamma_S) \right] \qquad (2.28)$$

A simple manipulation of the additional part in $J_r$ shows that

$$L_{rr'} - L_{r'r} = \frac{4(\hbar\omega_K)^2}{k_B}\,\exp(-\beta_o \hbar\omega_K) \times$$

$$\times \operatorname{Im}\left\{ \frac{B_r^{12}(\Lambda_{12} a_2^{r'} - \Lambda_{21}^{*} a_1^{r'}) - B_{r'}^{12}(\Lambda_{12} a_2^{r} - \Lambda_{21}^{*} a_1^{r})}{i\Delta m - \tfrac{1}{2}(\gamma_L + \gamma_S)} \right\} \qquad (2.29)$$

with

$$B_r^{12} = \pi\,N_r(\omega_K)\,\overline{(g_{rk}^{1}\,g_{rk}^{2*})}\Big|_{\omega_k = \omega_K} \qquad (2.30)$$

Clearly, if both T and CPT are conserved, this difference will vanish, since in this case $\Lambda_{12} = \Lambda_{21} = 0$ (Eqs. (2.11) - (2.12)). In all other cases, the violation of the Onsager relations is of the order $\varepsilon \exp(-\beta_o \hbar\omega_K)$.

## 5. The Arrows of Time

From the definition of $\varepsilon$ and $\delta$ it may be shown (Aharony, 1971 b, c), that under the transformation of time reversal T,

$$\varepsilon \xrightarrow{T} \varepsilon_T = -\varepsilon$$

$$\delta \xrightarrow{T} \delta_T = \delta \qquad (2.31)$$

Therefore, if we consider a time-inverted coordinate scheme, many of the experiments on the $K^\circ$-$\bar{K}^\circ$ system will give different results. In particular, the oscillation in the evolution of the H-function will have a different sign, (see Fig. 2.3), the fraction of $\bar{K}^\circ$'s in a beam of $K^\circ$'s etc., will have a different value (Aharony, 1970), etc.

Such an inverted coordinate scheme is suggested in the theories of an oscillating universe (Aharony and Ne'eman, 1970 a): In such theories if $\tau$ is the oscillation period, there must exist a complete identity of physical behaviour at time $-t$ and at time

($\tau - t$). Therefore, in a contracting phase one should observe different physical laws than in the expanding phase, due to the asymmetry in the microscopic behaviour discussed above.

In all the expressions we discussed the change of the initial state from $K°$ to $\bar{K}°$ results in a change in the signs of both $\varepsilon$ and $\delta$. Therefore, if CPT is conserved ($\delta = 0$), the behaviour of matter in a direct scheme will be the same as that of antimatter in an inverted scheme. In this case, the definition of the microscopic direction of time is not absolutely defined. But if $\delta \neq 0$, it is clear that no such ambiguity exists; the behaviour of a $K°$ decaying in a direct time scheme is clearly different (e.g. in the features of the entropy time dependence) from that of a $\bar{K}°$ decaying in an inverted scheme.

Thus we arrive at a clear connection between the microscopic, thermodynamical and cosmological arrows of time.

Lecture 3

QUASAR COUNTS AND THE LAGGING CORE MODEL ("WHITE HOLES")[1]

The morphological study of quasars lends plausibility to the following characteristics:

(a) quasars possess very dense cores (Greenstein and Schmidt, 1964);

(b) quasars are at cosmological distances (Bahcall and Bahcall, 1970; Gunn, 1971);

(c) they seem to fit in a general class of objects with dense nuclei (Burbidge, 1970);

(d) their space distribution displays an evolutionary effect, their number appearing to follow an exponential decrease (Schmidt, 1970).

We would like to point out that the observation (d) reinforces the arguments for quasars as Lagging Cores of the original expansion (Ne'eman 1965, Novikov, 1964) and would be difficult to reconcile with any other model.

Dense nuclei can be created in two ways:

A) through the gravitational collapse of matter in a galaxy or a protogalaxy. In this case it can have the following history:

1) it continues to collect matter and to grow;

2) at a certain point it undergoes further collapse and generates a black hole; in this case it either vanishes and thus has a finite lifetime, or it may go into a stationary state and is then equivalent to case 1).

B) dense nuclei represent lagging, as yet unexpanded, cores ("white" hole in presently fashionable nomenclature). At some stage in their history they finally expand and turn into "normal" distributions of matter, galaxies, etc. Members of this class of objects are thus only recognized as quasars in their initial stage.

In case A1, the number of quasars (and similar objects) should grow with time. The evolutionary effect should thus be the inverse of observation (d).

In case A2, with a finite lifetime for collapsed nuclei due to the formation of black holes, the number of quasars and dense nuclei should have at least reached a steady state at some early

---

1. Reprint of the article by Y. Ne'eman from the Proceedings of the IAU Symposium on Physics of Dense Matter, Boulder, Colorado Aug. 21-26, 1972 (Reidel, Pub., in press.)

stage in the formation of galaxies; alternatively, we would have some growth with time, though less than in case A1. Again, negative exponential behaviour does not fit this picture.

Now for case B. We assume that the total amount of matter in the universe is fixed, with much of it in unexpanded cores. Assuming a fixed probability $\lambda$ for any single core to start upon the final stages of its expansion in any given unit of time, we find for N the number of quasars at time t

$$N(t) = N_o e^{-\lambda t}$$

The time coordinate t here corresponds to some external observer, e.g. an observer linked to a universal space-averaged reference frame. We know that the expansion occurs very fast in the lagging core's own reference frame, but is extremely slow with respect to an outside observer. With the difference between the two frames becoming very large, it is natural to take a probabilistic view with respect to the occurrence of the final expansion for the outside time-coordinate.

Summing up, we see that observation (d) fits very well with the lagging core model.

Lecture 4

HYPERCOLLAPSED NUCLEAR MATTER[1]

1. Introduction

Physical theory has yielded in the past a series of predictions relating to new states of matter, or to novel types of material constituents. Successful predictions have included new chemical elements, previously unknown elementary particles, the world of antiparticles etc. and recently macroscopic nuclear matter in the form of neutron stars. As yet unfound are magnetic monopoles, quarks, W mesons, superheavy elements and collapsed material in "black holes" (or coming out of "white holes"). To this list we are now adding the hypothesis of the existence of hypercollapsed nuclear matter.

2. Repulsive Core and Attractive Heart

We were led to the first suggestions (Ne'eman, 1968 a, b) of the existence of such tightly bound nuclear isomers by considering the new evidence about a strongly attractive internuclear force at very short ranges, beyond the barrier of the repulsive core. It is this region that we shall name the "attractive heart" inside the repulsive core.

The evidence for the attractive heart comes from high energy physics. Internucleon forces, and not just in two-particle (four prong) processes, can be represented by the exchange of Regge trajectories (Collins, 1971). For forward and near-forward scattering these are meson trajectories; backward scattering is produced by the exchange of 2 nucleons. The meson trajectories appear in a Chew-Frautschi plot as straight lines: the spin is linear in $M^2$ for the various meson states lying at positive momentum-transfer squared t' at integer J, either even or odd.

The quark model suggests that there should be trajectories passing through states with (J = the spin; P = the space parity; C = the charge-parity of the neutral component):

---

1. Reprint of the article by Y. Ne'eman from the Proceedings of the IAU Symposium on Physics of Dense Matter, Boulder, Colorado, Aug. 21-26, 1972 (Reidel Pub., in press)

$$J^{PC} = 0^{-+} \quad ; \quad 1^{--}$$
$$\Downarrow \qquad \qquad \swarrow \downarrow \searrow$$
$$1^{+-} \qquad \quad 0^{++} \; 1^{++} \; 2^{++},$$

where the upper row corresponds to the two s-wave quark-antiquark combinations. The recent results of duality theory confirm the observation of exchange-degeneracy, which connects $0^{+-}$ to $1^{+-}$ and $1^{--}$ to $2^{++}$ (our double arrows).

Experimentally, all of these mesons are now known (Particle Data Group, 1972) except for the $0^{++}$ set which is as yet uncertain. For reasons of covariance, the $1^{++}$ set does not play an important role in the near-forward region. For zero-strangeness cases, we have two isoscalar and one isovector state for each $J^{PC}$; this would imply dealing with five states per $J^{PC}$, i.e., 30 states altogether (or 18 isospin x $J^{PC}$ multiplets). Actually, the isoscalar $0^{-+}$ is relatively weakly coupled to the nucleon, because of the peculiar F/D ratio in the SU(3) coupling. In the $1^{--}$, $0^{++}$, $1^{++}$ and $2^{++}$ the particular form of SU(3) breaking mixes isoscalars so that only one is coupled to the nucleon. This leaves us with the following, (masses are in parentheses) (Collins, 1971);

|       | $0^{-+}$ | $1^{+-}$ | $1^{--}$ | $0^{++}$ | $2^{++}$ |
|-------|----------|----------|----------|----------|----------|
| I = 1 | $\pi(140)$ | B(1235) | $\rho(765)$ | | $A_2(1310)$ |
| I = 0 | $\eta'(960)$ | | $\omega(785)$ | | f(1260) |

Phenomenologically, the couplings of the $2^{++}$ are about 5 times stronger (in $g^2/4\pi$) than the $1^{--}$ couplings (Michael). This implies an attractive heart which should have some chance of producing a second potential well, closer to the nucleon's center.

The existence of a tensor force in nuclei has been known for a long time. The new high-energy picture points to the possibility that aside from quadrupole moments, the longitudinal component of a $2^{++}$ exchange might create an attractive heart depending of course upon the particular many-body situations arising for various baryon numbers. This is plausible, since the elementary interactions, repulsive and attractive, are of the same order of magnitude, and it is only the variation in A and in configurations which can establish clear cut regions of preponderance.

3. Nuclear Isomers

Bodmer (1971) has studied the possible existence of hyper-collapsed nuclei in terms of the nuclear Hamiltonian. He assumes this Hamiltonian to contain in addition to the "normal" $H_A(N)$, a part $H_A(C)$ for the collapsed version corresponding to the same baryon number A.

Before going into any details, it is instructive to consider the analogy with molecular (nuclear) fusion and fission. Taking a deuterium molecule, (the N state) we know that it is fusion-favored, i.e., that given the right conditions it can "collapse" into a He atom (the C state); the D-D system is originally bound by $B_{2,2}(N) < B_{2,2}(C)$, the binding energy in He. On the other hand, at the other end of the periodic table we have fissioning nuclei (or "molecules"), i.e., with the fission products originally bound by $B_A(C) < B_A(N)$, the binding in a pseudo molecule made of the fission-product atoms.

The radius of the hypercollapsed state $R_c < R_N = r_o A^{1/3}$ and is either a constant ($\lesssim 0.5$ F) or a saturating radius $R_c \sim r_c A^{1/3}$ with $r_c < 0.4$ F. As in the above molecular examples, the most general possibility would correspond to a cross-over region $A_{crit}$, with

$B_A(C) < B_A(N)$ for $A < A_{crit}$

$B_A(C) > B_A(N)$ for $A > A_{crit}$

Bodmer estimates $A_{crit}$ to lie somewhere between A = 16-40. To reach this value, he treats the second case; assuming the lifetime of a "normal" stable nucleus to be $\tau_A(N) > 10^{31}$ sec (one collapse per mole per year), he estimates the penetrability of the saturation barrier (the hard core) $W(r)$ between normal and collapsed states. Comparing $\tau_A(N) \geq 10^{31}$ sec with the "period" $\tau$ corresponding to $R_A(N)$, i.e., about $10^{-22}$ sec, he gets (P is the penetration probability, $\tau^{-1}$ the number of penetration "attempts" per second)

$$P = 10^p \leq 10^{-22} / 10^{31} = 10^{-53}$$

Using various plausible values for the nuclear parameters does not modify p by more than 20%. P can now be related to the equation of state of nuclear matter which is of course part of our guess.

$$P = \exp\left\{-2(2M/\hbar^2)^{1/2} \int_{R_c}^{R_N} \left[W(r)\right]^{1/2} dr\right\}$$

where $M \simeq A\, m_N$ is the mass of $N_A$, the normal nucleus. It is by extrapolating smoothly from the region around $R_N$ where we know the equation, to the region of a hypothetical $R_c$, that Bodmer gets his estimate of $A_{crit}$.

For the binding energy per nucleon in the hypercollapsed state, we have to assume that it will stay in the hundreds of MeV, so as to avoid zero or negative energy nuclei.

4. Unitary Symmetry

The Fermi energy may rise to 0.5 $m_N$ or above. This will then lead to the creation of hyperons and mesons. In the extreme case, we might have the unitary spin equivalent of an $\alpha$ particle: $A = 16$, with all eight octet baryons appearing with spins up and down. The next "complete shell" would be at $A = 56$, with all components of 8 and 10 etc. The total strangeness (and electric charge) tends to be very low in general (and 0 in the "complete" cases). Bodmer gets similar results from a rough quark model calculation.

5. "Superbaryons" and Cosmology

In recent years, a thermodynamical approach to strong interactions (Hagedorn, 1970) has been used to predict results of high energy multiparticle collisions in accelerators and in cosmic rays, and to study the early stages of the expanding universe. This approach provides, it seems, an efficient way of accounting for the strong interactions through the insertion of an infinite spectrum of metastable ("elementary") states, i.e., there is a maximal temperature to hadrons. In this approach the development of concentrations (leading to the ultimate formation of galaxies) appears to be inescapably tied up with the existence of "superbaryons". A superbaryon is a quasi-elementary particle, with

$$10^{67} > A > 1$$

It is required (Hagedorn, 1972) that the decay of a galactic-mass state go through cascade emission of nucleons, with the large-A state appearing as a hadron. Ordinary nuclei do not fit the above description; hypercollapsed nuclear matter might be just what is required.

Note that as an energy source, if $B_A(C) \gg B_A(N)$, we might have a new mechanism in the gradual collapse of normal to hypercollapsed matter.

## 6. Observation

The search for hypercollapsed matter resembles the search for quarks, in that we are after unusual e/M ratios (here because of new M values instead of e values). Cosmic rays appear to be one possible place to search; present limits on nuclear isomers (Webber) do not get under 5%. Neutral collapsed matter would be more difficult to find. One could also search material on the earth's crust (though hypercollapsed matter might sink gravitationally).

## 7. Astrophysics

In the main, the existence of hypercollapsed nuclei would modify present estimates of the size of neutron stars etc. We would have a new region of stability, following white dwarfs and "normal" neutron stars. Neutron star material would also include dense "raisins" of hypercollapsed matter.

## REFERENCES

Achiman, Y., 1969, Lettere al Nuovo Cimento, $\underline{2}$, 301.

Aharony, A., 1970, Lettere al Nuovo Cimento, $\underline{3}$, 791.

Aharony, A., 1971 a, Ann. Phys., $\underline{62}$, 343.

Aharony, A., 1971 b, Ann. Phys., $\underline{67}$, 1.

Aharony, A., 1971 c, Ann. Phys., $\underline{68}$, 63.

Aharony, A., and Ne'eman, Y., 1970 a, Int. J. Theor. Phys., $\underline{3}$, 457.

Aharony, A., and Ne'eman, Y., 1970 b, Lettere al Nuovo Cimento, $\underline{4}$, 862.

Bahcall, J. N., and Bahcall, N. A., 1970, Publ. of the Astr. Soc. of the Pacific, $\underline{82}$, 487.

Bell, J. S., and Steinberger, J., 1965, Oxford International Conf. on Elementary Particles.

Bodmer, A. R., 1971, Phys. Rev., $\underline{D4}$, 1601.

Burbidge, G. R., 1970, Ann. Rev. Ast. and Astrophysics, $\underline{8}$, 369.

Cannata, F., Del Fabro, R., and Signore, O., 1970, preprint INFN/AE-70/6.

Casella, R. S., 1968, Physical Review Letters, $\underline{21}$, 1128.

Casella, R. S., 1969, Physical Review Letters, $\underline{22}$, 554.

Christenson, J. H., Cronin, J. W., Fitch, V. L., and Turlay, R., 1964, Physical Review Letters, $\underline{13}$, 138.

Collins, P. D. B., 1971, Phys. Reports, $\underline{1C}$, L03-234.

Crawford, F. S., 1965, Physical Review Letters, $\underline{15}$, 1045.

Dass, G. V., 1971, preprint TH-1373-CERN.

Greenstein, J. L., and Schmidt, M., 1954, Ap. J., $\underline{140}$, 1.

Gunn, J. E., 1971, Ap. J. Letters, $\underline{164}$, L113.

Hagedorn, R., 1970, Astronomy and Astrophysics, $\underline{5}$, 184.

Hagedorn, R., 1972, Ettore Majorana School of Astrophysics and Cosmology.

Lee, T. D., Oehme, R., and Yang, C. N., 1957, Physical Review, 106, 340.

Lee, T. D., and Wu, C. S., 1967, Annual Review of Nuclear Science, 17, 513.

Michael, C., "Regge Redisues", in Low Energy Hadron Interactions, Springer Tracts in Modern Physics, 55, 174-190.

Midiel, L., 1968, in "Proceedings of the VIII Nobel Symposium", (N. Suartholm ed.), p. 345, John Wiley, New York.

Ne'eman, Y., 1965, Ap. J., 141, 1303.

Ne'eman, Y., 1968 a, "the Strongest Force" in Science Year, World Book Science, Field Enterprises Educational Co., Chicago, p. 157-173.

Ne'eman, Y., 1968 b, "Nuclear Physics Implications of the Spin 2 Multiplet", in Symmetry Princimpes at High Energy, Proceedings of the Fifth Coral Gables Conference, A. Perlmutter et al., editors, W. A., Benjamin, Inc., New York, p. 149-151.

Ne'eman, Y., 1968 c, paper presented at the March, 1968 Session of the Israel Academy of Sciences.

Ne'eman, Y., 1970, International Journal of Theoretical Physics, Vol. 3, No. 1, p.1.

Novikov, I. D., 1964, Astron. Zh., 41, 1075 (trans. in 1965, Sov. Astron., 875).

Onsager, L., 1931 a, Phys. Rev., 37, 405.

Onsager, L., 1931 b, Phys. Rev., 38, 2265.

Particle Data Group, 1970, Rev. Mod. Phys., 42, 87.

Particle Data Group, 1972, Phys. Letters, 39B, 1.

Schmidt, M., 1970, Ap. J., 162, 371.

Schwartz, M., et al., 1967, Physical Review Letters, 19, 987.

Steinberger, J., 1969, in Proc. of the Topical Conf. on Weak Interactions (Geneva), CERN 69-7.

Wangness, K., and Bloch, F., 1953, Phys. Rev., 89, 728.

Webber, W., private communication to H. Kasha.

Zweig, G., 1967, Paper presented at the Conference on Decays of K Mesons, Princeton-Pennsylvania Accelerator, November 1967, Unpublished.

Structure and Dynamics of Galaxies

Lectures by

K.H. Prendergast

Columbia University

Notes by

K. Brecher, G. Cavallo

# Structure and Dynamics of Galaxies

Lecture 1

As you all know there is no such thing as a normal galaxy, and therefore I will be talking about something whose existence I will have to postulate. However, there is an atlas of such objects, the Hubble Atlas. As a reminder, I shall review the Hubble classification scheme or tuning fork diagram. Galaxies can be classified as ellipticals and spirals. The progression is shown from E → S in Fig. 1.1. In general, one tends to consider a morphological progression from E → S → Irregular. One must add one other box, labelled peculiar galaxies, which is the subject of most of this summer school. This is one of the most important boxes in any classification scheme. Nature cannot at first sight be made to fit a simple classification scheme. One needs a box where to put exceptions, and the box labelled "Arp" is it. Operational definition of a peculiar galaxy is that you cannot find it in the Hubble Atlas, but in Arp's Atlas. You can consider other criteria: HII regions, non-thermal emission, etc. There is a rather large choice, but these are not the subjects here. There are several features to note in Fig. 1.1.

The gas content increases from left to right, from no gas at all in ellipticals, to $\sim 10\%$ in irregulars. The degree of structure also increases in the same direction. By and large the spectral type also changes, tending to become "earlier" (younger, brighter stars) from E0 → Irr. This last property correlates better with the Morgan classification, where color correlates with properties of galactic nuclei (central concentration increases from Irr → E0). There is a certain amount of anonymity about ellipticals: given a picture of a random E0, unless an astronomer recognizes the star field he will not be able to identify it; in other words, ellipticals have no personality. On the other hand spirals and irregulars in the Hubble Atlas are all uniquely identified: they have personality.

I would like to say a few things about ellipticals, and the usual stellar dynamical models for them. The properties which any dynamical theory of elliptical galaxies must account for are:
(a) isophotes are similar ellipses
(b) surface brightness falls off smoothly with distance from the center of the galaxy (Fig. 1.2).

# Structure and Dynamics of Galaxies

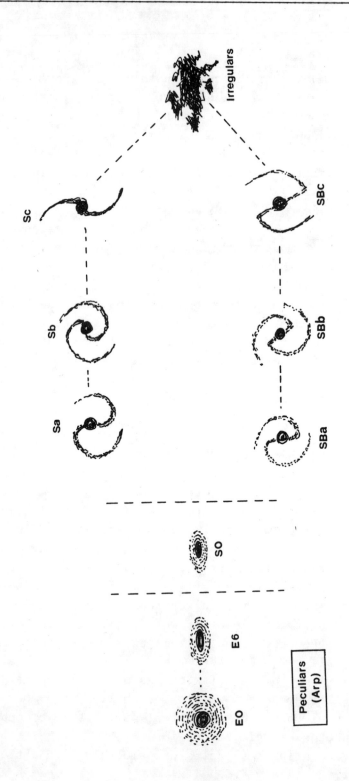

Figure 1.1 Hubble Classification Scheme

# Structure and Dynamics of Galaxies

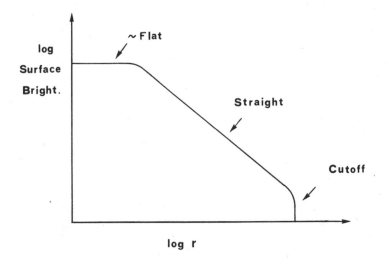

Figure 1.2

The plot in Fig. 1.2 fits either an E0 or E6. It is worth noting that an E6 edge on might appear as an E0. The existence of a class of galaxies of type E0 depends on the fact that there are too many spherical E galaxies to be accounted for on the basis of projection alone.

In general, these galaxies approximate very well the $10^{12}$ body problem. Theoretically they are very simple; observationally they are very dull. Let us, therefore, be simple and dull. This subject dates back to Eddington and Jeans. Consider an elliptical galaxy to be a system composed only of stars. The only question to be asked about them will concern the distribution function $f(\vec{r}, \vec{v}, t)$, which when multiplied by $d\vec{r}\, d\vec{v}$ tells one the number of stars per unit volume element in 6-dimensional phase space. Let us assume the following things (which we know to be untrue);

1) no stars are created or destroyed during the time interval under consideration;

2) each star moves independently in the "smoothed" gravitational field of all the others (neglect 2-body encounters, for example). We can then describe the behavior of f by the equation

$$\frac{\partial f}{\partial t} + u_i \frac{\partial f}{\partial x_i} - \frac{\partial \psi}{\partial x_j} \frac{\partial f}{\partial u_j} = \begin{cases} \left(\frac{\partial f}{\partial t}\right)_{\text{collision}} & \text{Boltzmann Eq.} \\ 0 & \text{Vlasov Eq.} \end{cases} \quad (1.1)$$

We shall use the Vlasov equation as an approximation to the truth. To check this, estimate the relaxation time. If one shoots a test star through the galaxy what will be its effect on the galaxy? Does it really move independent of individual stars, or does it test the "bumpiness" of the galaxy? A measure of this bumpiness is the time it takes for the energy integral to change. The relaxation time for n stars for unit volume of mass m and moving with average velocity $\vec{v}$ under the gravitational effect of all the other stars (G = gravitational constant) is given by

$$\tau_{relax.} = \frac{\vec{v}^3}{G^2 nm^2 \ln(\ )} \begin{cases} \gg 10^{10} \text{ years (for } 10^{12} \text{ stars)} \\ \simeq 10^{10} \quad " \quad \text{(for } 10^{6} \quad " \quad ) \\ \ll 10^{10} \quad " \quad \text{(for } 10^{3} \quad " \quad ) \end{cases} \quad (1.2)$$

Here the ln( ) term is the 2-body collision term. Thus for elliptical galaxies assumption 2) is true, though it is not for globular clusters, or even the very center of galaxies. Thus we shall use the Vlasov Equation. $\psi$ is determined from Poisson's equation.

$$\nabla^2 \psi = 4\pi G \rho \quad (1.3)$$

where $\rho$ is given by

$$\rho = m \int f(\vec{r}, \vec{u}, t) \, d\vec{u}. \quad (1.4)$$

We now have a closed problem, and one wants to solve (1.1) plus (1.3). We shall now assume that the galaxies are in a steady state, so that $\partial f/\partial t = 0$. This brings in another time scale, the dynamical time scale, defined as

$$\tau_{dynamical} \simeq (G\rho)^{-1/2} \quad (1.5)$$

where $\rho$ = nm. This is the time scale for gravitationally controlled pulsations of the system, etc., and is the time scale on which one expects f to vary. For an elliptical galaxy, $\tau_{dyn} \simeq$ $\simeq 10^7$ years $\ll H_o^{-1}$. Thus, on a time scale of $H_o^{-1} \simeq 10^{10}$ years, it makes sense to talk about a steady state problem. We then have a third assumption, that we are in a steady state.

Finally, we shall assume in the fourth place that elliptical galaxies have an axis of symmetry. This is less an assumption than an observational fact. Let us now try to solve this problem: to make a complete picture of a hypothetical elliptical galaxy.

One can solve the Vlasov equation in some simple case. Equation (1.1) says that f is conserved on a trajectory of a particle in phase space. For any one star, one can write

$$\varepsilon = \tfrac{1}{2}|\vec{u}|^2 + \psi(\vec{r}) = \text{const} \tag{1.6}$$

$$J_z = \vec{k} \cdot (\vec{r} \times \vec{u}) = \text{const}, \tag{1.7}$$

where $\vec{k}$ is along the direction of the axis of symmetry. Any function of these two integrals of the motion $f(\varepsilon, J_z)$ solves the Vlasov equation. Let us pick one: in particular, one that is easy to work with, and perhaps related to the real world. Pick a straight Maxwellian

$$f(\varepsilon, J_z) = \begin{cases} a \exp(-\alpha\varepsilon - \beta J_z) & \text{if } \varepsilon \leq \varepsilon_0 \leq 0 \\ 0 & \text{otherwise} \end{cases} \tag{1.8}$$

This is a truncated exponential, and is expected in thermodynamic equilibrium. If $\varepsilon_0 > 0$, stars have positive energy and escape; if $\varepsilon_0 = 0$, the system has infinite spatial extent. Then we choose $\varepsilon_0 < 0$, insuring a finite radius system. At this point, insert $f(\varepsilon, J_z)$ into (1.3), and compute $\rho = \left[\rho \; \beta r \sin\Theta, \psi(r, \Theta)\right]$. Absorbing constants into $f(f \to f')$ one wishes to solve

$$\nabla^2 \psi = f'(\beta r \sin\Theta, \psi) \tag{1.9}$$

with boundary conditions

$\nabla\psi = 0$ at origin (no mass singularity)

$\psi = 0$ at $r \to \infty$.

One wants to solve them for various values of $\beta$ and $\varepsilon_0$. Note that $\alpha$, $a$, and $G$ can be scaled out of the problem to make it dimensionless. This problem is similar to that of a polytropic model for a star, and is a self consistent, non-linear problem. It is like the Thomas-Fermi problem (or space charge problem). To solve it, one must devise a numerical algorithm. One writes

$$\nabla^2 \psi^{(n+1)} = f'(r\sin\theta, \psi^{(n)})$$

One starts by guessing a $\psi^{(n)}$, and when $\varepsilon \to \varepsilon_0$, $\rho = 0$; then one computes $\psi^{(n+1)}$, etc. This process diverges rapidly (say after 4 iterations), so one needs one more condition, that the total mass M is given. Thus there is a hidden eigen-value of the problem. One normalizes to a fixed mass, and iterates away. Does the distribution function at all resemble an elliptical galaxy? Are the isophotes similar ellipses? Does the intensity fall off with radius as in Fig. 1.2? The answer is yes. The projected mass density resembles the observations. This is a two parameter family of solutions: one parameter $\varepsilon_0$ is the cutoff, $\beta$ is the flattening. But note that any function of $\varepsilon$ and $J_z$ will work. Now the simple $f(\varepsilon, J_z)$ is reasonably good. How does an elliptical know which $f(\varepsilon, J_z)$ to have?

# Structure and Dynamics of Galaxies

## Lecture 2

The classical viewpoint makes some sense, but we must discuss the analytical form of the distribution function. We have already established that any functions of the total energy $\varepsilon$, and of the axial component of the angular momentum $J_z$ will do:

$$f(r, u) \stackrel{any}{=} f(\varepsilon, J_z)$$

However, one problem immediately arises: the motion of a particle in 3 dimensions requires 6 constants of motion, and we only know two. How does an elliptical galaxy know what to do? (It turns out that elliptical galaxies choose a limited range of parameters.)

There are many possibilities:

(i) The distribution function is <u>given at the outset</u> (but this possibility is not so pleasant).

(ii) The distribution function depends on the pre-stellar state, for example, a large diffuse cloud which then collapses. (This problem has not been tackled, and there is no answer so far.)

(iii) We must exclude collisions, because they make the system forget the initial conditions, like, for example, the gas in a room, which assumes a Maxwellian distribution no matter what were the initial conditions. In this case the medium will not forget its initial conditions: the $10^{12}$-body problem goes backward too. But can we make them non-important?

"VIOLENT RELAXATION" is the answer which has been proposed. At the initial instant the distribution function is $f_0(r, u, t=0)$, and the potential is $\psi_0(r, t=0)$. Let's assume no particular symmetry: both $f$ and $\psi$ change on the same timescale $\tau_{dyn} \simeq (G\rho)^{-1/2}$
Neither the total energy nor the z-component of the angular momentum for a single particle are conserved: only the total $\varepsilon$ and $J_z$.

One possible way is to demand that the 6N initial conditions are forgotten and only 4 quantities remain important, namely $\varepsilon$ and $\vec{J}$ (vector angular momentum). However, this problem is very complex and we do not see clearly at all our way through it. We shall therefore consider a much simpler situation.

Let's take the phase space of one-dimensional harmonic oscillator, and start particle motions at $t = 0$. We can predict the

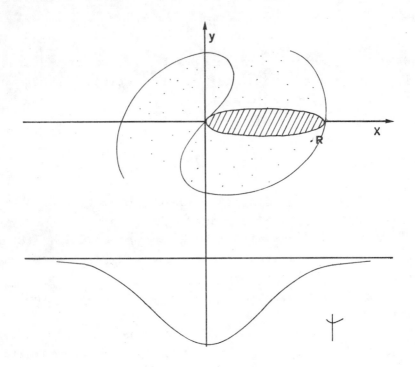

Figure 2.1

kind of configurations we shall find at the end of the process: let's put in particles in a region R at t = 0 with v = 0 (Fig. 2.1). At the end we shall have a configuration similar to a ring, or rather, a tightly wound, thin arm spiral. In other words, the system has mixed itself up without interacting at a time $t \gg t_{dyn}$.

We can consider $f(t)$ and $\psi(t)$ on the basis of some mechanism through which the system forgets the initial conditions. We are then dealing with a statistical mechanics system, of the kind extensively stated by Lynden-Bell. A sort of "exclusion principle" works for those systems, because the density of states is strictly conserved.

We cannot, however, test our notion of violent relaxation on a computer without extreme care. We shall now illustrate some of the horrors which present themselves while doing this problem on a machine. We want to discuss $f(r, u, t)$ which obeys the already quoted collisionless Boltzmann equation

$$\frac{\partial f}{\partial t} + u_i \frac{\partial f}{\partial x_i} - \frac{\partial \psi}{\partial x_j} \frac{\partial f}{\partial u_j} = 0$$

# Structure and Dynamics of Galaxies

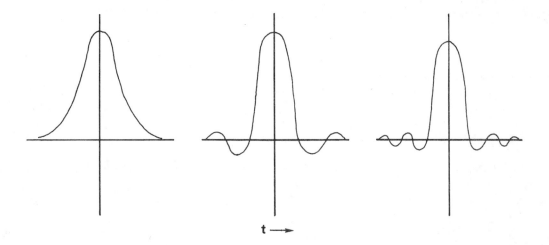

Figure 2.2

It is extremely important to notice that the right-hand side must be exactly 0, namely we do not want any truncation or round off error. As soon as we get a non-zero value, the system violently relaxes.

The distribution function will then evolve with time in the way illustrated in Fig. 2.2, with the catastrophic effect of becoming negative, a rather unpleasant feature for a distribution function.

Let's now go back to our problem: we want systems in which the initial conditions are unimportant. Such systems exist, and we shall quote two examples.

I Example:

Suppose we have slabs of uniform density $\rho$, and infinitesimal thickness, which extend all over the yz plane (Fig. 2.3a). The advantage of such a configuration is that the force is independent of the distance. The acceleration is therefore constant and the x(t) functions for the particles in such a system are parabolas. Let's call $t_0$ the time for the first collision to take place. At this time we start again (Fig. 2.3b), relabelling names of states. By repeating extensively the procedure, one can find that after a time T equal to a few tenths of the relaxation time the system appears like in Fig. 2.3c all spread over a ring.

This system is not time reversible by construction. Violent relaxation works in perhaps a few tenths of dynamical relaxation times to go to the configuration of Fig. 2.3c.

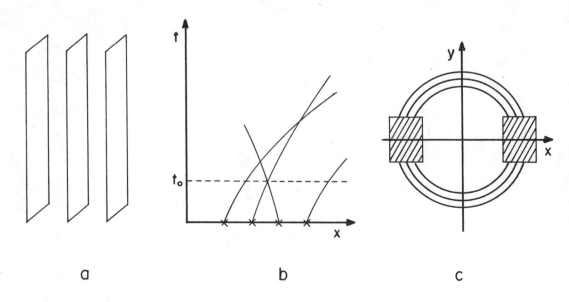

Figure 2.3

II Example

Another possibility is to take $10^5$ particles on a plane and to <u>quantize the phase space</u>. This fact of quantizing is not so exotic, since we always do it whenever we submit our problem to a computer. We can then construct a grid (Fig. 2.4), and play a game according to the following two rules:

(1) Starting from any point we like, (for example, $P_1$) shift to the right by the starting value of the y coordinate. We land in $P_1'$.

(2) Now shift up by the value of the x coordinate of $P_1'$, and find the new position $P_2$. In the case of the example, $P_1' \equiv P_2$. By performing the same two operations for a number of times, we will subsequently go into $P_3$, $P_4$, $P_5$, $P_6$ and we will finally be back to $P_1$.

This system is nothing but the usual harmonic oscillator in a disguise. The following few points should be mentioned:

(1) The orbit in the phase space is not quite an ellipse, and this is explained by our quantization procedure.

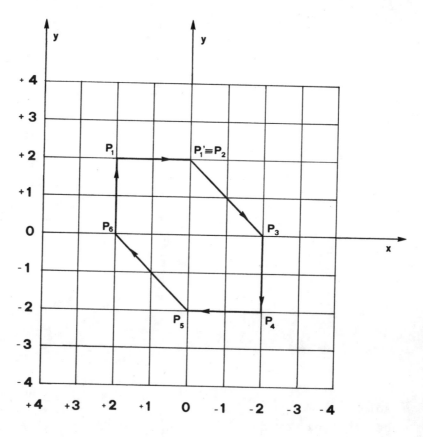

Figure 2.4

(2) Orbits which start at points different from $P_1$, $P_2$, $P_3$, $P_4$, $P_5$ have no points in common with these. (Particles don't interact.)

(3) The area in the phase space is conserved. (There is no diffusion.)

(4) The system is strictly time reversible. We have therefore a guaranteed solution of the Vlasov equation, which we can play with any potential provided that we also change the force. We can compute a smoothed force from a smoothed density.

What has been done is the following: $10^5$ particles have been considered in the space (x u v t). We wanted to see whether there was any violent relaxation to our equilibrium state, and what this equilibrium state precisely is. In this particular case a grid with 128 x 128 squares was considered, and the two dimensional Poisson equation with periodic boundary conditions

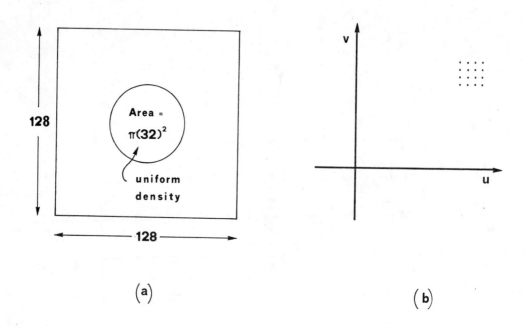

Figure 2.5

was solved.

The starting point was a distribution concentrated as shown in Fig. 2.5. We then turn on the gravitational interaction and give the system a mean velocity in order to balance the local gravitational field and to keep the system roughly in equilibrium.

We pack the representative points in the u-v plane as close as possible (Fig. 2.5b). Let the system be differentially rotating. The result of this computation is given by a series of plots of the density $\rho(x)$ at different times (Fig. 2.6)

The following features are worth noticing:

(1) Already in the second configuration we can see some fuzziness at the contour, because of the velocity dispersion we built in. The slower particles will fall in, the faster will go out. These particles which are left with less kinetic energy fall in within one dynamical time, $\tau_{dyn} \sim (G\rho)^{-1/2}$.

(2) After $2\tau_{dyn}$ a hole is formed in the middle.

(3) The system has an axial symmetry.

After 10 $\tau_{dyn}$ we reach the configuration (8): from then on there

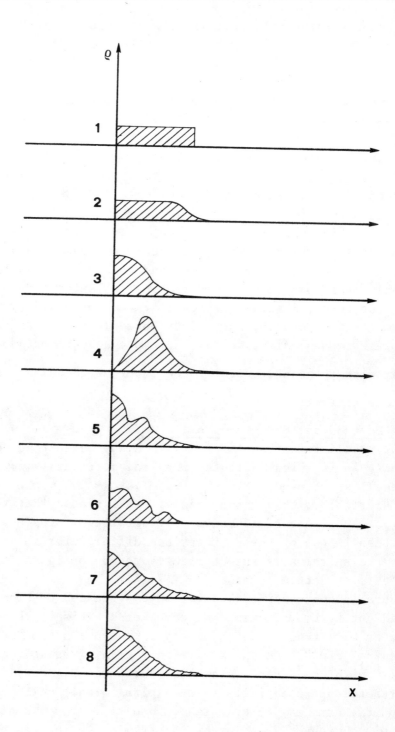

Figure 2.6

will be no substantial change in the function $\rho(r)$. A state of statistical equilibrium has been reached.

What do these results mean? Are they a real result or a product of round off? No. This is a time reversible calculation. We started with a rather simple mode of oscillation and we obtained a reasonable "squashed galaxy", in two dimensions. (We might do the same job in three dimensions, and it would just cost more money.) The obtained rotation curve (v vs r) is reasonable: in a sense we obtained a galaxy "without personality", i.e., the same object one would obtain from a galaxy if one were to leave only the population I stars (a slow, smooth distribution of old stars with high velocity dispersion). It would not be a good picture for our galaxy, which is not old enough. However, we see that it is the above process, and not two body collisions or initial conditions that make elliptical galaxies know what to do.

We can thus summarize our results: we get a slowly rotating, **axisymmetric** smooth, hot, distribution. Hot means that the tensor pressure gradient $<\rho \, \delta u_i \, \delta u_j>$ plays an important role together with the centrifrugal force in order to balance the gravitational attraction.

## Where Does Spiral Structure Come From?

The result is not satisfactory in a way because we do not see anything in it which indicates a spiral structure: we want population II stars, and not population I.

Historically, this was a puzzle: it is known that the rotation of a galaxy is not rigid body. If we allow a single piece to move, then, in two or three rotations, differential rotation will sheer the spiral arms into a tightly wound configuration. How does this structure persist?

There is one considerable way out, suggested by Linblad and worked out by Lin, Toomre, and coworkers. This is the so called "density wave theory" which attempts to explain the "eternity of the spiral arms". (A spiral arm in a galaxy is defined by the locally high gas, dust, star density; however, not all stars are distributed preferentially in the spiral arms.) Alternatively, the spiral arms may be transient, resulting, for example, from collisions with other galaxies, or from explosions in the nucleus. The problem can be complicated to the point of becoming almost impenetrable. Therefore we must rather simplify it: we can choose whether to deal with a disk of stars or (better) with a gaseous disk. We shall start from an axisymmetric state and shall need self-gravitation and pressure. The question then is whether it will go to pieces or not, and the density wave method is one way of dealing with it.

However, the main point of the problem, the cause of the density waves, remains unsolved. This is not a trivial point: the gas in a room does not clump, it does not self-gravitate, density waves are not produced. In a way, the gas is too hot, and the gas particles will quickly escape the density concentrations, and avoid their formation on a large scale. The same trouble happens in a rotating system. Here, the pressure suppresses the "small" concentrations, and the Coriolis force suppresses the "large" concentrations. Is there a scale of concentrations, in the middle, which is stable?

## Lecture 3

We have considered a <u>hot system</u>, i.e. one in which pressure or random velocities play a role in supporting the gravitational attraction in the sense that their contribution is of the same order as the rotational velocity. The conclusion has been that such a system is unlikely to produce spiral arms because reasonably long lasting condensations do not form.

We can start instead with a <u>cool population</u> (gas, globular clusters) with good hopes of forming spiral arms. First of all we shall consider the analytical approach, by giving a modest paraphrase of the difficult work done by Lin, Toomre and others. Can one consider spiral arms as self propelling self gravitating density waves in a cool population belonging to a galaxy? We may take as a starting point a thin, steady state, differentially rotating, cool, axisymmetric disk of gas (stars). The first question is: "Is it stable or unstable?" The point is that for such a system the normal modes contain a large family of spiral modes. The system will be characterized by polar coordinates, $\tilde{\omega}$ and $\phi$ (Fig. 3.1) and by a number of physical quantities, functions of $\tilde{\omega}$ and $\psi$, namely the pressure p, the density $\rho$, the rotational velocity v, and the gravitational potential energy $\psi$, whose unperturbed values at a given point $P(\tilde{\omega}, \phi)$ will be indicated by $p_0(\tilde{\omega})$, $\rho_0(\tilde{\omega})$, $v_0(\tilde{\omega})$, $\psi_0(\tilde{\omega})$. The corresponding (small) perturbations will be $\delta p$, $\delta \rho$, $\delta v$ (radial velocity), $\delta \psi$.

The usual technique is now to take the fundamental set of equations of the system, (i.e. the equations of hydrodynamics, continuity, Poisson, and assume the gas is a polytropic $\delta p = c^2 \delta\rho$, where $c^2$ is the speed of sound). Even in the linear approximation, the equations are so well entangled that there is no hope of getting an analytical solution. However, there is no need to despair, since we can reasonably guess the analytical expression of a normal mode. The time dependence will be expressed as $e^{ist}$, where s (to be later interpreted) can be complex ($s = \sigma + i\tau$); the polar dependence, for cylindrical symmetry, will be $e^{im\phi}$. So we are left with the problem of determining the radial expression. We will assume the very general form $\exp\left[if(\tilde{\omega})/\epsilon\right]$ where $\epsilon = 1/(\text{Mach number}) \simeq v_{rot}/c$, and its numerical value lies between 0.1 and 0.05.

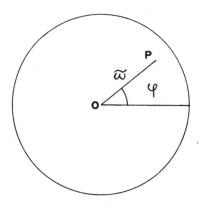

Figure 3.1

Symbolically we can then write the perturbations of all dynamical variables as:

$$\begin{pmatrix} \delta p \\ \delta \rho \\ \delta u \\ \delta v \\ \delta \psi \end{pmatrix} \sim e^{ist} \, e^{im\phi} \, e^{if(\tilde{\omega})/\varepsilon}$$

This problem can be solved via WKB method, so yielding the dispersion relation

$$\frac{df}{d\tilde{\omega}} = A^2 + (A^2 + B^2)^{1/2},$$

with appropriate expressions for A and B. What really matters is that the derivative is positive. We can now outline our attack on the linearized problem.

(1) <u>The structure of the pattern</u>: The shape of the arm (if we find one) is given by the position of the maxima of the function, which lie at

$$\psi = \frac{f(\tilde{\omega})}{\varepsilon m} \; (+ \, c)$$

If f is a monotonic function of $\phi$, we have a spiral arm. Now, the

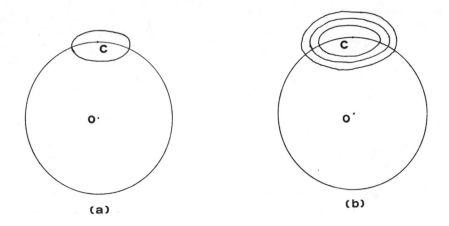

Figure 3.2

above expression of $df/d\tilde{\omega}$ is O.K., if f is increasing and, therefore, monotonic. The spiral, however, could be either leading or trailing.

(2) <u>The rotational speed of the pattern</u>: This is easily found to be

$$\Omega_{pattern} = \text{Re}(-\frac{s}{m}),$$

where m is an eigen-value, the number of arms. We can then conclude that the natural point of view of linear analysis automatically leads to spiral patterns, <u>if we find a normal mode</u>.

Only a local analysis has been performed: the theory does not tell anything about the eigen-value $s = \sigma + i\tau$. That has not been done, and this not because of laziness, but because of the true mathematical horrors of the problem, namely the presence of no less than 5 singularities. The singularity arises when the pattern rotates as fast as the gas (case of corotation), at $\omega_c$: this happens because a star at the corotation orbit gets accelerated in phase. Moreover, there are two so called "Lindblad resonances" (at $\tilde{\omega}_L$). Consider one star which moves on a nearly circular orbit: it will move on a <u>rosette</u> orbit. From the point

Structure and Dynamics of Galaxies       435

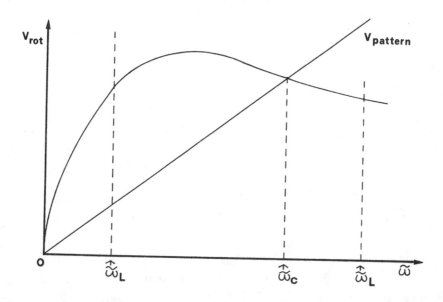

Figure 3.3

of view of a corotating observer G (moving at the average rate
of rotation about the center) there will be one star moving on a
circular orbit. All the other stars will describe elliptical
orbits with respect to him (Fig. 3.2). The period of this epicyclical motion will be $1/k$, where $k$ is the angular frequency.
When $k = \Omega$, the problem is again singular. The physical reason
is that a density wave, at a Lindblad singularity, hits the star,
at the same phase in the epicyclical motion: the star receives
more energy, and the initial ellipse broadens (Fig. 3.2b). We
can now ask whether the system is stable or unstable: the motion
will break up in normal modes; if a normal mode is unstable, its
growth rate is faster, and we only see that one.

The location of the singularities on the $\tilde{\omega}$ axis is sketched on
Fig. 3.3. The problem has been analyzed by Toomre: he has found
out that waves are dispersive for non-spiral normal modes. We
cannot go into that analysis in detail, because it is very involved. It can be taken as an example to prove that linear problems are not necessarily trivial. We shall then go back to the
computer. First, let us outline the classification system devised by Morgan for galaxies. It depends on the degree of concentration of light correlated with the radius. It so happens

Table 3.1

| Galaxy Type | Ellipticals | Spirals | Irregulars |
|---|---|---|---|
| Stellar Type | K (red) | G | A-F (blue) |
| Gas Distribution | Smooth | | Irregular |

that also the most common stellar types and the gas distributions are corollated with the thus constructed sequence of galaxies(Table 3.1). The fact that in elliptical galaxies stars are redder is a hint of what really happens.

We already know that if we have a system of $10^5$ points(purely a stellar system) self gravitating in a smooth field, we obtain a hot galaxy, and no trace of spiral arms. For spiral galaxies we shall need a cool subpopulation, otherwise, there will be an evolution to a statistical steady state, with no spiral structure. At any point of the system in the velocity space there will be a large spread in u (radial velocity) and v (tangential velocity). The first suggestion is then to cool the system, suddenly. But that's foolish, we know what will happen, namely a collapse, which will result in a smaller, hotter, steady equilibrium system, and therefore in a total failure. We can then think of cooling the system, but gradually, by slowly reducing the random velocities. The result is that we shall never get a galaxy, namely a coherent system, but rather a collection of lumps. It is not hard to see that the system is unstable: but what is it made of? Certainly not a stellar system. It is a model of a very dissipated system, like a gas with many internal degrees of freedom, to whom the energy is lost without being recuperated. As a third attempt we can finally do what we know that happens, namely form stars at a rate determined by $\rho$, the gas density

$$\frac{dN^*}{dt} = K\rho_{gas}^{\alpha}$$

Once we form stars, we get the most surprising result, i.e., we can get spiral galaxies. In fact, if we start with the wrong initial conditions, we can get also other things, like a small group of galaxies. It all depends on whether the objects produced in the initial collapse amalgamate or not. In order to see

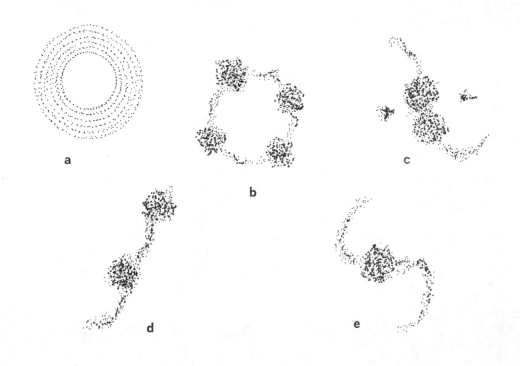

Figure 3.4

what kind of things happen, a computer movie has been produced. Since we do not have it, we will just give a rough idea (Fig. 3.4). The starting point is a UNIFORM DISK OF PURE GAS IN A STATE OF DIFFERENTIAL ROTATION. What then happens depends on the <u>rotation frequency k</u> and on the equation of state of the gas.

First of all it usually collapses into a differentially rotating ring (Fig. 3.4a) which is unstable and breaks up into a few lumps within 1 $\tau_{dyn}$ (dynamical time) (Fig. 3.4b). A cannibalistic process then takes place and the lumps reduce in number, while sort of spiral arms appear, (c). One lump generally sits at the center and another stays at the end of one arm, where it lasts for several $\tau_{dyn}$ (d). Finally, a spiral configuration appears with 2 trailing arms, most of the time.

Now a few comments:

First of all, if we look at the stars they do not show this picture: only the coolest among the stars show the same struc-

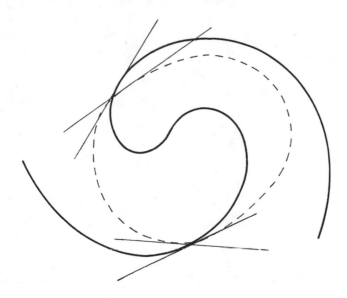

Figure 3.5

ture, although smoother than the gas. Secondly, if we follow the trajectory of a lump, we see that trajectories are refracted when crossing the spiral arms as we expect from shock waves (Fig. 3.5). Thirdly, we may ask whether we can also produce spiral galaxies with three arms (and the answer is yes). Finally, from Fourier analysis of the peak density we can find out that a density wave rotates like a rigid body and tends to wind up, but slower than the gas. The last point we want to mention is that we have to dissect the computations in order to know what among our initial assumptions is necessary and what isn't in order to get spiral arms.

We shall now list the modifications we introduce and what are the computer results:
(1) <u>Turn-off self-gravitation</u>: the pattern disappears fast.
(2) <u>Start with a bar or something like it and then stir</u>: we do not get a normal spiral; but a barred spiral.
(3) <u>Introduce self-gravitation and bar</u>: it works.
(4) <u>Leave only self-gravitation</u>: it works fine.
  The <u>conclusions</u> are open to discussion:
(I) There exist normal spirals.

(II)   We can make some sense of them.
(III)  It is good to use simulation.
(IV)   There is no wisdom in a computer.

Strong Interactions, Gravitation and Cosmology

Lectures by

A. Salam

Imperial College

Notes by

J. K. Lawrence

Strong Interactions, Gravitation and Cosmology          443

Lecture 1

ELEMENTARY PARTICLES AND GRAVITATION

I will speak as a particle physicist. There are few particle physicists who pay any attention to what gravitational people or cosmologists are doing. One of such groups is the one in Trieste and London which has been considering the possible interconnections. This talk will be a progress report on this work.

Let me start by assuming
1) That gravitational waves exist.
2) In the 21st century we will know that gravity is a quantum effect as are all other fields.

How can we ever discover the quantum nature of gravity, short of constructing very strong gravitational sources? The main reason particle physicists have never taken any notice of gravitation is the smallness of the coupling constant

$$G_n = \text{Newtonian constant} = 10^{-44} \, m_e^{-2}$$

The weakest coupling constant dealt with is the Fermi coupling wiich is $10^{-12} \, m_e^{-2}$. So graviton cross-sections will be smaller by a factor of $10^{32}$ or so from the presently very difficult weak interaction experiments.

The biggest effects of quantum gravity you can get are still in the three traditional tests of general relativity. To see the order of magnitude of quantum corrections, let's consider the bending of light in the sun's gravity field.

Classical Picture

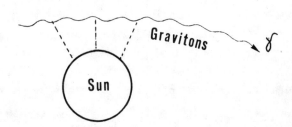

The biggest quantum correction arises when the graviton is involved in pair production in diagrams like

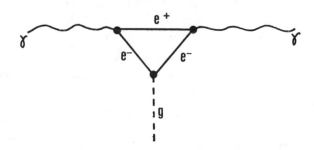

The correction to the Einstein value from this is
$\sim R_o^{-1} m_e^{-1} \sim 10^{-26}$.

In the Pound-Rebka redshift experiment, the effect is a little better, and the correction to the Einstein redshift is $\sim 10^{-11}$. So in any foreseeable time these experiments will be impossible.

Why are we interested in quantum gravity at all? There are some main reasons.

1) The old conjecture that gravity is the "civilizer" of all other forces. In calculating self masses and charges of particles you get infinities that have to be cut off externally. It has been conjectured that gravity would provide a built in cut-off, and so we have taken it very seriously. The amazing thing that happens is that $G_N$ does not appear in the calculations, but $|\log G_N| \sim 50$, a large number. This is the first new result I shall talk about.

2) The second result is that experimental particle physicists have discovered a <u>nonet</u> of spin 2⁺ particles. The two neutral ones have the same quantum numbers as the graviton. Thus the gravitons and f mesons must interconvert. A graviton must spend a small fraction of its time as an f meson (mass $\sim$ 1200 Mev). Therefore, all phenomena, including the final state of the universe, where singularities were expected to occur at radii of $10^{-33}$ cm must be revised. This radius is now $10^{-13}$ cm. This statement cannot be ignored. These regions are not of interest to particle physicists, but I assume they are to

at least some cosmologists. The "first stage" of collapse may occur when the universe becomes the size of a nucleon.

I am assuming that we may use the Einstein equations to describe the f-mesons. All phenomena deduced from them, like collapse, black holes, etc., in regions of strong curvature can be taken over into hadron physics. Now the geometrical language used by cosmologists to describe stars or the universe can be telescoped into regions the size of $10^{-13}$ cm. If this conjecture (this is all conjecture, though plausible) is correct, then laboratory energies will be sufficient to test some properties of solutions of the Einstein equations at the size of $10^{-13}$ cm.

3) C-P violations and the Klein-Kaluza theory (Thirring).

Of main interest to particle physicists is that gravity kills infinities in other interactions. Of main interest to cosmologists would be whether gravity kills its <u>own</u> infinities of the type discussed by Hawking and Penrose.

The problem of infinities has been with us since the beginning of particle physics. Lorentz ($\sim$ 1900) calculated the inertia of the electrons' electric field:

$$\delta m = \lim_{R \to 0} \frac{e^2}{R} = \text{linear } \infty.$$

This was later repeated by Weisskopf in 1935. He found that, when you take electron-positron pair production into account, and the spin ½ of the electron, you get only a logarithmic infinity

$$\delta m/m = \log 0.$$

This was considered a very great triumph. You still, however, cannot get a number (except infinity) for the self mass of an electron due to its electromagnetic forces.

### Outline of Weisskopf Calculation:

Electron emits photon and re-absorbs it

Photon propagator = $1/x^2$; electron propagator = $(\gamma \partial + m)(1/x^2)$.

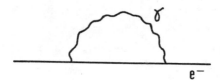

We compute the Fourier transform of the above diagram at the momentum $p^2 = m_e^2$:

$$\delta m/m = \alpha \int (1/x^2)^2 \, e^{ipx} \, d^4x = \alpha \ln 0$$

The infinity occurs at the light cone $x^2 = 0$.

What happens when you include gravity? We will mention three viewpoints.

1) Thirring: Quantum fluctuations in space-time smear out the light cone

$$1/x^2 \to 1/(x^2 + G_n),$$

thus killing the infinity.

2) Weisskopf: Below a certain wavelength, photons are captured within the Schwarzschild radius of the electron and can't get out to react back. This gives a cut-off in frequencies ∝ radius of the Schwarzschild sphere of the electron. This viewpoint translates the Schwarzschild radius into wavelengths.

3) Salam and Strathdee: This is the mathematically rigorous way of looking at it. Note that gravity theory has a non-polynomial Hamiltonian. The electrodynamical Hamiltonian has the polynomial form $\bar{\psi}\gamma_\mu \psi A^\mu$. This simple form is dictated by gauge invariance. The Einstein equations, however, are intrinsically non-polynomial. If we write $g_{\mu\nu} = \eta_{\mu\nu} + h_{\mu\nu}$ and take $h_{\mu\nu}$ as the fundamental physical field, than the contravariant metric $g^{\mu\nu}$ will be an infinite power series in h:

# Strong Interactions, Gravitation and Cosmology

$$g^{\mu\nu} = \eta^{\mu\nu} + (h)^{\mu\nu} + (h^2)^{\mu\nu} + \ldots$$

This will also be true of the affinities and the determinant $g \equiv \det g^{\mu\nu}$. Thus we see that Einstein's Lagrangian is intrinsically an infinite series in h. This is also true of the Einstein-Maxwell interaction Lagrangian

$$g^{\mu\nu} \frac{\bar{\psi}\gamma\psi A}{(-g)^{1/2}} = \bar{\psi}\gamma\psi A(1 + h + h^2 + \ldots)$$

(The rigorous argument for non-polynomiality is more subtle and does not depend on expanding $g = \det g^{\mu\nu}$ in terms of the physical field h.) This corresponds to an infinite sum of Feynman diagrams:

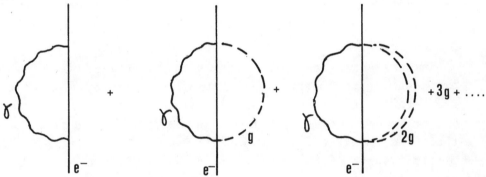

Looking at this as a perturbation expansion, you would think each term is smaller than the previous one by $G_n = 10^{-44}$ so only the 1st would be important:

$$e^2(1 + G + G^2 + \ldots) = e^2(1 + 10^{-44} + 10^{-88} + \ldots)$$

This is what has been done in particle physics. But now methods are available so we can sum this series exactly. The result is $e^2 \log G_n$. So, taking $G_n = 0$ has been the source of the infinity. Note that $\log G_n$ has an essential singularity at $G_n = 0$ and so cannot be expanded about that value.

This result is very desirable because it gives, for the electron self-mass

$$\delta m/m = \frac{\alpha}{5} \left| \log G_n m_e^2 \right| \simeq 1/6$$

This is a very reasonable result to lowest order in $\alpha$. This is perfectly believable in terms of the old dream of Lorentz that all mass should be due to interactions and no intrinsic bare mass. This is what is happening. The full series might be

$$\frac{\delta m}{m} = \alpha \log G_n m_e^2 + (\alpha \log G_n m_e^2)^2 + \ldots = 1$$

If we reverse the argument, and simply decree that $\delta m/m = 1$, then we get a relation between $\alpha$ and $\log G_n m_e^2$. Thus given that all mass is electromagnetic and given $\alpha$, I could compute $G_n$ and vice-versa. This method will remove all infinities in QED.

In our summing over an infinity of quantized gravitons we are assuming the metric is Minkowski at infinity. We couldn't even start without this. In some complicated Wheeler superspaces we don't even know how to quantize. Is there any sign that by summing an infinity of gravitons we can recover a familiar result? Consider the field produced by a very heavy static mass and a test particle. We want to get the Schwarzschild solution by summing an infinity of diagrams.

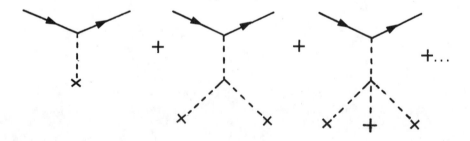

These are the $1/r$, $1/r^3$, $1/r^5$, etc., contributions produced by the source acting once, twice, etc. Duff, in London, has summed these to recover the exact Schwarzschild solution. This reassures us that what we are doing is not so stupid.

This leads us to the question of whether gravity will quench

# Strong Interactions, Gravitation and Cosmology 449

its own sigularities. It is our conjecture that this will happen, but we have no precise results towards this yet.

The second part of my talk concerns the f-mesons. This concerns the discovery in the last 5 or 6 years of the two $2^+$ neutral mesons f° at 1200 Mev and f°' at 1400 Mev. These are degenerate with the graviton and should interconvert with it. Probability of finding a graviton in the f state is $(G_n)^{1/2}$.

This interconversion business has been discussed even before we came to the gravitons. This arises in the vector dominance model of electrodynamics. The ρ-meson has the same quantum numbers as the photon. (ρ°, φ°, ω° have masses in the 700 mev - 1200 Mev region.) We are familiar with the photon no longer being the pure carrier of electrodynamics. In this picture, the photon interacts only with leptons directly, not protons. The photon turns into a ρ-meson which in turn interacts with the proton directly (but not leptons).

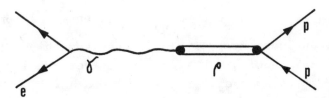

What we have proposed is to do exactly the same for gravitons and f-mesons. We let the gravitons interact directly only with leptons and with hadrons only through an f-meson.

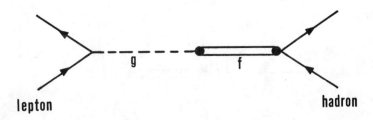

This will make no change in the equivalence principle or the Eötvös experiment, just as the various gauge invariances of electrodynamics require the electron and proton charges to be equal.

This does, however, in a certain approximation give a kind of metrical meaning to the f-meson. We can now write the Lagrangian, which was written in terms of one metric, g, in terms of two metrics, g and f. It is an extension of the Einstein Lagrangian:

$$\frac{1}{(-f)^{1/2}}\left[R(f) + T_{\mu\nu}(\text{hadrons, }f)\right] + \frac{1}{(-g)^{1/2}}\left[R(g) + T_{\mu\nu}(\text{leptons, }g)\right] + m_f^2(f-g)^2$$

The last term is the mixing term that turns a g into an f. In a certain approximation, we could regard g as flat and regard the entire f physics in geometrical terms. The only difference is that instead of the Newtonian constant $G_n \sim 10^{-44} m_e^{-2}$ we have the f constant $G_f \sim 1 m_{nucl}^{-2}$. Then in regions of strong curvature, nuclear physics is no longer something strange, now it's geometry. Nuclear physics is thus very analogous to gravity. There are some differences, of course.

For one thing the lepton-hadron attraction is changed to

$$V \simeq \frac{G_n}{G_n + G_f}\left[\frac{1}{r} - \frac{\exp(-m_f r)}{r}\right]$$

which will show up only for distances $r \sim m_f^{-1}$. Another effect is the so-called $M^{2/3}$ effect. If we take, say, a large neutron star, made up of hadronic matter, how will the graviton interact with it? It will interact via an f-meson if it has sufficient energy ($\nu \sim m_f$):

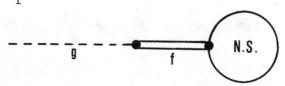

Because the f-meson has a very short range (Yukawa force), it never penetrates beyond $\sim 10^{-13}$ cm. The inside simply doesn't feel the effect, so you have a surface rather than a volume effect. Thus there is a weakening of the force to $M^{2/3}$. Something exactly similar should happen for photons interacting with large nuclei. There should be a $Z^{2/3}$ effect. Experiments at 18 Bev at SLAC seem to indicate a definite weakening in the electromagnetic potential (the experimental exponent is not 2/3 but 0.9). Anyway, we expect a similar weakening for gravity.

A final remark of relevance to the collapse people. When one is talking of the collapse of the universe to a radius $\sim 10^{-13}$ cm, one says that two stress-tensors always attract to second order. However, due to spin and velocity dependent effects, etc., we do not know what happens in <u>fourth order</u>. Normally, this is irrelevant because it is $10^{44}$ times smaller. But this is not irrelevant in the case of f gravity ($G_f \sim 1$) at this distance. We are not developing a repulsive core; we are changing gravity itself. The usual argument for unavoidable collapse:

pressure → increases $T_{\mu\nu}$ → increases gravitational attraction

no longer applies. The behavior of hadronic matter at distances of $10^{-13}$ cm. will be quite different from that of leptonic matter at $10^{-33}$ cm. The f meson dominant quantum region will be reached long before $10^{-33}$ cm. Very important for gravity as a whole.

Although the Einstein equation is the only pretty one for describing a spin 2 object, particle physicists won't look at it. We must convince them of advantages to themselves of looking at a geometrical picture of the f-meson physics (i.e. ordinary nuclear physics). Nuclear physics is misnamed and should be called "strong gravity".

A phenomenon peculiar to tensor forces is conservation of helicity. If spin is aligned in the direction of motion, it tends to remain so under a tensor interaction. This also seems to have been observed in elastic scattering of a certain number of elementary particles. This gives us the feeling we are on the right track and that there is a strong gravity dominance of nuclear forces.

One of the main problems in particle physics is that of quarks. They provide us with all possible symmetries. Is there some way we could have quarks inside an elementary particle, but not as free particles? None has ever been seen. Perhaps this could be done by borrowing some ideas from the black hole physics, i.e. by considering f-gravity black holes, combined with para-statistics. This might provide the necessary finding.

We have also taken seriously the Schwarzschild potential and considered scattering in a particle theory manner. We are trying to work out scattering cross-sections to see if Schwarzschild-like potentials are operative in nuclear physics.

Klein-Kaluza Theory and CP violation:

This is due to Thirring, and is the first use of the Klein-Kaluza theory or those theories that combine internal symmetries and space time. These theories require more dimensions (5, 7, etc., depending on the group you want to inbed in space-time). What happens to the displacements of the extra variables? You don't see them. According to Klein the universe is like a cylinder in those variables with a periodicity. Then, taking the periodicity into account, the eigenvalue $d/d\tau$ of the wave f is exactly equal to the charge. If you buy this argument, and include both curvature and torsion in the space, you get CP violation coming out of the usual, conventional theory. But to get right order of magnitudes, apparently f-gravity has to play a role.

Extragalactic Observational Astronomy

Lectures by

W. L. W. Sargent

California Institute of Technology

Notes by

J. Kormendy and G. Cavallo

Lecture 1

INTRODUCTORY REMARKS:  THE REDSHIFT CONTROVERSY

In 1929, Hubble published his now famous law relating the redshifts and distances of galaxies. For objects with $z \ll 1$, say with $z <$ several tenths, we can conveniently write this as:

$$d = \frac{c}{H_o} z , \qquad (1.1)$$

where $d$ = distance to the galaxy;

$z \equiv \dfrac{\lambda_{observed} - \lambda_{emitted}}{\lambda_{emitted}}$ is called the redshift of the galaxy;

$H_o \equiv$ Hubble constant at the present epoch (subscript $_o$), and

$c$ = speed of light.

Interpreting the redshift to be a Doppler shift due to a velocity of recession, Hubble then went on to study the expansion of the universe implied by the above relation. Later in life, though, he became doubtful that redshifts were always due to such an expansion. In fact, in 1947 (Hubble 1947) he listed the nature of the cosmological redshift as one of the three major unsolved problems in astrophysics. This forshadowed the present redshift controversy, in which two opposing schools of thought hold the following points of view:

A) Apart from velocity dispersions within galaxies or clusters of galaxies, the redshift is always a Doppler effect due to the expansion of the Universe. As a corollary, equation (1.1) gives the distance to any galaxy of known $z \ll 1$, subject to errors introduced by the velocity dispersions.

B) In at least some objects, such as perhaps QSO's or compact galaxies, a major part of the redshift is not due to the expansion of the Universe, but is due to an unknown mechanism internal to the object. Therefore, the true distance can be much less than that given by equation (1.1).

(B) will be dealt with extensively by other lecturers. A major purpose of these lectures will therefore be to present the conventional point of view (A). The second objective will be to review some observed properties of galaxies and clusters of galaxies to help set the stage for later theoretical discussions.

Hubble (1946) proposed the following test of the nature of redshifts, based on measurements of the surface brightness of some suitable kind of standard galaxy. It is easy to show that in an expanding universe, surface brightness is proportional to $(1 + z)^{-4}$. Alternatively, in any mechanism intrinsic to an object stationary in a non-expanding universe, surface brightness varies as $(1 + z)^{-1}$, by assumption. Measurements of various standard galaxies at different redshifts should be able to differentiate between the alternatives. However, this test has never been carried out. In fact, only a limited amount of data is available in support of (A), so that although a majority of astronomers subscribe to this viewpoint, the issue remains unsettled.

## SOME PROPERTIES OF NORMAL GALAXIES

A) Morphological Classification

1. Hubble Classes: The most commonly used classification system is due to Hubble (1926, 1936) and Sandage (1961), and is represented by the well-known "tuning fork" diagram (Fig. 1.1).

Elliptical galaxies are denoted by E, followed by an integer in the range 0 to 6, representing ten times the observed axial ratio.

For spiral galaxies, the classification criteria are:
i) the relative size of the nuclear bulge as compared to the disk, decreasing in the range a-c from being the dominant light source, to being almost invisible, and
ii) the appearance of the spiral arms, which are amorphous and of low pitch angle in Sa, and become more and more patchy and highly inclined in the sequence Sa-Sc. Spiral galaxies are further divided into those which have prominent bright bars across their nuclei, with the arms beginning from the radius of the end of the bar, and those which have no bars. In the latter case, the arms begin either at the nucleus or in a ring at some small radius (typically a few kpc).

Irregular galaxies of type I have a young stellar and gaseous content, similar to Sc and SBc galaxies. They are also

Extragalactic Observational Astronomy                                457

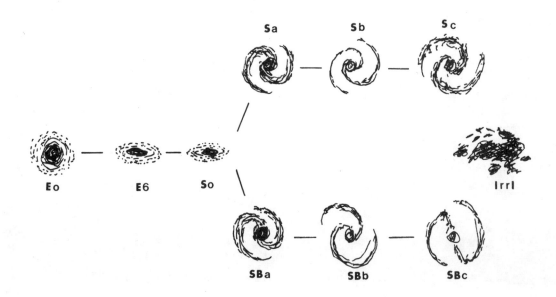

Figure 1.1  Tuning-fork Diagram of Hubble Classes

patchy, but are distinguished from spirals because there is no
overall organized structure.  Type II irregulars, such as M 82
and NGC 520 (see lectures by E. M. Burbidge and H. C. Arp),
are those galaxies too peculiar to fit into any of the other
Hubble classes, and therefore do not have a natural place in
Fig. 1.1, which attempts to portray a continuity of morphological properties.  Finally, from Hubble's unpublished notes,
Sandage (1961) added the class S0, for disk galaxies which were
generally very smooth and showed no gas or spiral structure.

   It should be noted that the two criteria for classifying
spirals do not always agree.  For example, there are galaxies
with smooth arms of small pitch angle, but with small nuclei.
Various classification schemes have emphasized different criteria.

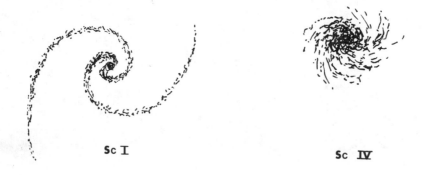

Figure 1.2  Variation of Morphology With Luminosity Class

Thus Morgan (1958) used only the size of the nuclear bulge, and Sandage used only the appearance of the arms when the criteria were in conflict.

2. Van den Bergh Luminosity Classes:

   Van den Bergh (1960 a, b) has developed a method by which the intrinsic luminosity of an Sb or Sc galaxy can be inferred from the regularity of its arms.  Bright galaxies (luminosity class I) have very regular, thin arms that cover a large range in azimuthal angle.  Faint galaxies (luminosity class V) have short, broad spiral arms, that cover only a small range in azimuth.  They have short spiral features, and many interarm branches, but little coherent spiral structure.  Fig. 1.2 shows this pictorially.

   This discovery and the subsequent calibration were made using the redshift-magnitude diagram (Fig. 1.3).  One interprets the spread in the points about a line of slope 5 as being due to a dispersion in intrinsic magnitudes.  Then the brightest galaxies toward the left side of the distribution in Fig. 1.3 had the characteristics subsequently named class I, and the

# Extragalactic Observational Astronomy

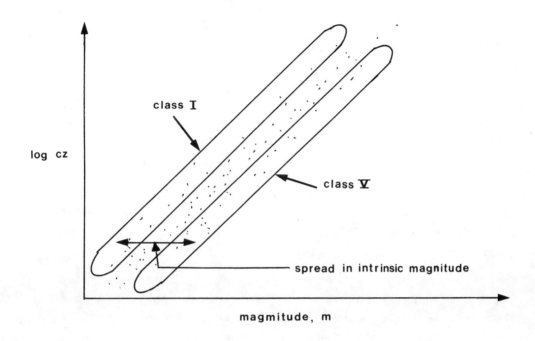

Figure 1.3 Luminosity Classes in the Redshift-Magnitude Diagram

faintest galaxies had class V appearance. This calibrates the luminosity classes. The original calibration, based on a Hubble constant of 100 km sec$^{-1}$ Mpc$^{-1}$ (giving the "indicative distances" advocated by Zwicky), is given in table 1.1, where colons indicate uncertainty.

Van den Bergh luminosity classes have proven to be very useful in our search for standard candles in cosmology, out to $z \simeq .1$, where morphology becomes difficult to observe. This will be discussed further in lecture II.

3. Important New Classes of Radio Sources:

In addition to Seyfert Galaxies, discussed in Margaret Burbidge's lectures, three new classes of radio galaxies require definition.

<u>N Galaxies</u> consist of a bright, starlike nucleus imbedded in a much fainter, small envelope. They are so distant that one cannot tell whether the galaxy is elliptical or spiral. Some

## Table 1.1

### CALIBRATION OF LUMINOSITY CLASSES

| Luminosity Class | Photographic Absolute Magnitude[a] ($M_{pg}$) |
|---|---|
| Sc I | −20.0 (± 0.5 approximately) |
| Sc II | −19.4    " |
| Sc III | −18.3    " |
| Sc IV | −17.3 :  " |
| Sc IV-V | −16.1 :  " |

(a) "Apparent magnitudes" are a logarithmic measurement of apparent brightness, with

(magnitude) = $-2.5 \log_{10}$ (intensity) + constant,

where the constant is roughly such that the brightest dozen stars are of first magnitude. The faintest apparent magnitude star visible to the naked eye is about 6. The subscript "pg" means that the magnitude was determined from a (blue-sensitive) photographic plate; that is, it measures the brightness in the blue part of the spectrum. Finally, the absolute magnitude M of an object is defined as its apparent magnitude m when the object is placed at a distance of 10 pc:

$M = m + 5 - 5 \log_{10}$ distance, distance measured in parsecs.

---

N galaxies are probably distant Seyferts.

D Galaxies are elliptical-like galaxies, with large faint envelopes. There appears to be a kink in the light distribution between the core and the halo. D galaxies are commonly found in rich clusters.

cD Galaxies are supergiant D's. They dominate some rich clusters, but have never been recognized outside such clusters.

B) Space Densities of Extragalactic Objects

Assuming a Hubble constant of 100 km sec$^{-1}$ Mpc$^{-1}$, the following number densities have been derived. For abnormal objects suspected of evolution, local densities are given representing distances less than a few hundred megaparsecs.

## Table 1.2

### NUMBER DENSITIES OF GALAXIES

| Type | Magnitude Range | Number Density Mpc$^{-3}$ |
|---|---|---|
| Normal field galaxies, locally | $M_B < -21 (10^{44}$ erg/sec optical) | $10^{-3.5}$ |
| " | $M_B < -20$ | $10^{-2.5}$ |
| " | $M_B < -19$ | $10^{-1.5}$ |
| " | $M_B < -18 (10^{43}$ erg/sec optical) | $10^{-1}$ |
| Rich clusters of galaxies, i.e. $\geq$ 100 galaxies within 3 mag. of the brightest galaxy | | $10^{-5}$ |
| Giant elliptical galaxies | | $10^{-4}$ |
| Seyfert galaxies, locally | nuclear $M_{pg} \sim -18$ | $10^{-3.5}$ |
| " | bright nuclei: $M_{pg} \sim -22$ | $10^{-6}$ |
| Very strong radio galaxies | $L \sim 10^{45}$ erg/sec radio | $10^{-10}$ |
| Radio galaxies (includ. Seyfert) | $L \sim 10^{44}$ erg/sec radio | $10^{-8.5}$ |
| " | $L \sim 10^{43}$ erg/sec radio | $10^{-7}$ |
| " | $L \sim 10^{42}$ erg/sec radio | $10^{-5.5}$ |
| QSO (radio and radio quiet) | $M_B = -26 (10^{46}$ erg/sec optical) | $10^{-8.5}$ |

Table 1.2 (cont'd.)

| Type | Magnitude Range | Number Density $M_{pc}^{-3}$ |
|---|---|---|
| QSO (radio and radio quiet) | ($10^{45}$ erg/sec optical) | $10^{-7}$ |
| " | $M_B = -21$ ($10^{44}$ erg/sec optical) | $10^{-6}$ |

We also note that about 90% of all normal galaxies near us are spiral, and about 10% are elliptical and irregular. This is not typical in the sense that most galaxies we observe are in rich, well-ordered clusters, and these consist mostly of E and S0 galaxies. No good estimates exist for what fraction of galaxies are in clusters and what fraction are in the field.

C) Mass and Light Distributions

The following methods are commonly used to measure galaxy masses.

1. Stellar Velocity Dispersion (for E galaxies): A spectrum is obtained by putting the spectrograph slit across the galaxy nucleus, giving the internal velocity dispersion from the absorption line widths. The range of observed values is from 60 km/sec in M 32, a dwarf elliptical, to 500 km/sec for the giant elliptical M 87. The mass distribution is then assumed to be proportional to the light distribution, giving the gravitational potential. Finally, the mass is determined using the Virial theorem:

$$2(\text{Potential Energy}) + (\text{Kinetic Energy}) = 0 \qquad (1.2)$$

in gravitational equilibrium. Masses obtained in this way range from $10^9$ to $10^{13}$ $M_\odot$.

2. Rotation Curves: This method is used for galaxies with emission lines, either in the optical, or at 21 cm in the radio spectrum. In particular, then, it is used for Sb and Sc galaxies, and Type I irregulars.

The rotation velocity as a function of distance from the galactic center can be obtained from spectra taken with various slit orientations (Fig. 1.4). One then models the mass distribution,

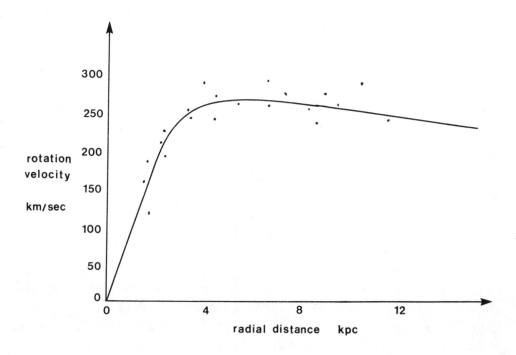

Figure 1.4  A Typical Galaxy Rotation Curve

for example as the sum of a number of homogeneous spheroids, and obtains the numerical values by fitting the rotation curve. It is assumed that the galaxy is in equilibrium, and that all motions are circular. Note that a rotation curve only gives that portion of the mass interior to the maximum radius at which the curve is observed. We often do not reach the decreasing portion of the curve, and therefore measure much less than the total mass of the galaxy.

The mass of our galaxy has been derived in this way (Schmidt, 1965).

3. Double Galaxies (Page, 1965): This method is particularly useful because it yields (statistical) total masses, and because it can be applied to ellipticals, which are otherwise very difficult to observe. The difference in recession velocity of two galaxies in a pair is found, and the galaxies are assumed to be gravitationally bound to each other. A statistical study can

then be made to allow for various orbital inclinations, eccentricities, and so on, giving mean masses of various kinds of galaxies.
4. Galaxy Magnitudes and Mass to Light Ratios: A very useful indicator of the stellar content of galaxies is the mass to light ratio:

$$\frac{M}{L} = \frac{\text{mass within some radius } R_o}{\text{light within radius } R_o}, \text{ in solar units} = \frac{(M/L)_{\text{Galaxy}}}{(M/L)_\odot}$$

Then spiral galaxies typically have $M/L \sim 5$, while ellipticals range in M/L from about 5 for dwarfs to about 40 for ordinary and giant ellipticals. Since the light output of a star varies as a large power (typically $\sim 4$) of the mass, this means that ellipticals are dominated by lower-mass stars (or other dark material) than spirals.

The above mass to light ratios of individual galaxies are inconsistent with the values of several hundreds obtained for clusters of galaxies using the Virial theorem. This suggests that there is a large amount of unobserved "missing mass", to make up the difference between the Virial masses and those obtained by adding up the observed galaxies. We will return to this problem in lecture 5.

To obtain mass to light ratios, we need to define some sort of standard magnitude. Since the surface brightness continues to fall off with radius as far out as we have measurements, an integrated magnitude is difficult to define. Two restricted magnitudes have been used:

i) the integrated magnitude within a standard isophote, such as the 25 mag per square arcsecond isophote, and

ii) the integrated magnitude within that radius at which the surface brightness has fallen to a standard fraction of the central brightness.

Since the next lecture will center on Sandage and Tammann's determination of the Hubble constant, we will describe here their method of measuring galaxy magnitudes. The first alternative is adopted, using the 25th magnitude isophote in the blue. (Note that the sky brightness is near 22 mag per square arcsec, some 16 times as bright as the cutoff brightness.) Sandage has found that one can conveniently see to a surface brightness of 22.6 mag/sq. arcsec on the Palomar Observatory Sky Survey. For E galaxies, the single accurate light distribution that reaches sufficiently faint is that of NGC 3379 (Dennison 1954, de Vaucouleurs 1959). This implies that the 25 mag isophote is 2.5 times as large as the 22.6 mag isophote. The procedure, then, is the following. The diameters of the galaxies to be observed are measured on the Sky Survey. These diameters are multiplied by 2.5 to obtain the size of the photometer aperture, $D_{25}$, which should be used to measure the galaxy. However, the aperture really has some fixed size $D_{phot}$, different from $D_{25}$. Therefore, the light distribution of NGC 3379 is used to tabulate magnitude corrections as a function of $D_{phot}/D_{25}$. The unknowns are then measured through the $D_{phot}$ aperture, and the magnitudes are corrected using the above table.

Of course, this is a somewhat crude procedure, because it assumes that the light distribution of NGC 3379 is universally applicable. This is probably not true, and in any case has been tested only at the bright end.

Lecture 2

DETERMINATION OF THE HUBBLE CONSTANT USING DIAMETERS OF H II REGIONS

A) Methods of Measuring Distances

The first value of $H_o$, published by Hubble in 1929, was 530 km sec$^{-1}$ Mpc$^{-1}$, and it was thought to be accurate to 15%. The latest Hubble constant is still believed to be accurate to 15%, but its value has decreased dramatically by a factor of 10. Sandage's recent recalibration, which led to the newest value of $55 \pm 7$ km sec$^{-1}$ Mpc$^{-1}$, is the subject of this lecture.

The problem of measuring $H_o$ reduces to that of finding the distance to objects sufficiently far away to make any local velocity anisotropies negligible. However, most kinds of things, whose properties we know, are too faint to be seen at great distances (m-M $\gtrsim$ 32).

i) For instance, the brightest stars in galaxies have absolute magnitudes M near -10. Working hard, it is possible to do photometry to 22nd magnitudes so that these stars can be used as standard candles to a distance modulus of m-M $\simeq$ 32. This is slightly farther than the Virgo cluster. But it assumes that one can recognize individual stars, whereas in practice stars are indistinguishable from clusters of stars. It is essentially this mistake which led to the early high values of $H_o$. In fact, individual stars can be used only at much smaller distances.

ii) Common novae, which occur quite frequently ($\sim$ 40/year in the Galaxy) have absolute magnitudes near -7 at maximum. If one could recognize a 22nd magnitude nova, this would take us out to a distance modulus of 29, but no such faint nova has ever been found.

iii) Cepheids: The best calibrated standard candles are undoubtedly cepheid variables. Henrietta Leavitt (1912) discovered a relation between the mean luminosity and the period of light variation, such that fainter cepheids had shorter periods. A 200-day cepheid has an absolute magnitude of -6. These stars are easily identified by their characteristic light curves. But they cannot be seen very far away; the greatest distance at which they have been used corresponds to a modulus of only 27 mag. (M 81 group: see Tammann and Sandage, 1968)

Extragalactic Observational Astronomy                                467

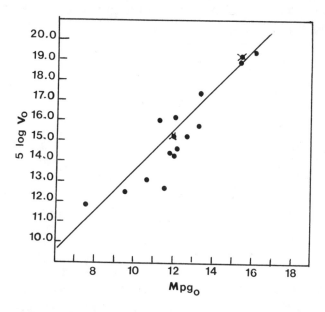

Figure 2.1  The redshift-magnitude relation for supernovae of
type I.  The dots refer to individual supernovae, and the crosses
represent averages for the Virgo and Coma clusters.

The cosmological recession velocity at this distance is only
200 km/sec, about equal to the random velocities between galaxies.
 iv) <u>Globular clusters</u> in the Galaxy have absolute magnitudes
of about −9.  However, there are indications that they are brighter
in brighter galaxies, since globulars in M 31 are brighter than
those in the Galaxy.  A dispersion in magnitudes leads to a
distance-dependent selection effect, since bright objects are
favoured at large distances.  In particular, in the Virgo cluster,
globular clusters are seen primarily around the brightest galaxies.
Thus globular cluster distances are open to considerable doubt,
even apart from the fact that the clusters cannot be seen very far
away.
  v) <u>Supernovae</u> of type I have the greatest potential for giving
a new calibration of the Hubble law.  They are routinely discovered at 19th apparent magnitude, and with absolute magnitudes of −18, this means they are useful to a distance modulus
of 37 mag.  Here the recession velocity is already 3700 km/sec,
much larger than the random velocities.  Furthermore, they are
quite good standard candles.  Figure 2.1 shows the redshift-
magnitude diagram for supernovae in galaxies of known redshift

(Kowal, 1968). The slope of the relation is very close to 5, so that the magnitude is independent of distance, and the deviation is only ± 0.6 mag, so that the dispersion is adequately small.

The problem with using supernovae as standard candles is that they are very rare. Thus calibration becomes very difficult. A typical SN rate is one per galaxy per 200 years. For instance, no SN has occurred in the Galaxy in modern scientific history. None has occurred in M 31 since 1885. In fact, only three SN as bright as 8.5 mag have occurred in modern times, two in NGC 5253 (1895 and 1972), and one well-studied example in IC 4182 (1937). These are nearby galaxies, so that the above techniques should be applicable, but accurate distances have not yet been measured.

At the present time, the calibration rests on Baade's (1945) analysis of Tycho's visual observations of the Galactic SN of 1572. Since Tycho made careful comparisons of the magnitudes and colours of the SN with those of Jupiter and Venus, a crude calibration is possible. A more accurate determination is clearly necessary, and such work is underway.

In the meantime, there is only one good measurement of $H_0$, that of Sandage and Tammann (to be published).

B) Sandage's Redetermination of the Hubble Constant

It was first suggested by Sersic (1960) that since the size of the largest, or the three largest H II regions in a galaxy varies little with Hubble type, H II region diameters might be useful as cosmological probes. H II regions have the added advantage of being very prominent and easy to discover on Hα plates, and of being measureable down to 2" apparent size. We will see that this corresponds to a large enough distance to calibrate our standard candles.

The problem has been that there is a small variation (about a factor of 2) in H II region size with galaxy type. This makes it necessary to choose a particular kind of galaxy to study. ScI galaxies have proven suitable, first because they are good standard candles, second because they are bright and easily recognizable at very large distances, and third because they have the largest H II regions.

The procedure used by Sandage and Tammann consists of the following eight steps.

1. __Determine the distance to the Hyades cluster to calibrate the zero-age main sequence.__

The __main sequence__ is a one-to-one relation between the absolute magnitudes and the colour or temperature of a star. It is satisfied by all stars during the major hydrogen-burning phases of their lives (see, e.g., Schwarzschild, 1958). The zero-age MS is this relation at the start of hydrogen burning (Fig. 3.1). An uncalibrated MS is easily observed in any cluster, where all stars are at the same distance from the sun. One finds that fainter stars are redder (or cooler). We need this relation in absolute form, i.e., with absolute instead of apparent magnitudes. The importance of the Hyades cluster then lies in the fact that we have a particularly trustworthy means of measuring its distance.

The classical moving cluster method is used. Motions in the sky, called proper motions, are measured for all the stars. These motions converge to a point, since the cluster as a whole is moving away from the sun at some angle $\eta$ to the line of sight. The angle $\eta$ is then given by the separation of the cluster and the convergent point. Finally, the distance is routinely derived from $\eta$, the proper motions, the radial velocities, and some simple geometry.

The distance then gives the distance modulus $m - M$, and a calibrated zero-age main sequence.

2. __Calibration of the period-luminosity law using cepheids in galactic clusters.__

The next step is to calibrate the period-luminosity law for cepheids in the Galaxy (Sandage and Tammann, 1968). About six cepheids are known in galactic clusters. The distances to these clusters are found by force-fitting the apparent magnitude-colour MS of the clusters to the known absolute magnitude-colour MS of the Hyades. This gives the distance modulus, determining the absolute magnitudes of the cepheids. However, only a small range in periods is represented among galactic cluster cepheids, so that the slope of the period-luminosity law is not given. Consequently, this slope is obtained from external galaxies, and the galactic cluster cepheids are used to set the zero point. Fig. 2.2 shows the result for our Galaxy and four other galaxies of the local group. The curves have been superimposed to minimize the scatter.

# Extragalactic Observational Astronomy

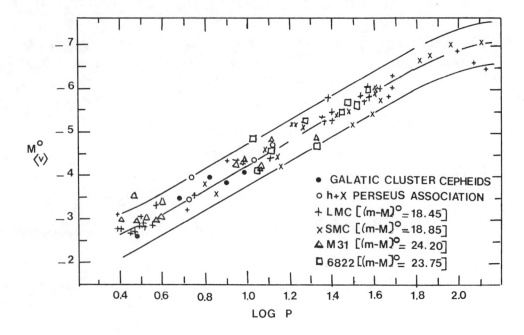

Figure 2.2 Period-luminosity relation for cepheids in galactic clusters, in an association in Perseus, and in four other galaxies in the local group.

3. **Use cepheids to obtain distances to galaxies containing H II regions, in the local group and the M 81 group.**

Hubble's first project on the 200" telescope was to try to find cepheids in M 81 and M 101. The latter turned out to be too far away: the cepheids are too faint. This is unfortunate, because M 101 is one of the nearest ScI galaxies, and the one we will use to calibrate H II region sizes. However, in the nearer groups many cepheids were found. The periods gave the absolute magnitudes via the period-luminosity law, and the apparent magnitudes then gave the distances. An excellent example can be found in Tammann and Sandage (1968), where 17 cepheids are used to derive the distance to NGC 2403 in the M 81 group.

4. **Calibrate the linear sizes D of H II regions as a function of luminosity class LC, or of $M_{pg}$, using galaxies from step 3.**

When hot stars are surrounded by clouds of hydrogen, they

ionize a volume around them whose size is determined by the
amount of ultraviolet stellar radiation, and the recombination
rate. The ionized volumes, called an H II region, have very
sharp boundaries, for the following reason (Kuiper et al.,
1937, p. 594-595). The ionizing stars have an approximately black
body spectrum, and the Lyman limit, below which photons can
ionize the gas, falls on the exponential part of the black body
curve. Higher-energy photons have more penetrating power, but
they are also much rarer. Thus H II regions are produced by
photons which are very near to but shortward of the Lyman
limit. They all have essentially the same penetrating power.
Now, the opacity of neutral hydrogen to these photons near the
Lyman limit is very high. Thus, consider what happens at the
radius at which the ultraviolet photons begin to be used up,
that is, where there are few enough so that recombinations begin
to dominate over photoionizations. This produces a little neutral gas, which strongly absorbs further ultraviolet photons.
Since the mean-free-path of such photons is so small, the region
in which there are both neutral hydrogen atoms and ultraviolet
photons must be very thin. A more mathematically rigorous
argument will be found in the above reference.

For our purpose, the sharp edges to H II regions mean that
diameters can be measured without involving any physics. The
problem of measuring diameters is purely geometrical, and it is
sufficient to show empirically that the largest H II region in
any galaxy of a particular type is always of the same size, so
that considerable experience is necessary before one can judge
what to measure.

The next step in obtaining $H_0$ is to measure the apparent
diameters of the largest H II regions in the galaxies used in
step 3. The cepheid distances are then used to convert apparent
to real linear diameters D, calibrating D as a function of luminosity class or absolute magnitude of the galaxy. Fig. 2.3
is this result.

A vital step in the above calibration is the extension to luminosity class I, which is not represented in the nearby clusters.
This is done using the dwarf companions of M 101, which have
been force fit to the nearer galaxies by sliding their curve
parallel to the "diameter" axis. Since M 101 is at the same
(previously unknown) distance as its companions, this locates
it in Fig. 2.3. The fact that it falls off the straight line

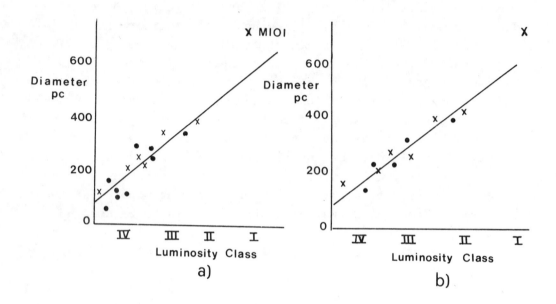

Figure 2.3 Mean H II region diameters as a function of luminosity class for 5 galaxies in the local group and 6 galaxies in the M 81 group. Galaxies in the M 101 group are shown as crosses, and have been force fit to the eleven nearer galaxies. 2.3(a) refers to the largest and 2.3(b) to the mean of the three largest H II regions in the galaxy. From unpublished work by Sandage and Tammann.

is of no significance, because there is no physical reason for any particular spacing of the purely morphological luminosity classes along the horizontal axis. Note that the large H II regions in M 101 are 600 pc. in diameter.

A plot of diameters versus absolute magnitudes of the galaxies can be constructed in the same way (Fig. 2.4). With this plot, we have a calibration that extends to $M_{pg} \simeq -21$. However, it is based on the single ScI galaxy M 101. This situation can be improved by going to the Virgo cluster, as we shall show below.
5. <u>Obtain distances to many field Sc and Irr galaxies in the range $28 < m - M < 32$ from the calibration of step 4.</u>

We can now find the distance to a large number of field Sc and Irr galaxies from their luminosity class and the apparent

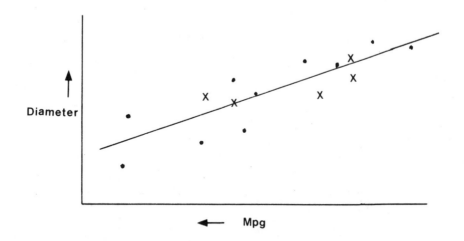

Figure 2.4 Sketch of mean H II region diameter as a function of the absolute magnitude of the galaxy. Dots refer to galaxies of the local and M 81 groups, and crosses refer to the M 101 group. From unpublished work by Sandage and Tammann.

size of their H II regions, using the calibration of step 4.
6. <u>From the galaxies in step 5, calibrate</u> $<M_{pg}>$ <u>as a function of Lc, for luminosity classes ScI - IrrV.</u>

For each galaxy, the distance gives the absolute photographic magnitude, resulting in a recalibration of the magnitudes of the various luminosity classes (Fig. 2.5, dots). Now the faint end of this curve is well known, including the zero point, but for the ScI galaxies the zero point still rests on the single galaxy M 101. This situation can be improved by going to the Virgo cluster, as follows.

The Virgo cluster, like the M 101 cluster, consists of a number of galaxies all at the same distance from us. The same procedure that placed M 101 in Fig. 2.3 in the first place can now be used to improve the calibration. The Virgo cluster yields a plot of apparent photographic magnitude against Lc. This is

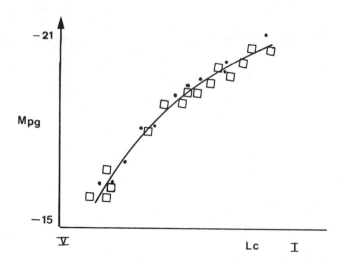

Figure 2.5 Recalibration of the photographic absolute magnitudes of the van den Bergh luminosity classes. Dots refer to field galaxies in the range $28 < m - M < 32$, and squares to galaxies in the Virgo cluster. From unpublished work by Sandage and Tammann.

the same kind of plot that is shown in Fig. 2.5 (dots), but with unknown zero-point. However, there are many ScI galaxies in this cluster, so the relative position of the ScI and other Lc galaxies is very well determined. The zero-point is found by sliding the curve vertically to bring it into coincidence with the dots in Fig. 2.5, producing the curve defined by the squares. This completes the recalibration of luminosity classes, the zero-point being defined locally, and the relative position of the ScI galaxies being determined by the Virgo cluster.

7. <u>Identify and measure the redshifts of many (58) ScI galaxies with cz >4000 km/sec, to get beyond any local anisotropy</u>.

This yields the Hubble diagram shown in Fig. 2.6. As we have already indicated, any class of objects is a standard candle if and only if it satisfies a line of slope 1/5 in the redshift-magnitude diagram. Fig. 2.6 shows that ScI galaxies are indeed standard candles: the observed slope is 0.2 to 1/2 standard deviation. The scatter about the line then determines the dispersion in magnitudes. Combining this with the mean magnitude

# Extragalactic Observational Astronomy

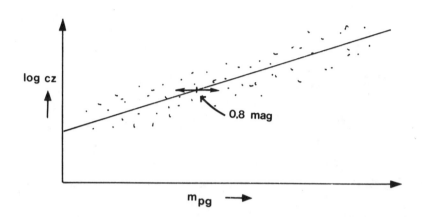

Figure 2.6  Redshift-magnitude diagram for 58 ScI galaxies. The line is the theoretical relation of slope 1/5. From unpublished work by Sandage and Tammann.

of ScI galaxies from step 6, we find that

$$<M_{pg}>_{ScI} = -21.2 \pm 0.4 \qquad (2.1)$$

This dispersion is surprisingly small, and results in a very well-defined value of the Hubble constant.

8. <u>Combine the Hubble diagram of step 7 and the calibration of $<M_{pg}>$ for ScI galaxies to obtain $H_o$.</u>

Recall the definition of distance modulus:

$m - M = 5 \log d - 5$, d the distance.

Then the definition of the Hubble constant (1.1) leads to :

$m - M = 5 \log(cz) - 5 \log H_o - 5$.

Thus the equation of the line in Fig. 2.6 is:

$$\log(cz) = \frac{1}{5} m_{pg} - \frac{1}{5} (<M_{pg}>_{ScI} - 5 \log H_o - 5). \qquad (2.2)$$

The intercept in Fig. 2.6 yields $<M_{pg}>_{ScI} - 5 \log H_0 - 5$, and (2.1) then gives the Hubble constant:

$$H_0 = 55 \pm 7 \text{ km sec}^{-1} \text{ Mpc}^{-1} . \tag{2.3}$$

The figure ± 7 in (2.3) represents the error in $H_0$ attributable to the magnitude scatter in the Hubble diagram (Fig. 2.6) and to the dispersion in the magnitudes of the nearby ScI galaxies used in the calibration. Because of the large number of steps involved in the above procedure, it is difficult to estimate how realistic this error estimate is likely to be.

The reciprocal of $H_0$, combined with the deceleration parameter $q_0$ (see Lecture 3), produces a new estimate of the age of the universe:

$$\tau \simeq 17 \times 10^9 \text{ years}$$

This is older than any kind of object for which we have independent age estimates. In particular, it is older than the oldest star clusters in the galaxy.

It is important to note that the Hubble diagram for the present (small!) sample of ScI galaxies shows no systematic difference between northern and southern hemispheres. This suggests that any local shear in the velocity field must be smaller than 100 km/sec. There is also no evidence for any systematic motion of the Galaxy with respect to the Virgo cluster. Now, there is no dynamical compulsion on the Local Group to revolve around Virgo, since the revolution time is approximately the age of the universe. Still, there is no a priori reason not to expect a difference in the peculiar velocities of the two clusters. Evidently this peculiar velocity must be very small. The observed velocity field seems to be surprisingly isotropic.

This concludes our discussion of Sandage and Tammann's derivation of $H_0$. At a considerably higher cost in telescope time, one improvement in the method might be to measure luminosities of H II regions rather than their diameters. This would presumably be more impersonal, and could be done out to much larger distances, where the apparent diameters are smaller than 2 arcseconds.

Extragalactic Observational Astronomy

Lecture 3

DETERMINATION OF THE DECELERATION PARAMETER

A. Introduction

The Hubble constant $H_0 = (\dot{R}/R)_0$ describes the rate at which the universe is expanding. Here R is the scale factor of the universe, which satisfies

$$\ddot{R} = -\frac{4}{3}\pi G \rho(R) R + \frac{\Lambda R}{3}, \qquad (3.1)$$

where $\rho(R)$ is the relativistic energy density,
    $\Lambda$ is the so-called "cosmological constant",
    and G is the gravitational constant.
The rate at which the expansion is decelerated by self-gravitation is defined by a second constant called the "deceleration parameter":

$$q_0 \equiv -\left(\frac{R\ddot{R}}{\dot{R}^2}\right)_0 = -\frac{\ddot{R}_0}{R_0}\frac{1}{H_0^2} \qquad (3.2)$$

The equation

$$3H_0^2 q_0 = 4\pi G \rho_0 \qquad (3.3)$$

relates $q_0$ to the mean matter density. Now, visible galaxies have a density of only about $7 \times 10^{-31}$ gm cm$^{-3}$ (Oort, 1958), leading to

$q_0 \simeq 0.03$.

But the universe is closed only if $q_0 \geq 1/2$. Many people prefer a closed universe for philosophical reasons, and so there have been numerous searches for enough invisible mass density to make $q_0 > 1/2$. However, there is no <u>a priori</u> reason to expect a closed universe, and the indications are growing that the above search will be fruitless. The local mass density supports an open universe.

A low value of $q_0$ has also been obtained by Gott and Gunn (1971), in a study of the X-ray emission due to the infall of matter into giant clusters of galaxies. A low value of the intercluster matter density is required. In fact all the observations seem to point to a value of $q_0$ which is near zero.

When the best available evolutionary corrections are applied, this includes the direct measurement of the deceleration of galaxies, described below.

The deceleration parameter can be found from the deviations of the magnitude-redshift relation from a straight line at large redshift. Sandage has been engaged in its derivation for many years, and has published a number of values (see, e.g., Sandage 1968). However, many large corrections must be applied to the observational data, which still make the determination of $q_o$ a somewhat implausible exercise. For instance, in his latest papers (Sandage, 1972 a, b), where he has published corrected as well as uncorrected data, we see that he arrives at a "formal value" of $q_o$: the value which is implied if his corrections are accurate. But he has necessarily omitted the most important (but most poorly known) correction, namely the effects of stellar evolution in the galaxies used.

In trying to determine $q_o$, we must confine ourselves to Friedmann models with cosmological constant $\Lambda = 0$. There is little enough hope to derive $q_o$, still less to find two parameters. In this case, the redshift-magnitude diagram satisfies the following equations (Mattig 1958, Sandage 1961).

For $q_o > 0$,

$$m = 5 \log \left\{ \frac{1}{q_o^2} \left[ q_o z + (q_o - 1) \left[ (1 + 2q_o z)^{1/2} - 1 \right] \right] \right\} + \text{constant}. \tag{3.4}$$

For $q_o = 0$,

$$m = 5 \log \left[ z(1 + \frac{z}{2}) \right] + \text{constant}, \tag{3.5}$$

and for $q_o = -1$,

$$m = 5 \log \left[ z(1 + z) \right] + \text{constant}. \tag{3.6}$$

The above expression (3.4) can be expanded for small redshifts to yield:

$$m = 5 \log z + 1.086 (1 - q_o)z + O(z^2) + \text{constant}, \tag{3.7}$$

for $q_o > 0$ and $z \ll 1$.

These are the theoretical relations to be fitted to the observations.

Two classes of objects have been used: radio galaxies, and the brightest galaxies in rich clusters. ScI galaxies, for instance, cannot be used, because we must go to $z \sim 0.5$ to see deviations from the Hubble law, and ScI galaxies cannot be distinguished at this distance. We will confine ourselves here to a discussion of Sandage's latest redeterminations of $q_o$, using brightest cluster galaxies.

B. Determination of $q_o$ Using Brightest Cluster Galaxies.
1. Corrections applied to the measurements

Most rich symmetrical clusters are dominated by a few E or D galaxies. The brightest of these is easy to pick out, and its B and V magnitudes are measured through several apertures. The redshift of the cluster is also found, by measuring a number of galaxies in it. One galaxy is not enough because velocity dispersions of order 1000 km/sec are common, and the redshift must be known more precisely than this.

The following corrections are then applied to the magnitudes:

i) <u>Aperture effect</u>: We require the magnitude to a standard radius (in kpc) from the center of the galaxy. But the relation between angular and linear diameters involves the unknown value of $q_o$. An iteration precedure must be set up, and it has been shown to converge. In the present case, it turned out that Sandage's original guess was sufficiently accurate.

ii) <u>K-correction</u>: The colours and magnitudes of galaxies vary with redshift. Increasing z moves their spectra to the redward under a fixed observing bandpass $\Delta\lambda$. At the same time, the spectra are compressed, so that more proper wavelengths are contained in $\Delta\lambda$. These are the two terms comprising the K-correction, as discussed, for instance, by Oke and Sandage (1968). To correct for these effects, we must know the spectral energy distribution of the emitted light; the usual way to proceed has been to use the spectra of nearby galaxies. The resulting corrections are quite large, amounting to about 0.5 magnitude.

iii) <u>Evolutionary correction</u>: We believe that star formations in E galaxies occurred very quickly just after their formation, and then stopped some $10^{10}$ years ago. Stellar evolution since that time has steadily decreased the maximum mass of stars still burning hydrogen. In particular, then, the distribution of stars in the colour-magnitude diagram varies with look-back time to galaxies of various z. For example, light from the most distant galaxies observed ($z = 0.46$) has been in transit for

$$\tau = z(1 + z)^{-1} H_o^{-1} \qquad (q_o = 0); \qquad (3.8)$$
$$\tau = 6 \times 10^9 \text{ years} \qquad (H_o = 55 \text{ km sec}^{-1} \text{ Mpc}^{-1}).$$

It is not clear how the spectra of E galaxies evolve in this time period, because we know neither the slope of the stellar luminosity function near one solar mass, nor the number of stars per unit mass along the main sequence. Nor do we know the evolutionary tracks well enough, for example, because we know too little about the composition. The situation is illustrated in Fig. 3.1, where the colour of a star is defined as the difference in magnitudes between its brightness B in a certain blue bandpass at $\sim$ 4400A°, and its brightness V in a certain yellow bandpass near 5500A°. Therefore, the redder the star, the larger is its B-V.

With so little knowledge of the stellar content, the evolutionary correction is very poorly determined. Recent calculations by Tinsley (1972) indicate that ellipticals may have been brighter in the past, but even this is not certain. All we are fairly sure of is that the correction is not negligible.

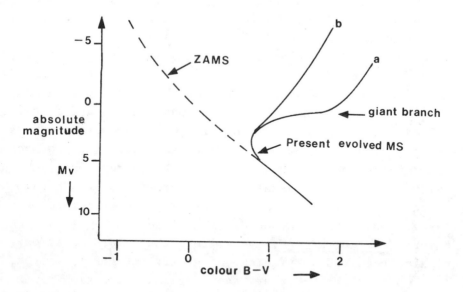

Figure 3.1 A schematic colour-magnitude diagram for an elliptical galaxy is shown. We cannot choose between the two alternative evolved stellar distributions (a) and (b). The former is similar to that obtained for metal-poor globular clusters in the Galaxy, and the latter resembles the curve for young metal-rich star clusters.

## 2. The Redshift-Magnitude Diagram

The sample of clusters used consists of the following:
  i)   33 clusters studied by Peterson (1970);
  ii)  10 clusters studied by Westerlund and Wall (1969);
  iii) 41 clusters with photoelectric photometry, worked on by Sandage. 28 of these had redshifts due to Humason, and 9 contain radio galaxies.

Except for Peterson's clusters, which have small redshift, the combined sample of 84 clusters is by no means random. Most of them had their redshifts measured long ago, before the Palomar Observatory Sky Survey enabled Abell (1958) to find a complete sample of clusters. The present clusters were in fact mostly found by accident, while investigating other objects. One could now take the Abell catalogue, which extends to $z \simeq .2$, supplement it with a search for fainter clusters, and select an unbiased sample using some random number technique. This has not been done. It is therefore conceivable that early researchers chose some special subclass of clusters in their attempts to measure redshifts. However, there are no indications at present that there is anything wrong with using the above sample.

Some of the clusters above contain radio galaxies, and one might wonder whether or not this has any effect. Sandage has looked into the problem and decided that, although their spectra show emission lines, most radio galaxies have the same range in B-V as the ordinary ellipticals. Any kinds of radio galaxies for which this is not true, such as N galaxies, have not been included.

Accordingly with the above sample, Sandage has constructed redshift magnitude diagrams as outlined in the previous section. Figures 3.2 and 3.3 show the results for the 41 and 84 clusters. In Fig. 3.2, the redshifts range from .0037 for the Virgo cluster to .461 for 3C 295. The gap between magnitudes 9 and 12 is presumably due to the scarcity of nearby rich clusters. (The clusters nearer than the gap are not very rich). A line of slope 1/5 has been least-squares fitted to the points in Fig. 3.2, giving

$$V_{corrected} = 5 \log cz - 6.76.$$

Let us now consider the scatter in the m - z diagram. We can adopt the following extreme points of view: (i) the magnitudes are infinitely accurate and the dispersion is only in

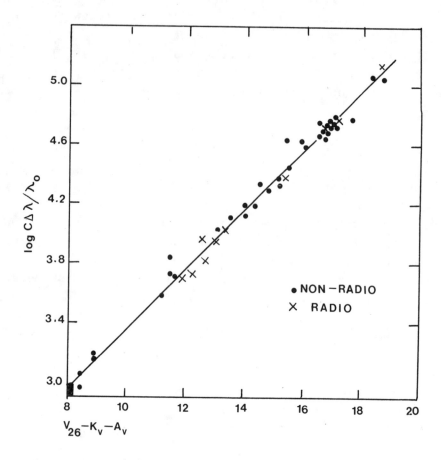

Figure 3.2 Redshift-magnitude diagram for the 41 clusters with photoelectric photometry. V magnitudes have been corrected for the aperture effect, K correction, and galactic absorption, but not for evolution. Radio galaxies are indicated by crosses. The redshift range covered by Hubble's (1929) original relation is shown by the box on the lower left.

log cz, and (ii) the dispersion is all in the magnitudes. In the first case, we find that at fixed m, $\Delta \log z$ is a constant, that is $\Delta z \propto z$. This is shown in Fig. 3.4. If we discard the two clusters which deviate most from the mean, then the histogram of $\Delta \log cz$ is approximately gaussian, and has a constant standard deviation at all redshifts:

# Extragalactic Observational Astronomy

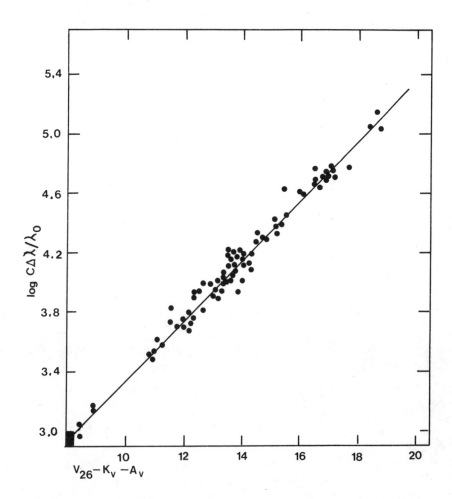

Figure 3.3 Same as Fig. 3.2, but for all 84 clusters. Note the larger scatter.

$$\sigma\left(\frac{\Delta z}{z}\right) = 0.063.$$

Incidentally, this places an upper bound on the motion of clusters such as Virgo ($cz = 1000$ km/sec). Their centre-of-mass velocities must be bounded by $\sigma(\Delta cz) \lesssim 100$ km/sec.

The problem with the relation $\Delta z \propto z$ is that it has no physical explanation. Sandage therefore discarded the alternative (i), and assumed the dispersion in Fig. 3.2 was due entirely to errors in the magnitudes. Then the two most deviant clusters are in-

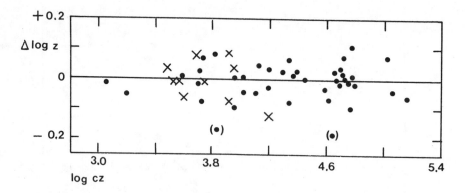

Figure 3.4   The dispersion in the logarithm of the redshift as a function of redshift. $\Delta \log z$ is independent of z.

terpreted as containing abnormally bright first-ranked galaxies, which is not surprising, since they are known to be cD systems. They are therefore discarded, giving for the sample of 41 - 2 = 39 clusters,

$$\sigma_{39}(\Delta M_V) \simeq 0.25 \text{ mag.}$$

This dispersion is small, showing that brightest cluster galaxies are excellent standard candles. If we go to the full sample of 82 clusters, we find the absolute magnitude of these galaxies to be

$$M_V = -23.3 \pm .32 \text{ (standard deviation).}$$

Due to the poorer photometry, the dispersion is larger, and the distribution is slightly unsymmetrical (Fig. 3.5).

A surprising property of both $M_V$ and $\sigma(M_V)$ is that they are independent of the richness of the cluster (Fig. 3.6). Richness class measures the number of cluster galaxies within 2 magnitudes of the brightest. If, for instance, first-ranked cluster galaxies were brighter in richer clusters, as envisaged by Scott (1957), a distance-dependent selection effect would result. At larger

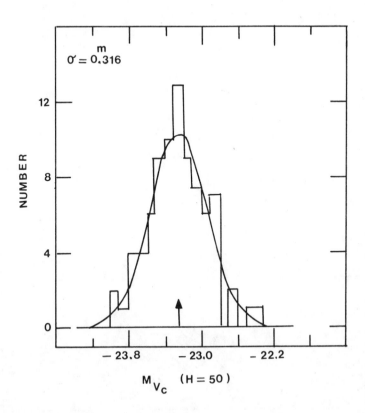

Figure 3.5 Histogram of magnitude deviations of 82 first-ranked cluster galaxies from a mean of $-23\overset{m}{.}3$.

distances, we would preferentially identify those clusters with brighter first-ranked galaxies. Of course, this would introduce curvature into the m - z relation, and thus give an incorrect value of $q_0$. The conclusion from Fig. 3.6 is that the effect does not exist. We note, however, that Peterson (1970) and Abell (1972) reached opposite conclusions in a similar study, based, perhaps, on poorer data.

3. The value of $q_0$:

Equation (3.7) can now be used to fit theoretical redshift-magnitude relations to the observations. As we can see from Fig. 3.7, values anywhere from -1 to +5 are about equally admissible. Sandage has plotted the R.M.S. deviation in magnitudes from the theoretical lines for various $q_0$ (Fig. 3.8),

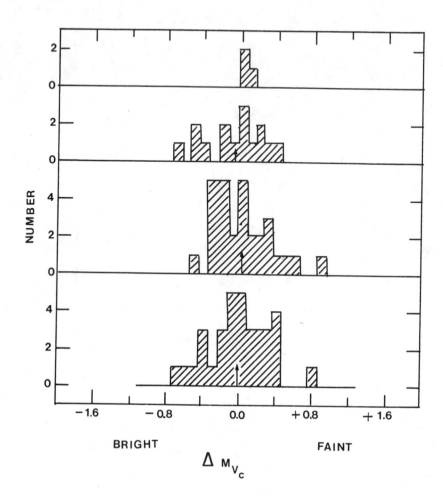

Figure 3.6 Histogram of magnitude dispersions of brightest galaxies in clusters of various richness classes. There is no evidence for a change in either $M_v$ or $\sigma(M_v)$ with richness class.

showing that there is a slight tendency to favour $q_0 = 1$. The homogeneous sample of 39 clusters, which seems to be the most accurate, gives

$q_0 = 0.96 \pm 0.4$          ($\Lambda = 0$).

Without including any evolutionary corrections, this is only a formal result. In fact, we can calculate the necessary evolution in $M_v$ to give $q_0 = 0$. The result is

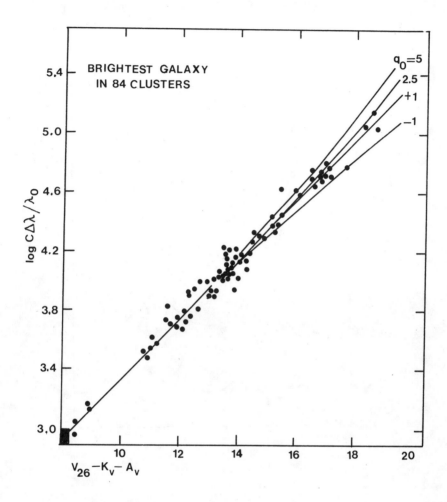

Figure 3.7 Redshift-magnitude diagram with various theoretical curves superimposed. Any value of $q_0$ in the range -1 to 5 appears possible.

$dM_V/dt = 1.09 (1 + z) H_0$ mag yr$^{-1}$.

This is about 0.1 mag in $10^9$ years, in the sense that the galaxies were brighter in the past. Tinsley's (1972) calculation gives a very similar value, so that the present observations favour $q_0 = 0$.

With modern observation techniques, the data in Fig. 3.7 can probably be extended by 2 magnitudes to $V \simeq 22$ mag. This would

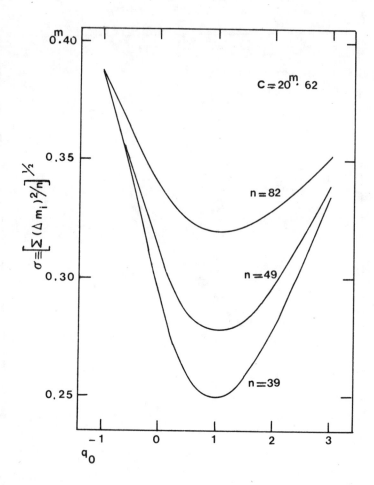

Figure 3.8 Root-mean-square deviations in magnitudes of brightest cluster galaxies from theoretical m-z relations, as a function of $q_0$. Curves are given for the three samples of 39, 49, and 82 clusters.

correspond to a redshift of $z \simeq 0.7$, and give us a much firmer value of the deceleration parameter.

C. On the Determination of $q_0$ Using Radio Galaxies

Radio galaxies can be used to determine $q_0$ in exactly the same way as brightest cluster galaxies. Fig. 3.9 is a Hubble diagram for 69 radio galaxies, mostly from the 3C catalogue (ie., with flux greater than $9 \times 10^{-26}$ watts m$^{-2}$ Hz$^{-1}$ $\equiv$

≡ 9 "flux units" at 178 MHz.) The usual K corrections and aperture corrections have been applied. We see that the points again fall on a line of slope 1/5, supporting the cosmological interpretation of their redshifts. The fit is particularly good if we exclude N galaxies, which have peculiar spectra. Then a Hubble constant of 50 km/sec Mpc gives

$M_V = -22.98 \pm .41$

for these galaxies. That is, they are slightly fainter than first-ranked cluster galaxies, and have a larger dispersion.

The magnitude distributions are compared schematically in Fig. 3.10. Radio galaxies are found to be distributed asymmetrically

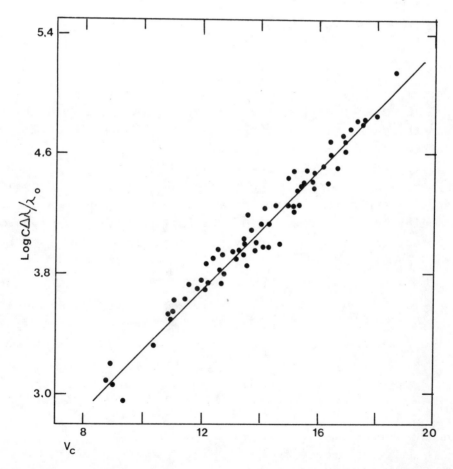

Figure 3.9 Hubble diagram for 69 radio galaxies. A line of slope 1/5 fits the points with good accuracy.

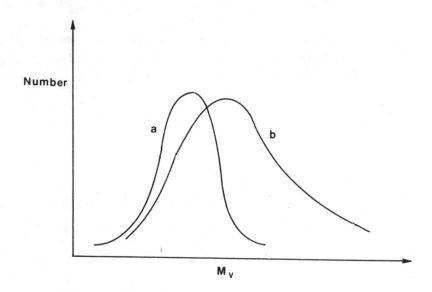

Figure 3.10 Magnitude distribution of brightest cluster galaxies (a), and the radio galaxies of Fig. 3.9 (b). Note the asymmetry and the greater width in curve (b).

in the sense that there are too many faint galaxies. The combinations of a larger dispersion and an asymmetric magnitude distribution unfortunately makes radio galaxies unsuitable as cosmological probes, leaving us with only the brightest cluster galaxies discussed above.

Lecture 4

EVIDENCE FOR THE COSMOLOGICAL INTERPRETATION OF QUASARS. THE LUMINOSITY - VOLUME TEST

A. Quasars in Clusters of Galaxies

When QSO were first being discovered, it was hoped that they could easily be used to firmly establish the values of $H_o$ and $q_o$, using the usual redshift-magnitude diagram. However, it was rapidly found that they do not fall on the usual line of slope 1/5 that is satisfied by radio galaxies. Rather, they are scattered throughout a large region predominantly above the galaxy line, as shown in Fig. 4.1. Conventionally, this scatter is interpreted as a dispersion in intrinsic luminosities, with the QSO being generally brighter than brightest cluster galaxies. There are other explanations, which usually involve a rejection of the cosmological interpretation of redshifts. Since such viewpoints are being discussed by other lecturers, we will review here the status of one test favouring the conventional view. This test consists of finding associations between QSO and clusters of galaxies of the same redshift. The redshift-distance relation for galaxies being rather well established, this is strong supporting evidence for using the same relation for quasars.

Two apparent associations have been found. Gunn (1971) showed that PKS 2251+11 lies in the line of sight of a sparse cluster of galaxies. He measured the redshift of one of the galaxies by fitting the continuum of a redshifted standard elliptical galaxy spectrum to the observed continuum of the galaxy. The redshift was $z = .33$, the same as that of the quasar. Supporting evidence was provided by a number of weak emission lines, but these were not confirmed by later spectra taken by Wampler. Apparently some light from the nearby quasar leaked into the diaphragm. In any case, the redshift from the continuum fitting suffices, and strongly implies that the quasar is associated with the galaxy, and hence is at its cosmological distance. Then its absolute magnitude is $M_B \simeq -25$, typical of bright QSO in the cosmological view, and brighter than any N galaxies. This fact is important to establish that we are dealing with a quasar, and not with a misclassified object of lower luminosity.

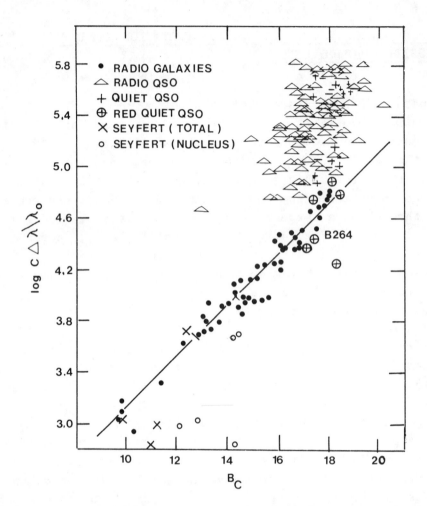

Figure 4.1 Hubble diagram for radio galaxies, quasars, and Seyferts in blue light. All magnitudes are photoelectric. Normal K corrections have been applied to the galaxies. No K corrections are applied to quasars because such corrections are small, and would be zero if $F(\lambda) \propto \lambda^{-1}$, which is nearly the case for QSS in the mean.

The second such identification (Oemler, et. al. 1972) involves 3C323.1, which has a redshift of $z = 0.28$. This is near a cluster of galaxies, at least one of which has the quasar's redshift. Again, the Hubble distance law implies that the quasar is brighter than any N galaxy.

Sargent and Hazard have searched for quasars in faint clusters

# Extragalactic Observational Astronomy

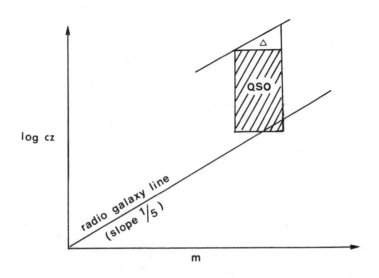

Figure 4.2 Schematic distribution of quasars in the Hubble diagram. See the text for discussion.

of galaxies. Four coincidences in position were found. However, the cluster redshifts could not be above about 0.3, from the magnitudes of the galaxies, whereas the quasar redshifts ranged from 0.68 to 1.27. This suggests that the QSO are background objects, with the clusters superimposed by chance. Now, a glance at any deep plate taken far from the galactic plane reveals numerous faint clusters, almost completely covering the plate. Hence this is not surprising. Nevertheless, the above view could only be established by some statistical list of the expected frequency of such coincidences, or perhaps by finding quasar absorption lines with the cluster redshift.

Thus the available direct evidence, although weak, is in favour of the cosmological viewpoint. According to Gunn (1972, private communication), many more candidates exist for possible identification, but the redshifts have not been measured. We can therefore look forward to a more conclusive result in the near future. Meanwhile, we will assume that redshifts are cosmological, as we have been doing throughout these lectures.

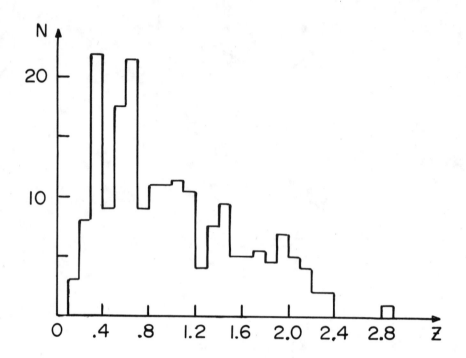

Figure 4.3  Redshift distribution of quasi-stellar sources.

B.  The Luminosity - Volume Test for Quasars
   1.  Observational Selection Effects in Quasar Identification
   The luminosity - volume test attempts to find any variation of quasar number densities with distance. In doing this, a thorough examination is necessary of the distance-dependent selection effects which go into the optical identifications. The following arguments show how important such effects can be (Sandage, 1972 b).
   First, consider again the distribution of quasars in the redshift-magnitude diagram, as schematically shown in Fig. 4.2. Three of the boundaries to the rectangular region occupied by quasars are fairly easily understood. We have few measurements of faint objects, accounting for the right-hand edge. This is a common sort of selection effect. Intrinsically bright quasars are very rare, creating the left-hand boundary. There is a lower limit on the redshifts near the line for bright ellipti-

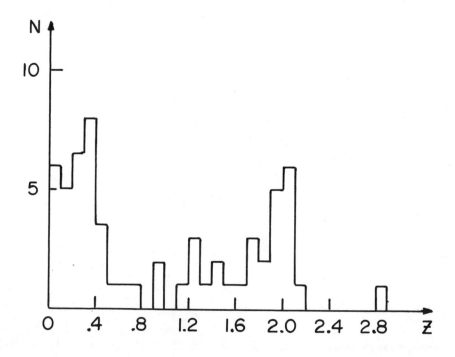

Figure 4.4  Redshift distribution of radio-quiet quasars.

cal galaxies, which Sandage (1971) has interpreted as evidence for quasars as nuclei of galaxies, so that only those objects which are at least as bright as galaxies are seen as stellar. Quasars which are faint enough to show the underlying galaxy have commonly been reclassified as, e.g., N galaxies (Arp, 1970). However, the cutoff at $z \simeq 2$ is not understood. For instance, if quasars have a maximum intrinsic luminosity, the triangle $\Delta$ in Fig. 4.2 should be populated. It could be that there is a strong selection effect which prevents our finding many quasars with larger redshifts. We will return to this question below.

Further selection effects are indicated by the redshift distribution of 208 quasi-stellar radio sources (Fig. 4.3), and of optically-identified QSO (Fig. 4.4). The distribution of the QSS is smooth, with a decrease near $z = 0$ due presumably to their rarity, and a decrease for large $z$ due to their being too faint to be easily observed. However, for radio-quiet QSO, the distribution shows a number of bumps whose origin is obscure. These objects are usually found by their excess emission in

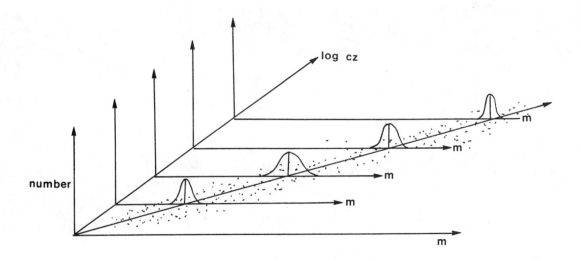

Figure 4.5 Derivation of the Density Function in the Ideal Case.
The density function is found from the number of objects between z and z + dz; that is, by integrating the n(m) curves at each z. The redshift distance and the n(m) curves also define the intrinsic luminosity function at each z, i.e., the distribution of quasars with luminosity.

the ultraviolet, as compared to ordinary stars. Perhaps we tend to select particular ranges of redshift when colours are used as search criteria. For instance, if a particularly strong emission line lies in the ultraviolet U bandpass ($\sim$ 3250 - 3950A°), then the object will have a larger UV excess, and will be discovered in a search for blue objects. At higher redshift this line would be shifted out of the bandpass, and the UV excess might largely disappear.

Incidentally, this may account for the cutoff in redshifts at z $\simeq$ 2. The Lyman limit enters the U bandpass near z = 2.5. If the gas surrounding the quasar is optically thick in the Lyman continuum, no light will be present blueward of the limit. The optical searches would then fail to find high-redshift quasars, because they no longer had UV excesses.

In view of the worse selection effects in optical identifications, only radio sources are used in the luminosity-volume test. This is unfortunate, since they are so much rarer than the radio-quiet quasars.

2. The Luminosity - Volume Test (Schmidt, 1968, 1972 a, b)

In the ideal case, in which one was certain that all sufficiently bright quasars were identified, one could look at the distribution of objects with magnitude at fixed redshift i.e., distance). When this is done for a number of redshifts, as shown in Fig. 4.5, a distance-dependent luminosity function as well as a density function are derived.

This cannot be done here, because there are two selection effects governing which QSO appear in a sample of optically-identified radio sources. First, there is a limiting radio flux $S_{min}$, below which the source will not appear in the radio catalogue. Second, given a radio source with $S > S_{min}$, there is a limiting optical magnitude $m_{max}$, such that fainter sources ($m > m_{max}$) will not have been optically identified. The following method has been devised by Schmidt (1968) for dealing with this situation. A value $q_o = 1$ will be assumed throughout.

First the distance R to the quasar is found from its redshift, properly accounting for space curvature. Next, the spectrum is used to correct the radio and optical fluxes to a standard emitted frequency or wavelength. The standards adopted were 500 Mhz and 2500A°, respectively. Two distances $R_{max, radio}$ and $R_{max, op}$ can now be computed. These are the maximum distances at which the source would still appear in the radio and optical catalogues. Thus, at $R_{max, radio}$ the source would have an observed flux of $S_{min}$. The volume $V_m$ interior to the smaller of the distances then gives the volume within which the quasar could have been seen. The volume V interior to its actual distance is given by R, and the ratio $V/V_m$ is computed for each quasar.

Now, if quasars are uniformly distributed within the volumes $V_m$ available for study, the $V/V_m$ distribution should be a constant function between 0 and 1, with a mean of 0.5. Since there are very little data, only this mean value is considered. Then if $<V/V_m> \neq 0.5$, the data indicate a variation of number

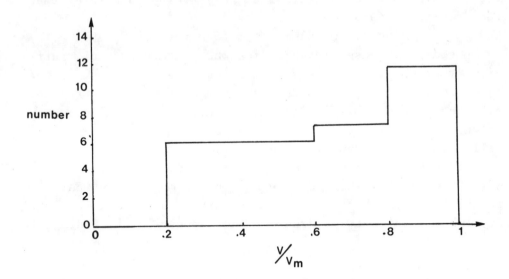

Figure 4.6 Frequency distribution of $V/V_m$ values for the sample of 4C quasars. Adapted from data given in Lynds and Wills (1972).

density with z. Note that the method is independent of any luminosity variations among quasars: $V$ and $V_m$ depend on luminosity, but $V/V_m$ does not.

Two different statistical samples of QSO have been studied; Schmidt (1968) used the identified 3C quasars, numbering 33 brighter than the assumed visual limit of 18.4 mag, and the radio limit of $S_{178} \geq 9$ flux units. It was found that

$$< V/V_m > = 0.70 \pm 0.05$$

This means that the number density increases with distance. In fact, Schmidt found that one could fit the increase with $a(1 + z)^6$ law in comoving coordinates. He emphasized, however, that this was only one of many possible functional forms, all of which reproduce the above dramatic rise in density.

The second sample consists of 31, 4C quasars in a .35 steradian area of sky, and was studied by Lynds and Wills (1972). In this case the radio flux limit was $S_{178} > 2$ f.u. and the optical limiting magnitude was $m_{vis, op} = 19.4$. Lynds and Wills

# Extragalactic Observational Astronomy

found that

$$\langle V/V_m \rangle = 0.67 \pm 0.05$$

in agreement with Schmidt's result. A histogram of the distribution is given in Fig. 4.6, showing again the strong increase with redshift. Thus, provided that quasar redshifts are cosmological in origin, the $V/V_m$ test implies an increase in number density with redshift that can be described by

No. density $\propto (1+z)^6$.

We can repeat the above discussion assuming that quasars are local. Space is then approximately flat, and the redshifts are used only to correct the fluxes to their values at the standard wavelengths. It is no longer necessary (or even possible) to know the distance; $V/V_m$ is computed from the fluxes alone.

$$V/V_m = \mathrm{Max}\left\{(S_{radio}/S_{max,radio})^{-3/2}, (S_{optical}/S_{max,optical})^{-3/2}\right\}$$

with obvious notation. This leads to:

$\langle V/V_m \rangle = 0.57 \pm 0.05$ for the 3C sample, and

$\langle V/V_m \rangle = 0.51 \pm 0.05$ for the 4C objects.

Thus, if quasars are local, the data are consistent with a uniform distribution with distance.

The $V/V_m$ test is related to the log N − log S curve in the following way. Suppose $S_0$ is the intrinsic flux of all the sources, so that the observed flux of a source at R is proportional to $S_0/R^2$. Since the number of sources with flux greater than S is $N(>S) \propto R^3$, we have

$$N(>S) \propto S^{-3/2} = S^{-\beta} \qquad (4.1)$$

This same relation holds if different sources have different flux. Thus the log N − log S curve should have a slope of −1.5. The observed value is near −1.8. This also suggests that the underlying assumption of uniform number density is incorrect.

In fact $\langle V/V_m \rangle$ can be related to $\beta$:

$$< V/V_m > = < (S/S_{limiting})^{-3/2} > .$$

But

$$< S^{-3/2} > = \frac{\int_{S_{limiting}}^{\infty} \frac{dN(S)}{dS} S^{-3/2} dS}{\int_{S_{limiting}}^{\infty} \frac{dN(S)}{dS} dS}$$

With $N(S) \propto S^{-\beta}$, this reduces routinely to

$$< V/V_m > = \beta/(\beta + \frac{3}{2}) . \qquad (4.2)$$

Note that if $< V/V_m > = 0.5$, $\beta = 1.5$ as required. If $< V/V_m > > 0.5$, then $\beta$ becomes greater than 1.5, as is observed. However, the amounts are not in agreement: $< V/V_m > = 0.7$ implies $\beta = 3.5$, much larger than the value observed. Finally, we note the following alternative interpretation of the quasar cutoff at $z = 2$ (Schmidt, 1972 b). Assume the QSO are cosmological, that they follow locally a density law proportional to $(1 + z)^6$, and that they are a stage in the evolution of bright elliptical galaxies. Then the density law must break down near $z = 2$, because at this distance the density of quasars has risen to equal that of elliptical galaxies. Since all ellipticals are then already quasars, no further increase with z is possible. According to Schmidt (1972 a, b), a dependence like $(1 + z)^0$ at $z \gtrsim 2.5$ is consistent with the observations. The major difficulty with this picture is that we must explain why elliptical galaxies all become quasars at the same time, long after galaxies were formed ($z \simeq 20$, see lectures given by K. Brecher).

# Extragalactic Observational Astronomy

Lecture 5

## PROBLEMS INVOLVING CLUSTERS AND GROUPS OF GALAXIES

The main problem in understanding clusters of galaxies has already been mentioned in Lecture 1, C) 4.: the virial mass-to-light ratios of clusters as a whole are larger than the mass-to-light ratios of their member galaxies. The discrepancy grows larger with the size of the cluster considered. No generally accepted theoretical explanation of this situation is available.

A. Rich Clusters of Galaxies

1. The Coma Cluster:

This is a well-known rich cluster of about a thousand galaxies, whose main concentration has a diameter of 1°. The redshift corresponds to a velocity of $\sim$ 6000 km/sec. The central region appears to be dominated by two giant elliptical galaxies, NGC 4869 and NGC 4874. The cluster is strongly centrally condensed. As in all rich clusters, the members of Coma are mostly E and S0 galaxies, and the distribution is rather smooth and symmetrical.

The mass-to-light ratio M/L is determined by assuming the cluster is in dynamical equilibrium, and applying the virial theorem. The kinetic energy is estimated by measuring a sample of radial velocities, there being far too many galaxies to allow a complete survey. In both the Coma and the Perseus clusters (see below), about 100 velocities are available. The dispersion $\sigma_R$ in the radial velocities is calculated, and amounts typically to $\sim$ 1000 km/sec. The kinetic energy is then:

$$<T> \simeq \frac{3}{2} M \sigma_R^2 ,$$

where M is the total mass of the cluster. The potential energy is obtained from the light distribution by assuming mass and light are proportional. Then

$$<\Omega> = f \frac{GM^2}{R_c} ,$$

where G is the gravitational constant, $R_c$ is some critical length at which the number density has fallen by a fixed factor, usually near 2 or 3, and f is a pure number, determined from the light distribution, which describes the central concentration.

The virial equation is

$$3\sigma_R^2 + f \frac{GM}{R_c} = 0,$$

which is easily solved for M.

For Coma we find a total mass of $10^{15}$ $M_\odot$, and a mass-to-light ratio of about 250 in solar units. The individual galaxies range in M/L from 50 to about 5. Here, as in all rich clusters, a discrepancy in the range 2 to 10 has occurred.

The first possible interpretation of this is that the cluster is bound, but that the main part of the mass is invisible. Among the possibilities proposed for the "missing mass" are black holes, intergalactic stars, and diffuse intergalactic matter such as dust or gas. Photometric scans across the two bright galaxies in Coma show that the light level does not fall to zero between them. Thus, either the halos overlap, or they share a large cloud of "intergalactic stars". Faint stars with large M/L may account for the discrepancy, at least in this case.

Alternatively, it is conceivable that the cluster is not bound, but is expanding at $\sim$ 1500 km/sec. This view suffers from the difficulty that the dispersion timescale is only $10^8$-$10^9$ years, much shorter than the age of the universe. If clusters dispersed with this speed, it would be very difficult to understand how so many of them could have survived to the present epoch.

It is not possible to invoke a sufficiently high mass for the dominating ellipticals to make them bind the cluster. Studies of a group of galaxies apparently bound gravitationally to NGC 4874 show that it has a mass of $\sim 10^{13}$ $M_\odot$. This is a normal value for a cD galaxy, and implies a normal mass-to-light ratio. We also obtain the same result if we assume the two central galaxies form a bound pair, and interpret their difference in velocity of $\sim$ 700 km/sec as orbital motion.

Thus we are left, in the Coma cluster, with a missing mass problem with no ready solution. This situation is typical of all rich clusters.

2. The Perseus Cluster:

The Perseus cluster is perhaps even more interesting than Coma, because we have more information about it, particularly in the X-ray and radio frequency range. It is rather irregular in shape, having a concentration of galaxies along a slightly curved line in the sky. Nevertheless, most of the galaxies are S0 or elliptical. The average distance between galaxies is of order 1 Mpc. At the point of highest galaxy concentration lies NGC 1275, a radio Seyfert which is also an infrared and X-ray source. It appears to be the seat of a violent explosion, as shown most directly by photographs due to Lynds (1970).

The Perseus cluster has been studied in detail by Chincarini and Rood (1971), and by Gunn and Sargent (unpublished). Their conclusions agree on the following main points. The velocity dispersion is approximately 1500 km/sec, implying a virial mass of $10^{15}$ $M_\odot$. This corresponds to $\sim 100$ giant elliptical galaxies, many more than the cluster is observed to contain.

There are other indications that the mass might be high. Radio maps by Ryle and Windram (1968) show that NGC 1265 and IC 310 have radio tails pointing away from NGC 1275. This suggests some sort of interaction, perhaps with energetic particles from the explosion in NGC 1275, and thus supports the view that NGC 1265 is a member of the cluster. But the velocity of NGC 1265 is 2700 km/sec higher than that of the cluster. If this galaxy were to be gravitationally bound, the cluster mass would have to be $\sim 10^{15}$ $M_\odot$.

Since the Perseus cluster is very elongated, we might ask whether the redshift distribution in space suggests the cluster is flying apart, or at least rotating very rapidly. However, there is no observed correlation between redshift and position of the galaxy.

Both Perseus and Coma are diffuse X-ray sources. Two emission mechanisms have been proposed:

i) <u>Inverse Compton Effect</u>: The clusters are diffuse emitters of low frequency radio radiation, which may be due to relativistic electrons ejected by the point radio sources. A **magnetic** field of $10^{-2}$ Γ would suffice to give the observed radio emission. The same electrons would then interact with the low-energy photons in the 2.7°K background, and produce X-rays by the inverse Compton effect. If this were true, it would lead to an enormous increase in our knowledge of the cluster magnetic field, and the spectrum and energy density of the relativistic electrons. But it would do nothing to alleviate the missing mass problem.

ii) The X-ray emission might be <u>thermal bremsstrahlung</u> in a gas of temperature $\sim 70 \times 10^6$ K (see, eg., Gunn and Gott, 1972). The required diffuse material might have fallen into the cluster from outside, or been expelled by explosion in cluster galaxies, or it might just be remnant material that never formed into galaxies. If it was not already hot, infall and the motion of galaxies at $\sim 2000$ km/sec through the gas would produce shock waves that would heat the gas rather rapidly. This might also account for the observation that relatively few galaxies in either the Coma or Perseus clusters show emission lines in their nuclei. In Perseus, only two galaxies in eighty, and in Coma, one in a hundred, show (O II) $\lambda$ 3727. In contrast, 17% of field galaxies have emission lines. In the above view, the galaxies might be stripped of this gas by ram pressure from the resisting medium through which they are moving. There is thus some slight supporting evidence for the thermal bremsstrahlung mechanism.

B) Small Groups of Galaxies

The mass-to-light ratios of small groups of galaxies present similar but less severe problems of missing mass. In this case, there are few enough galaxies so that all can be measured. If the group is in equilibrium, the usual virial theorem is applied. Otherwise, if the total energy $E = T + \Omega$ is positive, the group will disperse on a timescale of the order of its size divided by its velocity dispersion.

In applying the virial theorem, let M be the mass of the cluster, and $M_i = \mu_i M$ the mass of the $i^{th}$ galaxy. Then

$$T = \frac{1}{2} M \sum_i \mu_i v_i^2 ,$$

where $v_i$ is the $i^{th}$ velocity, and

$$\Omega = -GM^2 \sum_{i \neq j} \frac{\mu_i \mu_j}{r_{ij}} ,$$

$r_{ij}$ being the separation of galaxies i and j. However, we ob-

Extragalactic Observational Astronomy    505

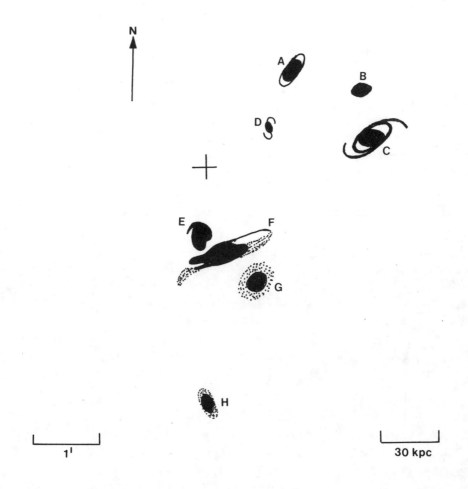

Figure 5.1  Sketch of VV 282 (Arp 320); see Arp (1966) for photograph.

serve the projected separation $r'_{ij}$, and the radial velocity $v_{Ri}$, not the quantities $r_{ij}$ and $v_i$. In the case of spherical symmetry,

$$T = \frac{3}{2} M \sum_i \mu_i v_{Ri}^2 \; , \text{ and}$$

Table 5.1

Mean M/L Ratios for Clusters of Various Sizes

| Sample | M/L |
|---|---|
| 6 rich clusters | 841 ± 168 |
| 9 poor clusters | 446 ± 102 |
| 29 small groups | 331 ± 96 |
| 11 triple systems | 85 ± 28 |

$$\langle 1/r_{ij} \rangle = \frac{2}{\pi} \langle 1/r'_{ij} \rangle .$$

The virial equation then becomes

$$3M \sum_i \mu_i v_{Ri}^2 - \frac{2G}{\pi} M^2 \sum_{i<j} \mu_i \mu_j \langle 1/r'_{ij} \rangle = 0. \tag{5.1}$$

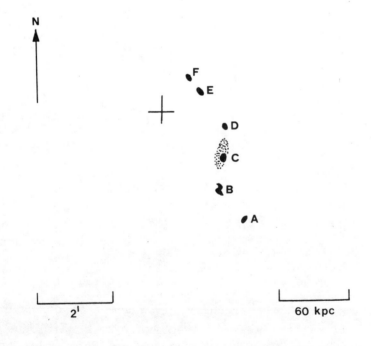

Figure 5.2  Sketch of Arp 330; see Arp (1966) for photograph.

Table 5.2

Redshifts of Galaxies in VV 282

| Galaxy[a] | $V_R$ (km/sec) | Type | $m_p$ | $M_p$ |
|---|---|---|---|---|
| A | 8913 | Sa | 15.2 | −20.4 |
| B | 9337 | So? | 15.5 | −20.1 |
| C | 9161 | SBb | 15.3 | −20.3 |
| D | −    | Sc  | 16.5 | −19.1 |
| E | 9048 | So? | 15.4 | −20.2 |
| F | 8608 | SBa? | 15.1 | −20.5 |
| G | 8935 | E   | 15.2 | −20.4 |
| H | 9515 | E   | 15.9 | −19.5 |

(a) the galaxies are identified in Fig. 5.1

---

Finally, the galaxies are weighted according to mass by assuming the same M/L for all, and assigning weights $\mu_i$ proportional to the apparent luminosity. Thus M is calculated from the measured luminosities, radial velocities and projected separations.

Table 5.1, due to Karachentsev (1966) shows that the M/L ratios decrease with decreasing richness of the cluster. But at the same time the member galaxies change from elliptical to spiral, and thus have much lower M/L themselves. Thus, while the difficulties become less acute as richness decreases, they still remain large.

Let us consider a few typical examples of small groups of galaxies.

1. VV 282[1] = Arp[1] 320: This group is sketched in Fig. 5.1, and

---

1. These designations refer respectively to a catalogue by Vorontsov-Velyaminov (1959), and to Arp's (1966) Atlas of Peculiar Galaxies.

Table 5.3

Redshifts and Magnitudes for Arp 330

| Galaxy[a] | $V_R$ (km. sec.$^{-1}$) | Type | $m_p$ | $M_p$ |
|---|---|---|---|---|
| A | 9093 | E   | (16)   | (−19.5) |
| B | 8237 | Sba | (16)   | (−19.5) |
| C | 8948 | E   | 15.5   | −20.0 |
| D | 9170 | E?  | (17)   | (−18.5) |
| E | 8831 | E?  | (17)   | (−18.5) |
| F | 7934 | E?  | (17.5) | (−18) |

(a) The galaxies are identified in Fig. 5.2

---

the galaxy redshifts are given in Table 5.2 (Burbidge and Sargent, 1971). The mass-to-light ratio is 74. The galaxies are mostly spirals, however, and so presumably have much smaller mass-to-light ratio.

2. VV 150 is a chain of four galaxies, having a mean radial velocity of 8080 km/sec, and a mass-to-light ratio of 38. Again, three of the four galaxies are spirals.

3. VV 169 is particularly interesting because it consists of one E galaxy and three smaller objects that seem to be strongly tidally interacting. Again the group is a chain, but assuming spherical symmetry leads to a M/L of 90.

4. Arp 330 is illustrated in Fig. 5.2. It was noted as being a prominent chain long before any redshifts were measured. The probability of the six galaxies forming such a configuration by chance was estimated to be one in $10^7$. Since then, the velocities have all been measured (Table 5.3), and yield a velocity dispersion too large for the system to be bound. For an edge-on plane distribution of matter, an M/L ratio of 3000 would be required to bind the galaxies. Since this is clearly unreasonable, the group is probably dispersing, on a timescale of $\sim 10^8$ years. The alternatives seem to be to blame another body, which has all the mass and none of the light, or to consider

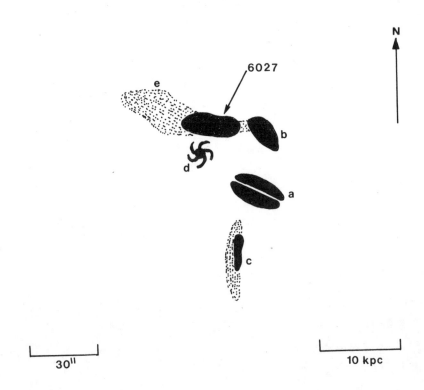

Figure 5.3  Sketch of Seyfert's Sextet, showing scale and identification of members.

galaxies B and F as being a separate pair closer than the other galaxies, or to invoke intrinsic redshifts, as Arp does. None of these are very attractive.

5. Seyfert's Sextet (Fig. 5.3) is one of the best examples of a very compact group of galaxies. Baade and others (Burbidge and Sargent 1971) have tried to decide on the basis of the visual appearance, and again before redshifts were measured, whether this could be a chance coincidence. One of the conclusions was that the probability of one of the galaxies being in the background was a part in a few hundred. Since then, galaxy d has been found to have a greatly-discrepant redshift, almost 20,000 km/sec as compared to an average of 4400 for the others. In view of the obvious gravitational interaction of the other galaxies, in which galaxy d does not take part, and because of its discordant redshift, galaxy d is suspected of being in the background. However, if it is cosmologically distant, its absolute magnitude

## Table 5.4

### Redshifts and Magnitudes for Seyfert's Sextet

|         | $V_R$ (km. sec.$^{-1}$) | Type       | $m_p$ | $M_p$       |
|---------|-------------------------|------------|-------|-------------|
| NGC 6027| 4468                    | Sa or So   | 14.7  | -19.4       |
| a       | 4141                    | Sa         | 15.1  | -19.0       |
| b       | 4430                    | Sa or So   | 15.3  | -18.8       |
| c       | 4581                    | Sbc        | 15.6  | -18.5       |
| d       | 19930                   | Sc         | 16.0  | -21.3 [a]   |
| e       | -                       | Irr. [b]   | 16.5  | -17.6       |

(a) Absolute magnitude calculated on the assumption that "d" has the distance corresponding to its redshift

(b) Component "e" is probably not a separate galaxy but matter tidally displaced from NGC 6027.

---

is -21.3, making it an ScI, contrary to the short and stubby appearance of its arms. The explanation of these observations, then is still not known.

The M/L ratio of Seyfert's Sextet, if galaxy d is rejected, is 49, not much different from that of the E galaxies.

Finally, the extreme compactness of the group is worthy of note. Using a redshift of $\sim$ 4400 km/sec, the mean three-dimensional separation of the galaxies is only 17 kpc. Thus the apparent tidal interaction is entirely to be expected.

Table 5.5, then, summarizes the results on mass-to-light ratios of small groups. Since these groups consist primarily of spiral galaxies with small M/L, the severity of the missing mass problem is amply illustrated.

## Table 5.5

### Groups and Chains Observed by Seyfert

| Name | Configuration (a) | $V_R$ km sec$^{-1}$ | Scale; pc. per second of arc | Mean 3-dim. separation (kpc) | Virial Theorem Mass (suns) | $M/L_p$ |
|---|---|---|---|---|---|---|
| NGC 833 group | G | 3885 | 248 | 89 | $4 \times 10^{11}$ | 10 |
| VV 161 | C | 8447 | 546 | - | - | - |
| VV 169 | C | 9087 | 626 | 37.5 | $4.0 \times 10^{12}$ | 90 |
| VV 144 | C | 6365 | 402 | - | $1.2 \times 10^{11}$ | (29) |
| VV 150 | C | 8080 | 525 | 34 | $1.3 \times 10^{12}$ | 38 |
| VV 282 | G | 9074 | 582 | 141 | $9 \times 10^{12}$ | 74 |
| VV 165 | C | 12742 | 812 | 70 | $2.4 \times 10^{12}$ | 42 |
| VV 159 | C | 10475 | 674 | - | - | - |
| Seyfert's Sextet | G | 4282 | 278 | 17 | $5.2 \times 10^{11}$ | 49 |
| Arp 330 | C | 8702 | 553 | 96 | $2.3 \times 10^{13}$ | 5560 |
| VV 197 | G | 11861 | 755 | 66 | $5.8 \times 10^{12}$ | 211 |
| VV 101 | G | 8165 | 510 | 104 | $1.2 \times 10^{13}$ | 87 |
| VV 208 | C | 7932 | 504 | 61 | $2.5 \times 10^{12}$ | 36 |

(a) G = Group; C = chain

Acknowledgements:
  It is a pleasure to thank Dr. Allan Sandage for discussions, and for furnishing most of the slides shown in these talks.

Appendix

SPECTRAL LINES IN GALAXIES

In this Appendix we will give a review of spectral lines observed in galaxies, expecially those used to determine redshifts. A similar discussion of QSO spectra is given by E. M. Burbidge.

A. Absorption Lines

Most galaxies are found to have absorption spectra. The lines are those commonly found in moderately cool stars, but have been broadened by stellar motions in the galaxy. These motions consist not only of random velocities, but also of an integration through the galaxy of the rotation velocity components in the line of sight of the slit.

The main lines are listed in Table A.1. The ionization state of the radiating atom is indicated by a roman numeral, equal to one plus the ionic charge. Normally these are all the lines that are seen, because of the Doppler broadening, and because most galaxies are so faint that low resolution spectrographs are necessary.

Table A.1

Major Absorption Lines in Galaxy Spectra

| Element and Line | Wavelength (A°) |
|---|---|
| Ca II  H and K | 3933, 3968 |
| Ca I | 4226 |
| G band, a blend of CH and CN | 4303 |
| $H\gamma$ | 4340 |
| $H\beta$ | 4861 |
| Mg I b lines, a blend of 3 | 5169 |
| Na I D lines | 5890, 5896 |
| $H\alpha$ | 6562 |

## Table A.2

### Strongest Emission Lines Observed in Galaxies

| LINE | WAVELENGTH (A°) |
|---|---|
| (O II) | 3727 |
| (Ne III) | 3869, 3968 |
| He I | 3889 |
| Hε | 3970 |
| (S II) | 4068, 4076 |
| Hδ | 4101 |
| Hγ | 4340 |
| (O III) | 4363 |
| He I | 4471 |
| Hβ | 4861 |
| (O III) | 4958 |
| (O III) | 5007 |
| He I | 5875 |
| Hα | 6562 |
| (N II) | 6548, 6583 |
| (S II) | 6716, 6730 |

However, in many cases, particularly in faint galaxies, only the very strong H and K lines of calcium are visible. Worse, at moderate velocities, one of these lines is redshifted onto the prominent λ4046 emission line due to the night sky. Then the redshift determination rests on only one line.

Nevertheless, the precision of most radial velocities is near ± 100 km/sec, entirely adequate when the velocity is thousands of km/sec.

B. Emission Lines

Emission lines are observed in those galaxies that contain H II regions. This includes especially the irregular galaxies,

## Table A.3

### Additional Lines Shown by Peculiar Galaxies

| LINE | WAVELENGTH (Å) |
|---|---|
| (Ne V) | 3444, 3425 |
| He II | 4686 |
| (O I) | 6300 |
| (Fe V) | ⎫ |
| (Fe VII) | ⎬ sometimes |
| (Fe X) | ⎭ |

and the arms of spiral galaxies. Table A.2 lists the strongest observed emission lines. Brackets indicate forbidden lines.

In measuring radial velocities, emission lines are always preferred over absorption lines, because they are narrower and more intense (more intense than the continuum, not less!).

### C. Peculiar Galaxies, Seyfert and Radio Galaxies

Many normal galaxies have weak (O II) and (N II) emission from gas in their nuclei. Much stronger emission is observed from the nuclei of Seyfert and radio galaxies, particularly N-type galaxies. Moreover, they often show a greater range in excitation, which cannot be attributed to a gas with a unique temperature, say of $T \sim 10^4$ °K. A range of temperatures is necessary.

These peculiar galaxies often show, in addition to the lines of Table A.2, lines listed in Table A.3. Most of these are due to highly ionized species.

The lines Hβ $\lambda 4861$ and (O III) $\lambda 4959, 5007$ form a characteristic and easily recognized pattern (Fig. A.1). This makes line identifications easy at low redshifts. But at intermediate z, they are moved out of the usual sensitivity range of the spectrogram. The strongest line left is then the $\lambda 3727$ line of (O II). Whenever a spectrum shows only one emission line, it is assumed to be (O II) $\lambda 3727$.

We have listed the major lines used in determining galaxy redshifts. The measurement then consists of finding the shift in position between the observed lines and their rest positions,

Extragalactic Observational Astronomy 515

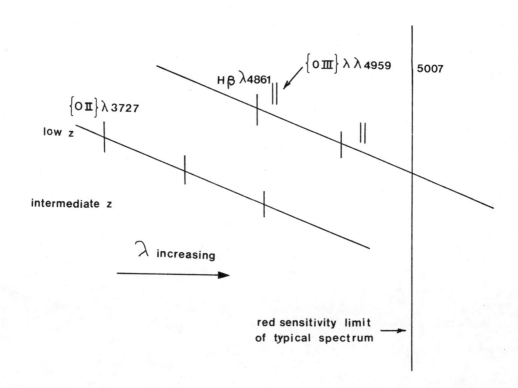

Figure A.1 Appearance of the Most Prominent Emission Lines as a Function of Redshift

using comparison spectra on the plate. This is routinely converted into a fractional wavelength shift z. Frequently only a few lines can be used. However, it should be noted that different observers almost always agree on the derived redshift.

## REFERENCES

Arp, H., 1966, Ap. J., Suppl., 14, 1.

Arp, H., 1970, Ap. J., 162, 811.

Abell, G. O., 1958, Ap. J., Suppl., 3, 211.

Abell, G. O., 1972, in "External Galaxies and Quasi-Stellar Objects", eds. D. S. Evans, D. Wills, B. J. Wills (Dortrecht-Holland: D. Reidel Pub. Co.) 341.

Baade, W., 1945, Ap. J., 102, 309.

Bergh, S. van den, 1960 a, Ap. J., 131, 215.

Bergh, S. van den, 1960 b, Ap. J., 131, 558.

Burbidge, E. M., Sargent, W. L. W., 1971, in "Study Week on the Nuclei of Galaxies", ed. D. J. K. O'Connell (Amsterdam: North-Holland Pub. Co.) 351.

Chincarini, G., Rood, H. J., 1971, Ap. J., 168, 321.

Dennison, E. W., 1954, unpublished PhD Thesis, Univ. of Michigan.

Gott, J., Gunn, J., 1971, Ap. J. (Letters), 169, L 13.

Gunn, J. E., 1971, Ap. J. (Letters), 164, L 113.

Gunn, J. E., Gott, J. R., 1972, Ap. J., 176, 1.

Hubble, E., 1926, Ap. J., 64, 321.

Hubble, E., 1929, Proc. Nat. Acad. Sci., 15, 168.

Hubble, E., 1936, "The Realm of the Nebulae" (New Haven: Yale Univ. Press).

Hubble, E., 1947, Pub. A. S. P., 59, 153.

Karachentsev, I. D., 1966, Astrophysics, 2, 39, (Astrofizika, 2, 81).

Kowal, C. T., 1968, A. J., 73, 1021.

Kuiper, G. P., Struve, O., Stromgren, B., 1937, Ap. J., 86, 570.

Leavitt, H., 1912, Harvard College Obs. Circ., No. 173.

Lynds, R., 1970, Ap. J. (Letters), 159, L 151.

Lynds, R., Wills, D., 1972, Ap. J., 172, 531.

Mattig, W., 1958, Astron. Nachr., 284, 109.

Morgan, W. W., 1958, Pub. A. S. P., 70, 364.

Oemler, A., Gunn, J. E., Oke, J. B., 1972, Ap. J. (Letters), 176, L 47.

Oke, J. B., Sandage, A., 1968, Ap. J., 154, 21.

Oort, J. H., 1958, Inst. de Phys. Solvay, Brussels, 2, 163.

Page, T., 1965, Smithsonian Inst. Astrophys. Obs., Special Report, No. 195.

Peterson, B. A., 1970, Ap. J., 159, 333.

Peterson, B. A., 1970, A. J., 75, 695.

Ryle, M., Windram, M.D., 1968, M. N. R. A. S., 138, 1.

Sandage, A., 1961, "The Hubble Atlas of Galaxies" (Washington: Carnegie Institution of Washington).

Sandage, A., 1961, Ap. J., 133, 355.

Sandage, A., 1968, Observatory, 88, 91.

Sandage, A., 1971, in "Study Week on the Nuclei of Galaxies", ed. D. J. K. O'Connell (Amsterdam: North-Holland Pub. Co.) 271.

Sandage, A., 1972 a, Ap. J., 178, 1.

Sandage, A., 1972 b, Ap. J., 178, 25.

Sandage, A., Tammann, G., 1968, Ap. J., 151, 531.

Schmidt, M., 1965, in "Galactic Structure", eds. A. Blaauw, M. Schmidt (Chicago: Univ. of Chicago Press) 513.

Schmidt, M., 1968, Ap. J., 151, 393.

Schmidt, M., 1972a, Ap. J., 176, 273.

Schmidt, M., 1972b, Ap. J., 176, 289.

Schmidt, M., 1972c, Ap. J., 176, 303.

Schwarzschild, M., 1958, "Structure and Evolution of the Stars" (Princeton: Princeton Univ. Press).

Scott, E. L., 1957, A. J., 62, 248.

Sèrsic, J. L., 1960, Z. fur Ap., 50, 168.

Tammann, G. A., Sandage, A., 1968, Ap. J., 151, 825.

Tinsley, B. M., 1972, Ap. J. (Letters), 173, L 93.

Vaucouleurs, G. de, 1959, Handb. der Physik, 53, 311.

Vorontsov-Velyaminov, B. A., 1959, "Atlas and Catalogue of Interacting Galaxies" (Moscow: Sternberg State Astron. Institute).

Westerlund, B. E., Wall, J. V., 1969, A. J., 74, 335.

General Relativity, Collapse and Singularities

Lectures by

J. A. Wheeler

Princeton University

Notes by

H. G. Hughes, N. Snyderman, J. K. Lawrence, R. K. P. Zia

(Lecturer unable to read and check these notes.)

## Lecture I

## GRAVITATIONAL COLLAPSE AND BLACK HOLE PHYSICS

Philosophical Introduction and Outline

In this series of lectures, I would like to talk about gravitational collapse at several levels. I do not know any issue in physics of our time or any other time which presents a greater crisis than gravitational collapse. One has the prediction, according to Einstein's 1915 theory of general relativity (field equations themselves plus Einstein's "boundary condition" that the universe is closed), that the universe starts from small dimensions, expands to a maximum, recontracts and collapses to a point where calculations are impossible. Yet, we are convinced that physics goes on even if the mathematics fails. So, the paradox is what comes after this collapse and what does this collapse have to do with physics as we know it. I believe that in this crisis, we are forced into the central issue of what the structure of matter (particles) is.

A Deeper Level of Structure:

Gravitational collapse proceeds at three levels: the universe, black holes, and quantum fluctuations in the geometry of space. Quantum fluctuations occur at the level of $10^{-33}$ cm. On the other hand, black holes of solar masses are of the order of 3 km in radius. Physics and astrophysics have taken us down to the dimensions of neutron stars ($\sim$ 30 km) so far, so that we have only one more magnitude of 10 to go before we get to black hole physics. Black holes will let us check out a number of the important ideas that we have about gravitational collapse.

Besides these mechanistic aspects of the question, we also have the "principles" side. Newton dealt with the problem of motion by cutting it into two pieces: one of which has to do with the nature of space and time, and the other, the equations of motion of particles. Although he was aware of the shortcomings of his notions of space and time, he was courageous enough to accept these and got on with the other half of the problem: describing motion of particles. However, his contemporaries were quite critical of the idea of absolute space and time, no one more so than Leibnitz.

This is the same Leibnitz who gave us the principle of least action, calculus, the idea that logic is at the center of physics, the concept of "the best of possible worlds", and the concept that everything at one moment of time is related to everything that has gone on before and afterwards. Of course, the principle of least action directly expresses the idea of going from A to B in terms of an integral. One speaks of comparing integrals along paths from A to B and the one history which extremizes this integral is "the best of all possible worlds" or the dynamically acceptable history. Further, the idea that that history also satisfies a local differential equation, a connection between a principle in the large and a principle in the small, is a central topic in physics today. The understanding of this principle of least action goes back to the quantum principle: the idea that for every history (H) there is a dynamical path length ($I_H$), which, when divided by t, gives us the phase of the partial probability amplitude to go from A to B along H. Although this idea of "democracy of all histories", that all histories have equal footing, disturbed Bohr, it is precisely this idea that is central to Bohr's exposition of the quantum theory through the famous two slit experiment. In the slit experiment, it is the democracy of two histories one going through one slit, one through the other slit. In the free motion of a particle (no slits!) from A to B, the sum over histories involves an infinite number of histories. The miracle is, of course, that histories which differ significantly from the classical path do not contribute, due to destructive interference. Only those histories which deviate little from the classical path ($I/\hbar \simeq 1$ radian) contribute significantly. So, here we have "the principle of best of all histories". The quantum mechanics of today has this close tie to the idea of Leibnitz.

The reason I bring Leibnitz into the story and tie him in to the two overarching principles of physics of the twentieth century, the quantum principle and general relativity, is because of his idea that logic is the primordial concept, lying ahead of any dynamical principles. For him, the categories of space and time, particle and motion are not the primordial categories but only approximations useful at the everyday level of analysis.

This idea has been restated by Sakharov. In dealing with the questions of gravitation physics, he said that it is a mistake to regard gravitation as a fundamental force. Instead, gravitation is no more than the metric elasticity of space. Let me be a little more specific.

If we look at a solid that is homogeneous and isotropic, then we need only two elastic constants to describe its properties. However, if we go down to one deeper layer of analysis, we discover that these elastic constants can be understood in terms of the hundred, or so different forces that act between atoms and molecules. On the other hand, the elastic constants reveal nothing about atoms and molecules. The idea of Sakharov is the same here. If we look at curved space geometry, we need only one constant, G, to describe the properties of gravitation. But, it is a misleading idea to regard curved space as everything. At a deeper level of analysis space is "stratified". If we "sock in" energy at a particular point in space, we cause particles to be produced. We see all the structure of the mass spectrum at this point. If we "sock in" energy at any other point in space or at any other moment of time, we get the same spectrum. Space really has much more structure than we can account for by a single constant! Sakharov points out that if we look at the sum of the zero point energy of these particles and fields, we get a divergent result which must be renormalized to 0. When we put the same particles and fields in a curved space, we alter all the momenta and energy. Now when we renormalize, we will no longer get zero but the result:

$$\delta I/\hbar \sim \left\{\int d^4 x \text{ (four-scalar curvature)}\right\}\left\{\int k dk\right\}$$

With a reasonable upper limit on the integral in the second factor $(1/\hbar^{1/2})$, the expression is finite. We obtain the action principle of general relativity, including the constant of gravitation. In a heuristic way, we have succeeded in producing gravitation out of particle physics, just as we can produce the elastic constants from atomic and molecular physics. So Sakharov reasons that gravitation and geometry are not primordial concepts in the construction of the universe.

Sakharov's point of view has further implications. The dream of Clifford and Einstein that somehow particles are to be made out of geometry can no longer be taken seriously. It is just as mistaken to try to build particles and fields out of geometry as build atoms and molecules from elasticity. This idea echoes Leibnitz, who feels that space and time are not the appropriate categories to use in describing physics. We should go to a deeper level.

In the case of atoms and molecules, deeper levels do exist: electrons, nuclei and Schrödinger's equation. Out of this simple foundation, we get all of chemistry. But we can study chemistry for a hundred years and not learn one thing about Schrodinger's equation! Each new feature of chemistry (van der Waal's forces, valence forces, ionic forces, homopolar forces) does not require for its description a new field or a new theory. All the richness of chemistry comes out of one single and very simple theory. In a similar sense, work hardening of metals, at the scale of 1 cm, can be understood in terms of dislocations, at the scale of $10^{-4}$ cm, which can in turn be understood as misalignment of atoms at the scale of $10^{-8}$ cm. One does not learn about dislocation from studying work hardening, nor atomic structure by examining dislocations under a microscope. The order of clarifications was the direct opposite: first atoms, then dislocations,then work hardening. And so, in the case of particles and fields, we need to go to a deeper level, which we will call pregeometry (Table 1.1) There should be some primordial structure, out of which we should derive and understand the nature of particles and fields. And then,following Sakharov, out of them, we understand gravity. But at this lowest level of discussion, "pregeometry", we expect to have to deal with an entity with <u>nothing</u> in it: no dimensionality, no space, no time! The traditional categories of physics are categories we can believe, that "lock us into" a wrong way of looking at things. To make progress, we have to find a deeper approach which goes back to something far more fundamental than such approximate ideas as space and time. To be concrete, Leibnitz was offering us long ago the calculus of propositions and logic as the fundamental starting point. At that level, there is no longer any division between mathematics, physics, and astrophysics.

Thoughts Toward Structure Without Structure:

The theme that I would like to speak on is gravitational collapse. The magnificent paradox confronting us is: What is

Table 1.1

| Elasticity of a solid | Work Hardening | "Metric Elasticity" (Sakharov) of Space |
|---|---|---|
| Two Elastic Constants | Work Hardening (1cm) | Spacetime (One "elastic constant": the constant of gravitation) |
| Atomic Bonds | Dislocations ($10^{-4}$ cm) | Zero Point Energy of Particles and Fields |
| Electrons, Nuclei and Schrödinger's Equation | Atoms ($10^{-8}$ cm) | Pregeometry |

---

the final state of gravitational collapse? What happens beyond the end of time? I certainly am not in any position to answer these questions. I simply want to put the issues before us. But I would like to propose one small step on the way toward sizing up what gravitational collapse has to tell us: the universe gets reprocessed from time to time after each collapse. In a new cycle of the universe, there will be a new set of constants, a new group of particles and a new nature of physics. The constants that we find ourselves within this cycle are the constants which permit life to occur.

Carter has examined the astrophysics of stars and finds that the luminosity of stars depends on $\alpha^{20}$ ($\alpha$ being the fine structure constant). Therefore a small change in this constant implies that a star like our sun would not exist. He offers us the idea that the particular value of $\alpha$ we have is one selected out of many cycles, each with a different value, by the requirement that we ourselves are here.

Dicke puts the idea in another way. To produce life, we need heavy elements, which depends on the nuclear reactions which take several billion years of cooking time in a star. According to general relativity, such a time scale demands a distance scale for the universe of several billion light years. So why is the universe as big as it is? Because we are here!

So, from these two very different viewpoints, we have this idea expressed of a universe so selected that we ourselves are present. This is very disturbing since we are used to the idea that we are observers of the universe "out there" rather than participants. But this disturbing feeling has shown itself before: when the beginning student first became acquainted with quantum mechanics. He learned that quantities such as position and momentum of, say, an atomic system <u>can not even be discussed</u> until he decides to measure one or the other. He eventually gets used to the idea that he is inescapably a participant: no participation, no physics! The idea here is simply that we ourselves are participants in, and have effects on, the entire universe; and there is no escape from this. We are saying that our participation is central in determining such numbers as $\alpha$ and the size of the universe. But once we started on this track, there's no stopping: even gravitation, electromagnetism, and other "fundamental" features and properties of the universe are structures organized into the universe by the very fact that we are here to look at them. This is the theme then: structure without structure.

Since we must agree that we simply don't know how to give a complete account along these lines today, we must preface this title and call our discussion "Thoughts toward structure without structure".

Superspace Description of the Dynamics of Geometry

How much and what can we do toward spelling out this line of thought ? Although we sound as if we have been very radical, the whole point is exactly the opposite: to hold fast to the principles which we know are sound. These are the quantum principle and general relativity.

But if we pursue the consequences of these principles systematically, we find that it is simply wrong to use the idea of space-time. This is the shock: "space" makes sense, "space-time" doesn't. These are only approximate ideas, classical ones, the use of which is not compatible with the quantum principle.

In brief, the argument is as follows: to predict the future of a dynamically evolving geometry describing the expanding universe, we have to have a present configuration of the 3-geometry and the present time rate of change of that configuration. But the uncertainty principle tells one that if one

knows the present configuration, it is absolutely excluded for
one to know the time rate of change of that configuration.
Only in a certain approximation, the classical one, can one do
this. So, it is wrong to speak of space-time and time. One is
excluded from knowing simultaneously, both the geometry and the
time rate of change of it.

The line of development that leads us into this is known
under the word of "superspace". To unite the two principles,
quantum and relativity, we have the superspace description of
the dynamics of geometry. The pay off is not so much the immediate astrophysical consequences as the larger point of view
that the idea of space-time with which we are familiar is simply
not the right idea. It is only an approximate one.

## Black Holes

Now that I have run through the general theme so that one can
see what the point of view is and why the subjects are being
taken up, let me get down to the subject originally listed:
black holes. We will summarize briefly the key points in black
hole physics.

There are two very different scenarios of production of black
holes: the individual collapse of a star and the coalescence of
stars in a galactic nucleus. In the first case, we are speaking
of a black hole with a mass of the order of one to a hundred
solar masses. On the second case, we are talking about $10^6$ to
$10^9$ solar masses.

The subject of black hole physics, in a sense, is a very old
one. The first person to propose the notion of a black hole
was Laplace, in 1798. He pointed out that the critical velocity
required for escape from a massive body increases with the mass
and decreases with radius. So, to get larger escape velocities,
we could either go to larger masses or smaller radii. Or, if
we want to consider fixed density, we could go to larger radii:

$(\text{Escape velocity})^2 \sim (\text{Mass})/(\text{Radius}) \sim (\text{Density})(\text{Radius})^2.$

Laplace noted that if we take a star with the same density as
the sun but with a radius as large as the radius of the orbit of
earth, then the escape velocity is greater than the speed of
light. Therefore no light can escape from such an object. The
details of a calculation on such an object are due to Oppenheimer
and Snyder (1939). If it takes as long to discover a neutron

star (1968) after its proposal (Baade and Zwicky, 1934), as to discover a black hole, we should see it soon!

The best opportunities to see the effects of a black hole seem to be: (i) gravitational radiation given off at the time of formation, (ii) X-rays from the accretion of matter onto the black hole after its formation, and (iii) the activity associated with a "live black hole". According to Bardeen, we must distinguish between two types of black holes: a live one or a dead one. A dead black holes is one which is not rotating and has no disposible energy. A live black hole is rotating and can give up energy to its surroundings.

In the future lectures, I would like to say something about these three methods of looking for black holes.

## Lecture 2

The general theme of structure without structure appears in many ways to be the lesson of gravitational collapse. The object of our lectures is to see a collapse itself and along the way to get some flavor of what this idea might be, structure without structure, because I meant the concept to be a kind of Jason's fleece for which we are still searching and have not yet found.

I don't know any better way of giving a first impression of what this idea is than to talk about the structure of the gravitational field equations themselves in terms of the idea that the boundary of a boundary is zero. This is the most central idea that one knows how to use in formulating General Relativity. What I would like to do is first to trace out a little bit what this idea is, then to use it to give an impression of what the gravitational field equations are and what they mean and then to go on to applications to gravitational collapse. I shall indicate the highlights and I think one will see the subject in perspective.

Let me recall the great difference between electromagnetism and gravitation. In the case of electromagnetism everything goes back to the Lorentz equation of motion for a charged particle; the acceleration of this particle is governed by the charge to mass ratio, by the velocity of the particle, and by the field, and as we know this Lorentz force equation provides our most compact way of defining what we mean by the electromagnetic field. In the case of relativity, it's been a long time to learn that relativity deals not with the notion of one particle, but with the relative motion of two particles. One particle and its acceleration has, as we know, zero value. The whole idea of the equivalence principle is this: all particles fall at the same rate, therefore, there is no acceleration to be talked about. In a local Lorentz frame the acceleration is zero. However, if we take above the earth a railway car, and take two particles which have a certain vertical separation one from the other, then we know one particle falls more rapidly, with a greater acceleration than the other. We describe this in terms of a tide producing force given by the derivative of the gravitational field. We know that the whole

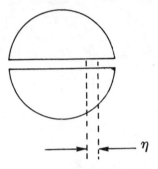

Figure 2.1

theme of Einstein's work was the idea that the simple way to do physics is locally, not to do physics in terms of long distance effects. Therefore, the wrong thing to talk about is the acceleration of this or that particle relative to the center of the earth; the right thing to talk about is the relative acceleration of one of these particles relative to another.

Instead of looking at accelerations outside the earth, it is simpler for our purposes to consider a hole bored through the interior of the earth (Fig. 2.1) and particles allowed to oscillate back and forth there, especially if we idealize to an earth of uniform density, because then we know the motion is a simple harmonic oscillation. We know that the mass inside a sphere of radius r is given by the third power of r, and that the force goes inversely as the square of the distance, and so the force itself goes as the first power of the distance. So force being proportional to distance we have simple harmonic motion, so that if we look at the world lines of these two test particles we see a simple harmonic motion. It doesn't matter if we talk about the displacement of either particle relative to the center of the earth or switch to Einstein's way of talking about the acceleration of one particle relative to the other; in either case we have simple harmonic motion, governed by an equation of this form:

$$\frac{d^2\eta}{dt^2} + \frac{4}{3}\pi\rho\, \eta = 0,$$

where the acceleration term is measured directly by the density. This equation for the relative motion of these two test particles, in this very simplest of examples, is generalized in relativity to an equation of this form:

$$\frac{D^2\eta^\alpha}{Dt^2} + R^\alpha{}_{\beta\gamma\delta} \frac{dx^\beta}{dt} \eta^\gamma \frac{dx^\delta}{dt} = 0,$$

the so called "standard equation of geodesic deviation". We think of these two world lines as having a certain separation. This separation of the two points is described by a 4-vector, and as time goes on this 4-vector changes its value. The magnitude of the separation grows compared to what it would otherwise be. In order to describe that acceleration we need more components to the equation than we had in this elementary example. We have an equation which is the natural generalization of this equation: four components for the 4-vector instead of the one component of the separation variable we had here, and we have the 4-velocity coming in as well. Under normal conditions, slowly moving particles, the 4-velocities of the particles have only one significant component, the time component, so all the terms go out in this equation except the terms of the form, if we are talking about the space components,

$$\frac{d^2\eta^i}{dt^2} + R^i{}_{ojo}\, \eta^j = 0,$$

so this equation greatly simplifies.

We tie these ideas, that go back to the elementary physics of what gravity is and what it means, into the ideas of differential geometry in a simple way. In a non-Euclidean space, if a vector is translated parallel to itself along a close circuit, it doesn't coincide with the initial vector. One is already familiar with this idea (change in a vector when it is carried parallel to itself around a close circuit) when he goes to the motion on the surface of the earth. If we take a vector that points north

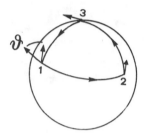

Figure 2.2

(at 1) on the equator (Fig. 2.2) and carry it to another point (2) along the equator,, and then we carry the vector parallel to itself to the north pole, and from here back down to the equator (at 1), it is now turned by 90° with respect to the direction it originally had. The angle of turn more generally which the vector experiences in going around a circuit on the surface of the earth is given by the product of two factors, of which one is the area circumnavigated and the other is the curvature of the earth. To be more explicit, in this example we just have taken, the area is one eighth the area of the surface of the earth, the curvature is given by one over the square of the radius of the earth, and the angle the vector has turned through is $\pi/2$.

$$\frac{\pi}{2} = \frac{1}{a^2} \frac{1}{8} 4\pi a^2$$

If we translate back from this elementary example to the case of true 4-dimensional geometry we have a general vector $A^\alpha$ and the vector undergoes a change on being translated parallel to itself around a circuit, and this change in its components is governed by the curvature tensor, by the original vector, and by the area of the region circumnavigated. If we talk about a small local region, a local parallelogram, we can describe that parallelogram around which we have taken the vector in terms of the length of the two principle sides of the parallelogram $\Delta x^\gamma$ and $\Delta x^\delta$:

$$\delta A^\alpha = R^\alpha{}_{\beta\gamma\delta} A^\beta \Delta x^\gamma \Delta x^\delta.$$

So we have these two ways to describe curvature, one in terms of geodesic deviation, and the other in terms of rotation of a vector in parallel transport. This is therefore the geometrical animal which is the center of attention in general relativity, and one turns now to asking what kind of equation this geometrical animal will satisfy. In order to introduce the point of view that is useful in talking about this topic, it is appropriate to go back as always to electromagnetism as a kind of guide.

If we think of space and time, and think of a single point P in space and time, we know that one of the central components of Maxwell's equations at that point is a statement that magnetic lines of force are neither created nor distroyed in an elementary volume element. That is to say, the divergence of the magnetic field is zero. That is a statement which applies to the physics on a spacelike hypersurface through that point P. Nobody can keep us from drawing another spacelike hypersurface through that very same point P with a different tilt to it, that is to say a spacelike hypersurface corresponding to an observer moving with a different velocity. If we write down the same statement that the divergence of the magnetic field is zero there, and compare its content with the content of the original equation, we find they are not the same statement: they are different pieces of information, and the difference between the two gives another piece of information which says that the curl of the electric field is given by the time rate of change of the magnetic field (or at least the x-component of that, if we are talking of a displacement here which is a Lorentz transformation in the x-direction). In other words, the content of Maxwell's equations is contained in the statement that magnetic lines of force never end. Of course we have another equation we have to take into account, namely, the statement that the divergence of the electric field is given by the charge density. The two equations, magnetic lines of force never end and electric lines of force never end except on charge, therefore, comprise the whole of electromagnetism. One can get all of Maxwell's equations from those elementary statements!

We can do the same thing in gravitation physics, and if this provides a simple way to look at electromagnetism, the corresponding way to look at gravitation is therefore also a simplification. Let me write down the answer at once and then come back and interpret it, because this is the central principle of General Relativity. If in space-time we consider any hypersurface cutting through space-time and any point P cut through by that hypersurface, then on this spacelike hypersurface there are two kinds of curvature that are relevant. One of them is the curvature of the 3-dimensional geometry intrinsic to this hypersurface, the other the curvature that tells how fast this geometry is changing with time in the enveloping 4-dimensional geometry, how fast geometrical figures which were present on the original surface are being expanded. One of them is intrinsic curvature, the other is extrinsic curvature, and the statement is that the sum of the two kinds of curvature is given by the energy density!

$$\begin{pmatrix} \text{Intrinsic} \\ \text{Curvature} \end{pmatrix} + \begin{pmatrix} \text{Extrinsic} \\ \text{Curvature} \end{pmatrix} = 16\pi \begin{pmatrix} \text{Energy} \\ \text{Density} \end{pmatrix}$$

Example:

$$6/a^2 + 6/a^2 (da/dt)^2 = 6a_0/a^3$$

All of General Relativity is contained in this one equation! If once again we take different spacelike hypersurfaces through a given point and consider the difference between the results we get in the different ways, and subtract the two, we get all ten of Einstein's field equations out of this one equation! So this is the center of the subject. Let me, by the way, just for purposes of background, say that if we are talking about a closed universe with uniform curvature (a universe of radius a), then the first term (intrinsic curvature) has a value $6/a^2$. The second term (extrinsic curvature) has a value governed by the time rate of change of the curvature. And if in addition this universe is described by energy which is associated with dust, so that the amount of energy is simply governed by the dilution over a certain volume given by the cube of the radius, then nobody can keep us from writing this term in the form of a certain constant $a_0$, which has the meaning of the radius at the maximum expansion divided by the cube of the radius. ($16\pi\rho =$ $= 16\pi\, m_0/(4/3)\pi a^3 = 6(2m_0)/a^3 = 6\, a_0/a^3$) The whole theory of

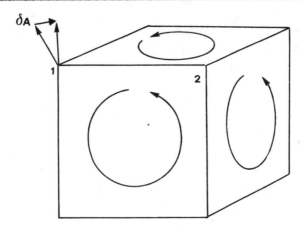

Figure 2.3

the universe is contained in this one equation, so it really is a compact way of stating the whole subject! And we'll see in a moment that also the Schwarzschild geometry lends itself to statement in terms of this way of looking at things.

But our original objective here was to see what the meaning of this equation is, to see it in deeper terms, and for this we go to the point of view developed by Cartan in his book on Riemannian geometry. It was his marvelous geometric insight which gave us the mathematical machinery which today is universally adopted. It took a long time to make its way in the world, because people were not so accustomed to thinking in so geometrical a fashion as he was. And unhappily his marvelous way of looking at General Relativity has not made its way in the world, and it needs really to be brought out. The general idea is the following: Consider at a given point on a spacelike hypersurface a small 3-dimensional volume element which can be, for example, a 3-dimensional cube (Fig. 2.3). Take a vector $A^\alpha$ and transport that vector parallel to itself around the top surface of the cube. We are dealing with the curvature of space and the vector does not come back to its original direction. Its departure from its original direction is a measure of the curvature associated with that face of the little cube. Once again, to recognize we are not talking just abstract mathematics, but real physics, we know we can give a meaning to this thing in terms of the ideas of geodesic deviation, the separation between test particles. Of course in the discussion we are going to be carrying out we will have different faces, and if we wanted to give it all a detailed description in terms of test particles we would need quite a large number of particles. If we now carry the vector around another face (particle) we get a different alteration of the vector $A^\alpha$. It might seem a little difficult to transport the vector around still another face except for the following fact, that we can carry the vector from point (1) over to point (2), carry it around this surface (right surface), and then bring it back, so that really it has a meaning to talk of what happens to the vector being carried around any face. If we add up the total changes in this vector brought about by carrying the vector around all six faces, we can see without doing any work at all that the result must be zero, and this for a very simple reason: the vector is carried once along an edge in one direction, as it goes around on a face, and once in the opposite direction around the opposite face. So each edge is traversed twice, or an even number of times, and the contributions cancel. This is the statement of Bianchi, the so called Bianchi Identity, which Cartan gives in this beautiful geometric form: the sum of rotations associated with all six faces is zero.

This reminds us of a problem in mechanics, so let's take a minute off and look at a familiar problem, the equilibrium of a solid body under the action of forces. We know that one condition for equilibrium is that the sum of forces acting on the body should be zero. But we also know this is not enough for equilibrium. There is an additional requirement, the requirement that the sum of moments should be zero. Moments about what? Well, happily it does not make any difference about which point we take moments, because the difference in the total moment about one point or another point is governed simply by the separation of the two points and by the sum of the forces. But the sum of the forces is already zero and therefore it makes no difference about what point we take the moment.

The only reason for bringing up this problem of mechanics is to give us motivation for considering what Cartan next considers. He considers not the sum of rotations associated with all these faces, but the moments of these rotations. Take any point, for example some point within the interior of this cube, or, if you want, a point outside of the cube, and take the vector running from that point to the center of the given face, and mul-

# General Relativity, Collapse and Singularities

tiply that vector by the rotation which one has associated with that face. This is a geometrical animal which is known technically as a time vector. The main point is, however, that if we add up the time vectors that measure the moment of rotation associated with all six faces of this cube we get what we call a moment of rotation, the total moment of rotation associated with the cube as a whole. This we call Cartan's moment of rotation. If we take this totalized moment of rotation, then that is equal to the energy density! This is the content of General Relativity!

Why is this an important idea? Why should we consider this totalized moment of rotation? Here we come to the central theme and idea that the boundary of the boundary is zero. I am now trying to give the background of why nature should like to use a law like this. I am not going to trace out and prove the equivalence of this way of doing business with that before seen (intrinsic plus extrinsic curvature), but this statement and that statement can be shown to be the same kind of statement. Come back to the idea that the boundary of a boundary is zero. If we look at a 3-dimensional figure such as Fig. 2.4, each boundary is made up of 2-dimensional faces, so the total boundary of the 3-dimensional cube (if we want to write it in mathematical form) is the sum of the six faces of the cube:

$$\partial V = \sum_6 F$$

If we take any one of those faces, then we can say that the

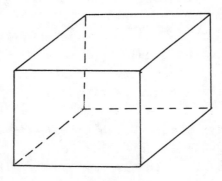

Figure 2.4

boundary of that face is the line that runs around the edge of it which is the sum of four shorter lines:

$$\partial F = L = \sum_4 1,$$

each endowed with a certain direction. So each line appears once as the boundary of a face, then the same line appears with the opposite direction, as the boundary of another face so that we can say that when we look at the total boundary, the sum of all six faces, and we take the boundary of all six faces, and we add all those little edges up, we count each edge twice, once with one sign and once with the other sign, and we get zero. But this sum of all the six faces was the boundary of the 3-dimensional volume, so we end up with the statement that the boundary of the boundary of the volume is automatically zero, which our mathematical colleagues summarize in the statement that the boundary operator applied twice gives always an identically zero value:

$$\partial \partial V = 0, \text{ or just } \partial^2 = 0.$$

This statement is a statement free of any appeal to any dimensionality. It rises above dimensionality and this is what we're looking for in waiting to see the deeper content of our subject in the most penetrating form possible. But now let's turn back to the world of the here and the now and the simple 4-dimensional story.

Think of yourself as the designer of the world and, wanting to start the world out right, decide to make sure some things in the world are conserved, so that in any given region of space and time that quantity you're talking about is neither increased nor decreased. That is to say the production of that something in this volume element should be zero. What that something is that you'll want to be conserved is different in the two cases of electromagnetism and gravitation. In the case of electromagnetism it is charge, but we go immediately to the discussion of General Relativity, where the thing we're conserving is energy. The bottom of this 4-dimensional cube (Fig. 2.5) is a little 3-dimensional cube $V_1$. In that we have a certain amount of energy. In the top cube $V_2$ we'll have a different amount of energy. In the cubes on the side we'll have what we call an energy flow; these cubes on the side have three dimensions, of which two are spacelike dimensions and the third is a timelike dimension. On such a side face we are talking about an area and about the energy that flows across that area in the available time. So the side faces represent energy that has flowed in, and the law of conservation of energy says that the sum of the energies associated with all those faces add up to zero. The statement that the energy change between the lower surface and the upper surface shall be equal to the energy that has flowed in (law of energy conservation), then, is to show up as a statement about the sum of these quantities representing energies in each of these 3-dimensional faces. So we want the idea to come out automatically, how are we going to wire up the world, how are we going to connect up our surfaces? We want to make this law of conservation of energy come out automatically as a consequence of the way we wire up the world. In brief, the appropriate way to wire up the world is to have it come out so that the energy associated with any face is given by the moment of rotation associated with that face. Why is this the right way to do business? The reason is the following: if we look at the face at the bottom, we know that face at the bottom had a lot of 2-dimensional faces associated with it, the amount of energy in that bottom cube is given by the sum of the moments of rotation associated with all the faces of that cube. But now the cube on the side has 2-dimensional faces too, and where the 2-dimensional faces meet we have there a 2-dimensional face that is counted with one side in one case and one side in the other case. So when we add up the contributions from all the 2-dimensional faces, the moments of rotation, everything automatically goes out. You get conservation automatically as a consequence of the idea that the boundary of a boundary is zero! So this is the marvelous way General Relativity is designed to capitalize on this principle, to wire up the source to the surroundings in such a way that conservation shall automatically be upheld. We could go through electromagnetism and show how the structure of electromagnetism capitalizes in another way on this principle that the boundary of a boundary is zero. This is just to give an idea how we are gradually, as we understand physics better and better, seeing that the ideas that come into it are simpler and simpler than we realize.

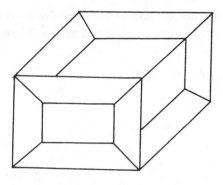

Figure 2.5

Now let's turn to gravitational collapse, because that is the central theme that directs our attention into why we should be worried about this aspect of physics, and let's look at the most elementary of problems, apart from the problem of the Friedmann Universe, that we have already sketched out here. You'll notice that a central point of our discussion is to go from a 4-dimensional description to a 3-dimensional description, because there one sees things most simply and the physics along a spacelike hypersurface is particularly simple in the case when this is a time symmetric surface, that is to say, one in which the physics that happened before this moment is the mirror image, timewise, of the physics that happened after that moment. If we consider situations, from the point of view in which the extrinsic curvature is zero, because the spacelike hypersurface is neither expanding nor contracting, the simplest among such situations is that in which the geometry is static for all time. So let's look in particular at a situation which is not only static but also spherically symmetrical:

$$ds^2 = (1 - 2m(r)/r)^{-1} dr^2 + r^2(d\theta^2 + \sin^2\theta d\phi^2)$$

Nobody can keep us from writing a spherically symmetrical geometry in this form. It's true that the coefficient that appears out here in front seems to have a rather elaborate form in this way of writing it, but nobody can keep us from writing any crazy function of r in this form. The advantage of writing that crazy function of r in this particular form is the following: when we do so, the equation that intrinsic curvature equals energy density, the central content of relativity, boils down to the statement that the function m is given by

$$m(r) = \int 4\pi r^2 \rho(r) dr$$

Of course the simplest of all situations, if we keep on simplifying (first we've got time symmetry, then we've got spherical symmetry, and one additional simplification is to take the case where there is no matter present, or a region of space surrounding a region where matter is constant) is then $\rho = 0$ and m = constant. So we have the standard Schwarzschild geometry. This is the simplest case of gravitational collapse to discuss, but first we should really get a geometrical description and interpretation of the geometry, and for this purpose it is appropriate to consider the 3-dimensional world that we are talking about embedded in a 4-dimensional world in which the fourth dimension actually has nothing to do with time. The fourth dimension is purely a matter of convenience in allowing us to see what is going on. We are going to consider this curved 3-dimensional geometry in a flat 4-dimensional spacelike geometry. This flat 4-dimensional spacelike geometry has a fourth coordinate in addition to the x y z of the spacial coordinates. We're going to find that this geometry we are just dealing with here can be described in terms of a slice through this 4-dimensional geometry of this fashion (Fig. 2.6), where we are going to take $\Theta$, for example, to be measured around in this direction. Let's supress one dimension here for simplicity in order to visualize things a bit more easily. So we are taking the geometry

$$(1 - 2m/r)^{-1} dr^2 + r^2 d\Theta^2$$

and we are expressing that geometry as a geometry $dx^2 + dy^2 + dw^2$, where the coordinate w here is to be thought of as depending upon the choice of x and y, $w = w(x, y)$, or, since we are dealing with spherical symmetry, $w = w(r)$. So $dx^2 + dy^2 = dr^2 + r^2 d\Theta^2$ and $dw^2 = (dw/dr)^2 dr^2$. We have the following

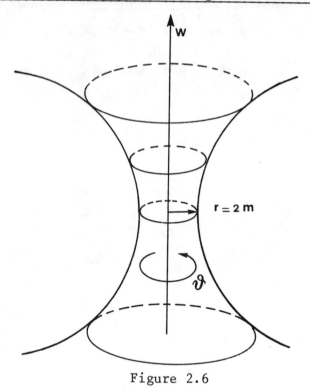

Figure 2.6

statement: the angle terms in the two ways of writing the geometry are equal to each other, the radial part in the one way of writing the geometry has to be equal to the radial part in the other way of writing the geometry:

$$dr^2 + (dw/dr)^2 dr^2 = (1 - 2m/r)^{-1} dr^2$$

This is a very simple equation which tells us how to draw the surface in order to reproduce this geometry. We can check out for ourselves that this is a parabola. So what we are looking at here is a 3-dimensional geometry which represents a solution of Einstein's equations at a particular moment of time, the Schwarzschild geometry. If we wanted to distinguish this from the Schwarzschild geometry surrounding a star, we can say what we have here is a complete Schwarzschild geometry, whereas when we are dealing with a star, we are dealing with a set of conditions under which this geometry stops at a certain point, namely the surface of the star, and is replaced interior to that region by a geometry which does not have this trough on it. There is no simpler case to consider than the case of uniform density, and in the case of uniform density we already know the answer. We know

# General Relativity, Collapse and Singularities 543

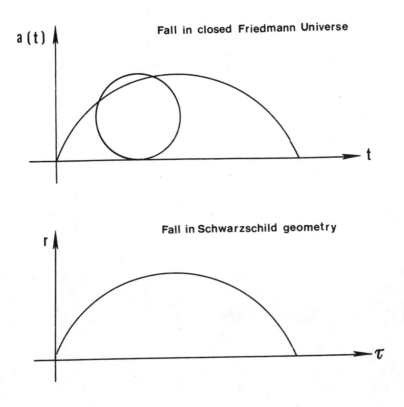

Figure 2.7

that the geometry inside is part of a 3-dimensional sphere of a radius of curvature that is governed by the density inside. So if we want to turn to the problem of gravitational collapse of a cloud of matter, we have all the ingredients here with which to begin the problem. If we want to discuss the final state, we also have the story. Let me summarize now the statement, let me try to collect the conclusions about these two very different problems of gravitational collapse that we are treating at one and the same time.

Problem number one: a cloud of matter of uniform density which starts out at the time t = 0, and starts falling inward. One could be perturbed about the back pressure of the material as it is compressing and this would certainly be a justified reaction, if we were considering at the moment a star. If, however, we are considering a sufficiently dilute cloud of dust, then it is perfectly clear that the interaction of one dust particle with another has no part in the procedure. We are talking about a pressure free problem, and the interesting thing is that the collapse, if the dust cloud is big enough, already carries the cloud inside of the Schwarzschild radius long before the dust particles have made contact, one with the other. So the issue of gravitational collapse is one which poses itself free of these issues of the back pressure of matter. The behavior of this cloud of dust is easy to discuss because the geometry here is the same as the geometry of a closed universe, and we know that, if we solve the equation for the radius as a function of time, it gives us a curve which is a cycloid curve for radius as a function of time, a cycloid curve obtained by rolling a circle along a plane. This collapse of the matter inside the cloud poses a real problem for us because we say to ourselves: look, what happens at the interface, at the region where there is matter and where there is no matter? Happily, it is possible to consider what goes on there by thinking of a particle at this point. This particle can be looked at in two points of view, and the results of the analysis have to be the same. If we look at it from the point of view of a particle falling freely in the complete Schwarzschild geometry we get one answer, where radius, r coordinate, as a function of proper time, follows a curve like that shown in Fig. 2.7. It's a remarkable fact that the General Relativity problem of free fall leads to the very same

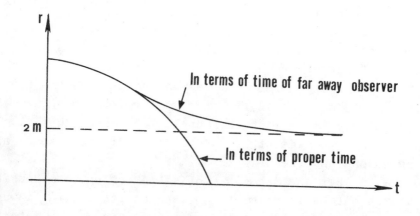

Figure 2.8

# General Relativity, Collapse and Singularities

Figure 2.9

equation which one has in the Newtonian theory of gravitation for a mass particle falling under the field of an inverse square center of attraction! The differential equation is the same and the solution is the same, and the cycloid is the same. This cycloid was already known long ago to our colleagues in the field of gravitation physics. If we compare this curve, which is the curve we deduce from the fall in the Schwarzschild geometry, with the curve we deduce from the properties of the closed Friedmann universe, we have the same result. The particle does indeed fall according to the same law viewed in both ways. Of course what this means is the following: it means that the radius of this geometry is all the time getting smaller, the closed universe is collapsing!

Now it's worthwhile to look at this question of fall also from the point of view of a time coordinate based on the time as seen by a far away observer. If we think of the mass started far away then, if we analyse in terms of proper time, we have the cycloid curve which we've just looked at. If we think of the r coordinate in terms of the time appropriate to a far away observer, then the curve in the beginning looks much the same, but it separates and the radius never gets closer than the point $r = 2m$ (Fig. 2.8). This curve is important in considering the issue of gravitational collapse.

If we already have a star that has undergone collapse, and then we add something in addition, one more particle being dropped in from outside, then as seen from outside the particle goes faster and faster accelerating: lower velocity, higher velocity, still higher velocity, but then it seems to go slower and slower (Fig. 2.9) and "ooze" up to the final point, This enormous difference between the physics as seen from far away and seen close up is of course one of the features that's most disturbing from the point of view of anyone, because it seems to force one to take a stand: am I going to follow the thing in, in which case I go into this catastrophe in a finite time, or am I going to stay far away, in which case indeed there is no catastrophe, but I don't get to see what is happening there? It's somehow reminiscent, although it doesn't have the slightest connection with quantum mechanics, in the sense that one has to make up his mind, am I going to measure position or am I going to measure momentum, I can't measure both simultaneously.

The example of the fall of this particle toward this "black hole", toward this completely collapsed object, is useful as an example of the most extreme form of asymmetry that one can very well imagine. Many times the question has been raised, will gravitational collapse still happen if the distribution of mass with which we are dealing is not spherically symmetrical? Here we've let a spherically symmetrical collection of matter collapse and then we've added in the most asymmetrical thing that one can imagine, namely a hair sticking out, if you will, a prong sticking out, and this prong smooths off, and we end up with the smoothness of a polished billiard ball at the end. This feature that the details of an object disappear in gravitational collapse is extremely important in giving a new insight into the laws of conservation of baryon number, lepton number and charge. One has discovered through the work of many people, among them Hartle,

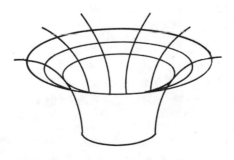

Figure 2.10

# General Relativity, Collapse and Singularities    547

Teitelbaum, Beckenstein, Wald, that the possibilities of measuring the number of baryons or leptons in an object disappear in the final stages of gravitational collapse! There is no possibility to give those ideas a meaningful content. If one draws a black hole and drops in the greatest variety of objects, in the end one comes to the conclusion that the only quantities that are left that have meaning for the black hole are angular momentum, mass, and charge. Baryon number and lepton number have disappeared! We might think that there would be enormously more detail preserved because we would say: look, if this has a positive charge on it, and over here a negative charge, we can arrange a pattern of positive and negative charges that we drop in over a big sphere that spell out some marvelous message in terms of the information content in that sphere. But this is a completely mistaken result and concept because in point of fact if just one charge goes in at one point, the line of force redistribute themselves in such a way that the electric lines of force appear eventually all to be emerging from the center, even though the particle has ended up elsewhere. So that observations from the outside offer no prospect whatever of getting at these details.

Well, I'd like to go to the possibility, next time, of observing a black hole, the physical methods of seeing such an object.

Lecture 3

DYNAMICS OF A PARTICLE IN GEOMETRY

If it is true that no-one has yet observed a black hole, yet it is also true that no one knows how to prevent a black hole from being formed. So it seems inevitable that somewhere, and perhaps in many places, there are black holes. The central reason for our ignorance on this point and for our inability to make strong predictions is our ignorance of a central point of astrophysics, namely the course of normal burning of a star. If we think about the evolution of a star in terms of temperature and density and its track on the T-ρ diagram (Fig. 3.1), then there is no more central issue than the question of loss of mass. We do not know whether a small fraction of stars, a negligible fraction of stars, or a very large fraction of stars in the course of evolution reach a point where collapse to a black hole is possible.

Let us discuss a star which has developed a white dwarf core, with a large atmosphere around it (Fig. 3.2). This core has some angular momentum, of course, so we have two important parameters in the problem, the mass and the angular momentum of the core. We want to talk about the development of this core as it becomes unstable against gravitational collapse. As it collapses we have, because of the conservation of angular momentum, a great increase in angular velocity, and the resulting neutron star will in some cases have the shape of a pancake, for example, the one shown in Fig. 3.3.

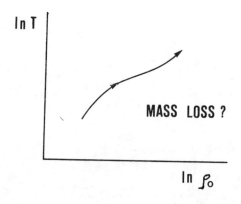

Figure 3.1

# General Relativity, Collapse and Singularities 549

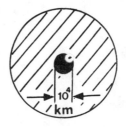

Figure 3.2

Now if we change the parameters, the situation can be entirely different. For example, we can get coagulation as illustrated in Fig. 3.4. It is worth noting that the masses here are the direct opposite of what they appear to be, because as a neutron star loses mass, it grows in size, so the larger pieces here are the less massive, and the smaller pieces more massive.

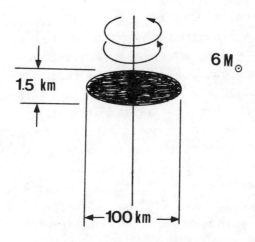

Figure 3.3

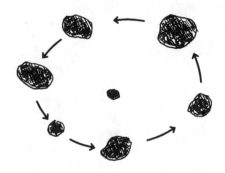

Figure 3.4

This system will lose angular momentum by gravitational radiation on a time scale of the order of minutes or hours, so one by one these objects will fall together, and we will have a black hole. If we look at the most obvious signal from this process, we will see sudden changes of quadrupole moment as the initial pancake forms, as it breaks up, as various pieces fall together and as they all collapse down to a central black hole. This very characteristic signal (Fig. 3.5) is something we would very much like to look for in detecting gravitational waves from these events, especially since these collapse events are surrounded by clouds of debris, so that light signals cannot get through, though of course gravitational waves can. This gives us the chance of detecting sources of gravitational waves in our own galaxy, and it is not necessary to go through the arguments about energetics to say that even events in nearby galaxies may be detected in this way.

I would like to take a moment to state my position on Joseph Weber's experiment. I have followed Weber's work since he first became interested in gravitational radiation in 1956, and I have great respect for his energy, persistence and faith. I have kept careful track of his data, and no one has done more than I to understand his effects from other causes, lightning

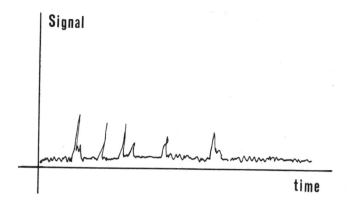

Figure 3.5

storms, solar flares, radio signals for communicating with submarines at sea, etc.: one after another he shows that these effects are excluded. On the other hand, the history of physics is full of apparent effects that have been shown to be due to causes quite different from what anyone had previously thought of. So one must be careful in interpretation, especially until his events are confirmed with independent equipment.

We have now seen that gravitational radiation created at the time of formation of a black hole is one way of detecting the existence of black holes. Another way is the accretion of matter onto a black hole that is already formed. The flow of matter onto a black hole would be given by the formula

$$\begin{pmatrix}\text{Surface area}\\ \text{of a sphere}\end{pmatrix} \times \begin{pmatrix}\text{Rate of inward}\\ \text{flow}\end{pmatrix} \times \begin{pmatrix}\text{Density of}\\ \text{matter}\end{pmatrix}$$

$$4\pi r^2 \qquad\qquad v_r \qquad\qquad \rho \quad = F,$$

and in the simplest view one could take the radial velocity comes from the kinetic energy, which goes as $1/r$, so that $v_r \sim 1/r^{1/2}$, and then $4\pi r^2 v_r$ goes as $r^{3/2}$, and so the density must go as $r^{-3/2}$. One is therefore dealing with a compression of the incoming gas, which results in a heating of the gas, up to $T = 10^{10} - 10^{11}$ °K according the Zel'dovich and Novikov. Such temperatures are sufficient to produce X-rays. Thus X-rays too are a possible tool for the diagnosis of black holes.

Incidentally I should point out that the X-rays do not come from inside the black hole. They come from the traffic jam of gas that develops on the way to the black hole. Thus we may also expect X-rays to be produced from gas accreting on the surface of a neutron star. Also the intensity of the X-rays depends on the density of matter surrounding the object, a number which is not at all well known. Also there will be a cloud of debris around if the black hole is recently formed. Moreover there is the issue of the speed of the black hole moving through the surrounding matter. Thus there are great complications in determining the shape of the spectra produced, and in sum, it looks like the last prospect in the world of distinguishing between a black hole and a neutron star.

There is another point about this situation: we have a gas coming in against the pressure of radiation coming out, and this is exactly the sort of situation that leads to Taylor instability, so that instead of a steady flow of material in, we may expect great irregularities. It is like mercury on the top of water, an unstable situation. The mercury will come down in globs and clumps. We might call this the "chug-chug phenomenon".

If we have a double star system in which one component is compact, massive, invisible, and cannot be interpreted as a normal star or a white dwarf, then we have reason to think that it may be a black hole. What is the critical limiting mass above which we could say that such an object cannot be a neutron star? If a neutron star has no angular momentum, then the point of instability is reached somewhere between $\frac{1}{2}M_\odot$ and $3M_\odot$, depending on the equation of state. However, we can get quite large stable neutron stars if we allow differential rotation. So the situation is not clear and a firm theory of rotating neutron stars would be of great help in the problem of identifying massive invisible components of binary systems.

A third field of study of the black hole has to do with the question of activity. It is perhaps easier to approach the question of the activity of a live black hole (a rotating one) by first looking at the motion of particles in the field, of a dead (nonrotating) black hole, the traditional Schwarzschild field. We shall go into this problem in some detail, because the same techniques will be useful to us in the study of superspace and the gravitational physics of the universe itself.

The point of view that shortens the gap between classical and quantum physics to a minimum is that of Hamilton and Jacobi. The Schroedinger equation for the wave function

$$\psi(x, t) = A(x, t) \exp\left[\frac{i}{\hbar} S(x, t)\right]$$

can with suitable approximation be translated into the Hamilton-Jacobi equation for the phase of the wave function:

$$-\frac{\partial S}{\partial t} = \frac{1}{2m}\left(\frac{\partial S}{\partial x}\right)^2 + V$$

We may call this a dispersion relation for matter waves. We note that it does no good at all to solve this equation by separation of variables, because that separates the space from the time dependence, and as we know these waves are spread out over all space and time. In Fig. 3.6 are drawn the surfaces of constant phase.

We know that to describe motion we must talk about a wave

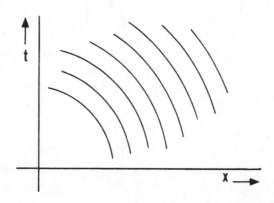

Figure 3.6

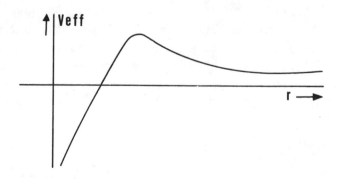

Figure 3.7

packet which we do symbolically by putting two terms here

$$\psi \sim A_E \exp\left(\frac{i}{\hbar} S_E\right) + A_{E+\Delta E} \exp\left(\frac{i}{\hbar} S_{E+\Delta E}\right)$$

for two different energies E and E + $\Delta$E. We know that the region where the two different sets of wavefronts agree with each other is the region of constructive interference, and this is the classical world line. So one gets to motion by looking at the idea of constructive interference. One looks for values of x and t where the phases are equal: $S_E(x, t) = S_{E+\Delta E}(x, t)$.
Thus we try to find values of x and t where the derivative $\partial S/\partial E$ vanishes.

$$\frac{\partial S_E}{\partial E}(x, t) = 0.$$

Now if we write $S_E(x, t)$ explicitly

$$S_E(x, t) = -Et + \int^x \left[2m(E-V)\right]^{1/2} dx$$

and take the derivative, we find

$$0 = -t + \int^{x} \left[(2/m)(E-V)\right]^{-1/2} dx$$

which is the classical time required to reach a point: the integral of distance divided by velocity.

If we now go to the problem of motion in the Schwarzschild geometry, there is a great simplification: there is no potential. The particle moves freely, in free fall, and the only source of change of its motion is the geometry, not a potential. So the equation contains only the quadratic terms in the particle's momentum and the square of the rest mass, which we may take to be unity:

$$g^{\alpha\beta} \frac{\partial S}{\partial x^{\alpha}} \frac{\partial S}{\partial x^{\beta}} + 1 = 0.$$

Written out explicitly for the Schwarzschild metric, this equation gives

$$0 = \frac{-1}{1 - 2m/r} \left(\frac{\partial S}{\partial t}\right)^2 + (1 - \frac{2m}{r}) \left(\frac{\partial S}{\partial r}\right)^2 + \frac{1}{r^2} \left(\frac{\partial S}{\partial \theta}\right)^2 +$$
$$+ \frac{1}{r^2 \sin^2 \theta} \left(\frac{\partial S}{\partial \psi}\right)^2 + 1$$

As usual, we interpret the angular parts of the equation in terms of an angular momentum.

If we solve by separation of variables, we see that the time part is related to the energy, and we get the energy in terms of radial and angular momenta:

$$E^2 = (1 - \frac{2m}{r})(1 + \frac{L^2}{r^2}) + (1 - \frac{2m}{r})^2 \left(\frac{\partial S}{\partial r}\right)^2.$$

One may ask oneself where are the turning points, where is the radial momentum going to be zero? We may look at the motion in terms of an effective potential, plotted in Fig. 3.7. It is the energy at which one comes to a turning point at a given distance. This effective potential depends on the mass of the black hole and the angular momentum of the test particle. For

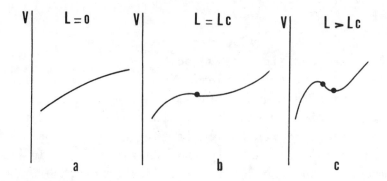

Figure 3.8

zero angular momentum, (Fig. 3.8a) the curve just falls as you go in. As you increase the angular momentum to a critical minimum value (Fig. 3.8b), you finally get a level spot, and as you go above that critical value for L (Fig. 3.9c), you get both a peak and a minimum. The latter corresponds to the Kepler orbit at that radius and the former to the barrier that must be surmounted to get into the black hole. When you go to large angular momenta the curve is like the curve you get for Newtonian gravity, and the new features that come with general relativity arise from the "pit in the potential" which is responsible for the lengthening of scattering times which is responsible for bending of light by the sun, the precession of the perihelion of planets, etc. It is a remarkable fact that Kepler's laws still holds exactly for the circular orbits in the Schwarzschild metric.

I might mention a technical point. The equations involve not dr itself, but $dr/(1 - 2m/r)$ and so a useful radial coordinate to adopt is $r^*$, where

$$+r^* = \int (1 - 2m/r)^{-1} dr = r + 2m \ln (r/2m - 1)$$

I like to call this the "tortoise coordinate" because it correctly

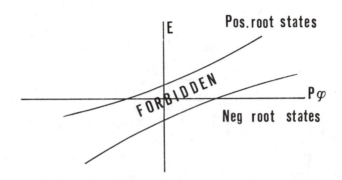

Figure 3.9 Diagram drawn for r = 1.4 M. Only the positive root states are allowed, but they can have E < 0.

represents the distant observer's view that the accreting particle never reaches the horizon, since r = 2m corresponds to $r^* = -\infty$. (It is also called the Regge-Wheeler coordinate.)

Now let us turn to the topic of rotating black holes. Brandon Carter first made the analysis that leads to the equation

$$E^2 (\hat{r}^3 + \hat{r} + 2) - 4E\hat{p}_\phi - \hat{r}(\hat{r} - 1)^2 + (2 - \hat{r})\hat{p}_\phi^2 = 0.$$

Here enter the energy E and angular momentum $\hat{p}_\phi$ of the particle and the distance $\hat{r}$ measured in units of the gravitational radius: $\hat{r} = r/m$. The case under consideration here is the extreme (maximally rotating) Kerr black hole.

A feature of the previous solution that we did not mention was the negative root solution for the energy. We simply discarded this solution before, but if we look at the corresponding case for the Kerr metric, we find that <u>both</u> solutions for E can be negative, for certain values of the angular momentum (Fig. 3.9). Thus a particle can be in an allowed state outside a black hole and yet have a negative energy. This can only happen for particles which remain within the <u>ergoshpere</u>: $M \leq r \leq 2M$. (For extreme Kerr, the horizon is at r = M.)

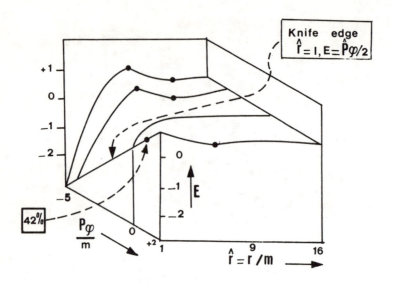

Figure 3.10

Now to a locally co-moving observer, there is nothing strange about this negative energy particle, but in terms of the energy measured by an observer at spacial infinity, the energy is negative. This means that the particle can never escape to infinity.

Let me try to illustrate the situation with a diagram applicable to the case of maximal angular momentum (Fig. 3.10). As you see the effective potential for the zero-angular-momentum particle decreases steadily until the horizon at $\hat{r} = 1$ is reached, so such a particle encounters no boundary and falls straight into the black hole. If we consider particles with sufficient negative angular momentum, then we once again get potential barriers and potential minima, the latter corresponding to stable circular orbits. If we take a particle in such a circular orbit and give it sufficient radial momentum, then its orbit will no longer be circular, and it may be able to surmount the barrier and fall into the black hole. For particles of positive angular momentum, we also get potential minima corresponding to stable circular orbits. As you make the angular momentum smaller and smaller, the location of the stable orbit grows closer and closer to the horizon of the black hole, and eventually, for the orbit which is just in contact with the horizon, 42% of the particle's energy has been extracted by the time it has reached this configuration.

But the most striking feature of this plot is the knife edge where the positive and negative energy states (which I have not drawn) come together. This knife edge is located at the surface of the black hole $r = M$, and at the energy-angular momentum points for which $E = \frac{1}{2}\hat{P}_\phi$.

I think all of us have heard the story of Roger Penrose coming into London on the train for one of his weekly seminars and suddenly realizing that he had no speaker. Under pressure to think of a topic to talk about, he began thinking about the process of a particle coming into the vicinity of a rotating black hole, and how this might serve as a source of energy for a particle. If we consider a particle coming into the ergosphere, fissioning there, and putting one fragment into one of these negative energy states, then the particle that comes out must have an energy greater than the mass-energy of the incoming particle, because the accretion of the negative energy fragment diminishes the mass of the black hole. Thus the conservation of energy demands an increase of energy for the particle that escapes.

Thus we have a mechanism to extract energy from a live black hole. The effective potential diagram together with the equations

$$E_{\text{primary}} = E_{\text{sacrifice particle}} + E_{\text{escape particle}}$$

$$\hat{P}_{\phi\,\text{primary}} = \hat{P}_{\phi\,\text{sacrifice}} + \hat{P}_{\phi\,\text{escape}}$$

provide all the tools one needs to study this process of Penrose. Christodoulou has looked at this process and asked the question, what happens to a black hole if you use the Penrose mechanism in the most efficient way, which is never to waste any of the sacrifice energy in radial momentum. He discovered that in these circumstances the energy and angular momentum of the black hole will be decreased according to a certain curve which he calculated, considering infinitesimal changes. He finds that the most efficient series of changes result in the following formula

$$m^2 = m_{ir}^2 + \frac{J^2}{4m_{ir}^2}$$

in which $m_{ir}$ is the <u>irreducible mass</u>, the mass which cannot be taken away by this process, no matter how efficiently you perform the changes. Later it was explained that this formula is related to the surface of the black hole. Now one might ask, could nature provide a mechanism by which this most efficient series of fissions could take place? Mashoon has studied the action of the tidal forces of the black hole on a star which comes in, to see whether the right kind of breakup could occur so that fragments could emerge with relativistic velocities, even if the original star had quite nonrelativistic velocities. I should now distinguish between small black holes of a few solar masses and the very large black holes that we speculate may exist in the centers of some galaxies, which could have masses of $10^6$-$10^9$ $M_\odot$, hence a large fraction, in some cases, of the total mass of the galaxy. It is only for these latter cases that the work of Mashoon is applicable, since only in these cases is the size of the incoming star negligible compared to the size of the black hole whose effect he calculates. He finds that there are indeed enormous stresses and turbulences generated in the incoming star, and we might well expect from this that fission does occur and that fragments can be drawn off with relativistic velocity.

So, in summary, we have in this activity of a black hole a third testable consequence of black hole physics, so that it seems that in a reasonable time, given good luck in the number of stars, we may be able to test the predictions of gravitational collapse at the level of a single star. And this then brings us to the topic of the gravitational collapse of the universe itself, and to the topic of superspace, which is the subject of the last lecture.

## Lecture 4

In a time when progress is being made on so many concrete fronts, nobody can have more qualms than I about speaking about a general topic such as the problem of gravitational collapse. My one excuse is that I hope that what I shall be doing is more to raise questions than to answer questions. I believe that in this issue, which is the most important issue physics has faced, the kind of answers we are going to get will depend very strongly on the kind of questions we ask. It is important to realize how far reaching the issues are and to look at them in the widest possible terms.

We have reviewed the principle issues, and compared them with those faced in the 1910's in Table 4.1.

How can we see into these things without a great deal of formalism? The central ideas are straightforward. Consider a harmonic oscillator:

Ground state wave is $\psi \sim \exp\left(-\frac{m\omega}{2\hbar} x^2\right)$. Zero point fluctuations $\Delta x \sim (\hbar/m\omega)^{1/2}$ must give up the idea of a deterministic world line through ordinary space.

In the case of geometry, the dynamics goes on in the arena of SUPERSPACE. Each point in superspace describes the whole shape of a three dimensional geometry. Between the arena of a particle moving in ordinary space and geometry developing in superspace, the easiest intermediate point is the dynamics of the electromagnetic field. The electromagnetic field may be described as a collection of harmonic oscillators. The ground state of each oscillator is described by a Gaussian function. The individual oscillators, in turn, govern the magnitude of the magnetic field throughout space. We cannot know both E and B simultaneously; E is governed by the time rates of change of the amplitudes. Thus we confine our discussion to B.

Just as the magnitude of B throughout space is governed by the amplitudes of the different field oscillators, so conversely we can find the different field oscillator amplitudes from the magnetic field. When we express the oscillators in terms of B, we get the following expression:

Table 4.1

|  | 1910's | 1970's |
|---|---|---|
| Dynamical Object: | Electrons | Geometry |
| Collapse | Should have collapsed in $10^{-15}$ sec into the nucleus; $\infty$ K.E. | Collapse to infinite compaction in finite time. Einstein equations + assumption of closed universe → collapse of universe. Big Bang represents same problem in time reversed picture. Radio astronomers hope to detect the focusing effect of a closed universe. |
| 1st try at escape | Give up Coulomb's law. Nothing better found. | Give up the Einstein equations. Nothing better found. |
| 2nd try at escape | Electron orbits without radiating. Causality problems. | Only allow limited compressibility of matter; but then $v_{sound} > c$ for high pressures: $$v^2 < \frac{dp}{d\rho} < c^2$$ |
| Way out via the Quantum principle | $\Delta p \sim \frac{\hbar}{\Delta x}$<br>Led to chemistry of atoms and molecules. | $\Delta \pi \sim \frac{h}{\Delta G}$<br>Not possible to give deterministic prediction of time development of geometry because can't know both geometry and its time rate of change simultaneously. |

$$\exp\left(-\frac{1}{32\pi^3 \hbar c} \int \frac{B_1 B_2}{v_{12}^2} d^3x_1 d^3x_2\right)$$

This tells us the probability of a distribution of magnetic field throughout all space. The most probable field is everywhere zero, but small departures from 0 are not much less likely, and there are so many ways to make these departures, that some fluctuation is much more probable than $B = 0$.

To get an order of magnitude estimate, what is the probability of a fluctuation $\Delta B$ from 0 in a region of four-space $L^4$? The integral is $\sim (\Delta B)^2 L^4$; we then set the exponent to 1 and solve for

$$\Delta B \sim (\hbar c)^{1/2}/L^2.$$

This is an estimate of the probability of a field fluctuation of magnitude $\Delta B$ in a region $L^2$. The smaller the region considered, the larger the fluctuation expected. The electromagnetic potential fluctuation goes like

$$\Delta A \sim (\hbar c)^{1/2}/L.$$

About the most important development since World War II was the prediction and confirmation of the effect of these fluctuations on the motion of the electron: on the theoretical side Tomonaga, Schwinger, Feynman, Dyson, Bethe; on the experimental side Lamb and Rutherford, who determined the shift of the energy levels of the hydrogen atom. The electron felt not only the field of the nucleus, but also the fluctuation field throughout all space, which produced an energy level shift given by the spread in the position of the electron produced by the field

$$\langle \Delta E \rangle \sim (\nabla^2 V_{nuclear})(\Delta x)^2$$

Since $V_{nuc.}$ is known, one can measure $\langle \Delta E \rangle$ and find $\Delta x$ to compare with theoretical predictions. The results checked. The great lesson of physics since World War II has been the inescapability of these fluctuations throughout all space.

We use the same ideas for gravity:

$$\Delta g \sim \frac{(\hbar G/c^2)^{1/2}}{L} = \frac{1.6 \times 10^{-33}(\text{Planck length})}{L}$$

The fluctuations get bigger the closer you look, like an aviator looking at the surface of the ocean.

There is another way to look at what we have just said here. We go back to the story of superspace and recognize that the dynamic development of geometry with time (expansion, recontraction) is a classical picture. Quantum mechanically there is a wave associated with this history and this gives an indeterminism to the whole picture.

Although we could never go to a different history of the universe, it is an interesting point of principle, that there is no difference between these quantum fluctuations off the classical history and a point on some quite different history of the universe. The essential point is that we do not have a deterministic track through superspace.

To discuss in detail the motion in superspace, from a quantum point of view, we apply the same set of ideas that we applied in Lecture 2 to discussing the details of a particle moving in a potential. The point of view which most closely unites the classical points of view is the Hamilton-Jacobi one. The phase of the particle wave is the Hamilton-Jacobi function of the Classical trajectory.

The General Relativity analogue of the dispersion relation for geometry is

$$\begin{pmatrix}\text{Extrinsic}\\ \text{Curvature}\end{pmatrix} + \begin{pmatrix}\text{Intrinsic}\\ \text{Curvature}\end{pmatrix} = \begin{pmatrix}\text{Energy}\\ \text{Density}\end{pmatrix}$$

The extrinsic curvature is related to the time rate of change of geometry and is the momentum. In the particle case we let $p \to \partial S/\partial x$. Here we write for the extrinsic curvature

$$\frac{1}{2g}(g_{ik}g_{jl} + g_{il}g_{jk} - g_{ij}g_{kl})\frac{\delta S}{\delta g_{ij}}\frac{\delta S}{\delta g_{kl}}$$

This tells us the whole story of the propagation of these waves in superspace at the Hamilton-Jacobi level of discussion. A nice paper by Gerlach in the Physical Review about 4 years ago shows how one can work back from this equation to recover general relativity. One can require that propagation via different

routes in superspace should give consistent results. (Must be
related somehow to the boundary of a boundary = 0 discussion.)
Kuchar has shown recently that one can, by the condition of
consistency of propagation alone, derive our Hamilton-Jacobi
equation (and hence General Relativity). So General Relativity has some great simplicities in it, that we are in the process of learning about.

Note that the phase of the probability amplitude is

$$S = S\left[{}^{(3)}G\right]$$

where ${}^{(3)}G$ = the equivalence class of all 3 geometries connected by coordinate transformations, independent of coordinate choice.

What is the physics of all this? What has it to do with the central, underlying issue, the final state of gravitational collapse?

Although there are similarities, there are also great differences between the motion of particles and geometries. Although in the Schrodinger theory the proton potential is singular, we can go ahead and treat Rutherford scattering and get a probabilistic description of the final scattering states. The superspace formalism, however, gives us no information on what final states we might expect after the singularity of collapse. It is as if the situation is trying to tell us we are not thinking about things in a deep enough form. Maybe we are under an illusion that there is such a thing as three dimensional space and thus are unable to get past the singularity.

Nevertheless, the most natural thing to think of is probabilistic scattering to alternate universes. When one does so, he is forced to recognize that all the dimensionless numbers in physics are things we have been trying for years. Have we not been giving the right answers to the wrong question? Maybe these are nothing more than initial value parameters appropriate for this particular cycle of the universe, and might have different values in a different cycle.

Then we come to the point of view, already briefly touched upon, brought forward by Dicke and Carter, the idea that most cycles of the universe have physical constants incompatible with the existence of life and that the physical constants we see in this cycle are very special in this respect. One could call this the principle of biological selection of physical constants.

This concept of the physical constants not being constant opens one's eyes to a number of issues. It is normally believed that the process which gives rise to matter in the universe gives equal amounts of matter and anti-matter. But one of the truly impressive features of black hole physics is that the details of what falls into the black hole get washed out. All that remains are total charge, angular momentum and mass, which are represented by infinite range forces (Gauss' theorem). But such things as baryon or lepton number have no long-range fields associated with them, so a black hole has no measurable baryon or lepton number. This might be different if, say, baryon number were associated with a long-range field, but none has ever been seen, and there are arguments that none can exist. There is no possible way to distinguish between a matter or an anti-matter black hole. Baryon and lepton numbers are not destroyed by the act of falling into the black hole, but the concept of the usefullness of their conservation is transcended. If a black hole existed in the early universe, one could, in principle, drop in all the anti-matter and leave only matter. Thus there would be no need for them to be equal.

There has been some discussion of the concept of a white hole as contrasted to a black hole. I would like to give a set of arguments why I think it is unreasonable to think of white holes.

When an electron passes by a nucleus it can accelerate, radiate energy to infinity, thus losing energy. Another solution of Maxwell's equations would allow just the right radiation, perfectly timed, to arrive from infinity, produce the same acceleration and increase the electron energy; nothing in microphysics rules this out; we must rely upon our experience: it is just like heat never flowing from a cold to a hot body. In many ways the surface of a black hole is like infinity. We expect radiation to go in, not out. No physics requires this to be so; it is just a generalization of our experience of the reversibility of physics. The existence of a white hole in nature would be contrary to every generalization of statistical mechanics.

But the first days of the big bang universe is the exact opposite of a black hole. If we want to talk about a white hole,

# General Relativity, Collapse and Singularities

this is the biggest one ever, with everything coming out. At this one time all the laws of statistical mechanics are violated. Just as conservation of baryon and lepton number are transcended by a black hole, we can think of them as transcended in the big bang, with more matter coming out than anti-matter.

1972 is the 500th anniversary of the conception of Copernicus, who freed physics from the idea that everything is centered on man (or the Earth). What kind of a step backward from that is it for us to be talking of the biological selection of physical constants, of physical constants being tuned to the existence of man? We could be unhappy with this, considering the kind of crazy theological discussions that went on before we got physics on a straight track. Isn't it an enormous step backwards? But we simply have to face each new issue on the road of physics. If we are going to be discussing the problem of gravitational collapse as the most central problem of the physics of our time, or that physics has ever faced, then we must re-examine all of our pre-suppositions of the past. One of these presuppositions is that the chances are overwhelming that there is life elsewhere, just as there is on Earth. But now we are talking about a completely different point of view toward the universe; we are talking of the cheapest universe we can get by with and still have life present in it. Thus the probabilities would be strongly against having life elsewhere. This prediction is decisively different from the common view.

In the same spirit, imagine a camera with a trigger mounted at some height above the ocean

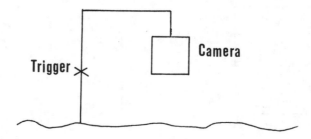

When a bigger than usual wave comes along, the camera will take a picture which looks like this:

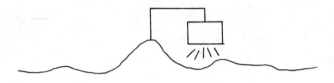

When we try to explain why the picture looks like this, the answer is that we selected it. In the same way, we, by our existence, have selected a particular universe.

What kind of picture does this open up to us for the future? It would be a great mistake to think that this picture of the dynamics of geometry is the end of the road. It is only the beginning of the road. There is nothing more impressive among the recent developments than what Sakharov was saying in his picture that geometry is only the metric elasticity of space (Fig.

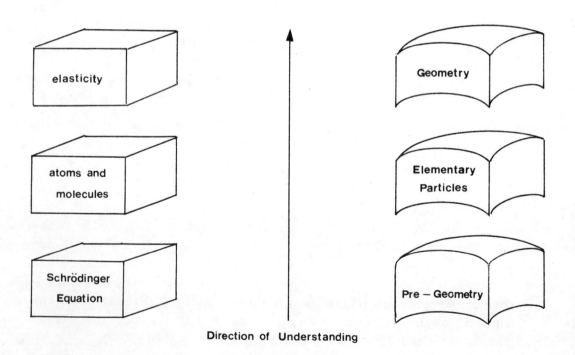

Figure 4.1

# General Relativity, Collapse and Singularities

4.1). We believe that under elementary particles is a simpler underlying structure, called Pre-Geometry, or whatever. We must understand this before we can understand elementary particles, just as one cannot understand the richness of chemistry without the Schrodinger equation.

In earlier history, it was possible to do experiments that worked down; in the present crisis, no one has ever seen a way to penetrate beneath elementary particles to pre-geometry by experiments, whether we express the difficulties in terms of the distances involved or the energies necessary to reach the distances. It would appear that the only hope we have to penetrate to this level of structure is by a leap of the imagination. One recalls the person who, of all people, made the greatest step forward by a leap of the imagination, guided by a central philosophical principle, the principle of equivalence, Einstein, and his great statement: "What really interests me is whether God had any choice in the creation of the Universe."

What magic quality is there that translates from equations into reality. There must be a principle so compelling, so simple, so strange, so beautiful that, if one once saw it, he would realize immediately that there could be absolutely no alternative to it. So one is invited to explore this avenue among others, and to ask himself, what is the unique guiding principle?

This has been at the center of the thoughts of many philosophers over many decades. It is therefore interesting to think back to Leibniz, mentioned in the first lecture, and his concept that logic was the central be-all and end-all of all discussions. The simplest item of structure that we have at all is the choice between yes-no, true-false, up-down. The simplest of all structures that one knows how to build or that elementary building block is the calculus of propositions. If we consider short propositions and long propositions and the statistics of large numbers of propositions, we can ask ourselves whether out of this we can get anything which, in a statistical limit, is as different from the elementary mechanics of collisions of 2 and 3 particles. These are propositions about nothing, like a switching circuit. It has no purpose except as it is put to use.

We are talking of the idea in the sense of Leibniz that the loop of analysing structure where we penetrate more and more deeply at last comes back to the observer himself.

Far from saying that the universe collapses and starts over

a new cycle that's like the present cycle, we say the direct opposite. The universe has a well defined structure only on the statistical average. When it has collapsed it is as structureless as wisps of hay blowing around in the wind both before and afterward. The very production of the organization here is a part of and connected with the existence of ourselves.

This may be an impossible way of looking at the situation, but it is helpful to be guided by people who have spoken in the past:

Wigner: "As we consider situations in which consciousness is more and more relevant, the necessity for modifications of the regularities obtained for inanimate objects will be more and more apparent."

In quantum mechanics, we ourselves must participate to make things happen; whether we are going to observe momentum or energy, etc. A single slogan for the whole story I am trying to spread out here would be: OUR PARTICIPATORY UNIVERSE. This concept that we ourselves are bringing about what we see is so upsetting that we recall Schrodinger's words at the Copenhagen Seminar, when he heard about the uncertainty principle and such things, "I wish I had never had anything to do with it."

Other phrases out of writings of the past: "The shaping power of the mind," and "what we choose is what we are". The concept of Leibniz is expressed in the words of Carr: "The mind experiences the universe as opposition felt within to an activity developing outward." Schelling in Weyl's "The Philosophy of Natural Science": "Thus there ought to be discernable in experience something which, without being in space would be the principle of all spatiality." and Thomas Mann: "As in a dream, where our own will unconsciously appears as inexorable objective destiny, everything in it proceeding out of ourselves and each of us being the secret theater manager of our own dreams, so also in reality the great dream, which is single essence the will itself dreams with us all. Our fate may be the product of our inmost selves, of our wills and we are actually bringing about what seems to be happening to us."

The central ideas, then, are this Leibniz loop, on the one hand, and this concept that there is a structure down underneath that is far simpler and far stranger and far more beautiful than anything we have had to deal with before, which somehow furnishes the mathematical framework, the calculus of proposition as a first working hypothesis for a pre-geometry which produces the framework within which the loop is set up and operates.

# Index

Abnormal galaxy, 111, 122, 237
  criteria for, 117–121, 122–123
  filamentary structure of, 122
Absorption, 358, 362, 372
  advanced wave, 361, 362, 363
  complete, 358, 380
  perfect, 363, 370, 380
  retarded wave, 361, 363
Action-at-a-distance theory, 348, 374
$\alpha$-particle, 410
Ambartsumian knot, 34
Angular size, 89, 90, 237, 241, 242, 246
  measurement of, 252
Anisotropy, 101, 474
  initial, 104
  and shear, 117
Antenna system, 65
Antimatter, 97
  and matter, 392, 393, 567
Antineutrino, 195
Antinucleus production, 278
Antiparticle, 265, 337, 407
Arp hypothesis, 178
Arrow of time, 389
  cosmological, 389, 391–393, 395
  direction of, 392, 395
  microscopic, 391–393, 394, 395
Atmospheric absorption, 204
Attractive heart, 407–409. See also Repulsive core

Balmer limit, 137
Balmer series, 116, 138, 146
Baryon, 198, 288, 410
  production of, 198
Baryon number, 259, 280, 287, 288, 409, 547, 566
  conservation of, 546, 567
Big Bang cosmology. See Cosmology; Universe
Binary star, 204, 215, 216, 219, 221, 223, 225, 228, 552. See also Star, double

Binary star (Continued)
  closed, 229
  double spectroscopic, 227
  eclipse of, 221, 223, 226
  and evolution, 232–233
  period of occultation of, 223, 225, 228
  radius of orbit, 224
  short-period, 204
  variation in period of, 224
  as x-ray emitter, 204, 231, 233
Black body radiation, 195, 205, 238, 259, 267, 364
Black body radiation curve, 337
Black body radiation field, 97
Black body spectrum, 175, 196, 337, 471
Black hole, 185, 196, 217, 219, 252, 285, 405, 407, 445, 502, 521, 527–528, 546–548, 553, 556, 558, 559, 566, 567
  antimatter, 560
  dead (nonrotating), 528, 553
  detection of, 252, 552
  energy of, 559–560
  f-gravity, 452
  formation of, 528, 551
  live (rotating), 528, 553, 557, 559
  Kerr, 557
  mass of, 527, 555
  tidal force of, 560
Black hole physics, 321
Blank field, 21, 23, 25, 40
Blueshift, 115, 166, 167, 308
  in QSOs, 185
  in Seyfert galaxy, 166
Boson, 272–273
Bremsstrahlung, 192, 205, 239
  thermal, 192, 205, 215, 227, 248

Carbon-to-hydrogen ratio, 156
Cartan's moment of rotation, 535, 536. See also Geometry

Cascade, 205, 411
Centaurus cluster, 238, 241.
    See also Clusters; Galaxy,
    cluster of
Cepheid. See Variable star
CFN paper, 280, 281, 284, 287,
    288
Classical action principle,
    350-353
Closed loop, 378-384
Clusters, 117, 249. See also
    Galaxy, cluster of;
    Globular cluster; Star,
    cluster of
  galactic, 117
    Cepheids in, 469
  kinematic center of, 244
  mass of galaxies in, 247-248
  spectra of, 246
  star, 476
  velocity of, 117
  x-ray emitting, 244, 246,
    247
Collisional damping, 363, 371
Color-magnitude diagram, 479,
    480
Coma cluster, 49, 61, 241,
    244, 501-502. See also
    Galaxy, cluster of
Compact galaxy. See N galaxy
Compact nonthermal sources,
    data for (table), 168-169
Companion galaxy, 16, 37, 53,
    61, 63, 68, 72
  compact, 71
  high-redshift, 42, 69
  table, 17
Core, 405
  lagging, 405, 406
  of quasar, 405
  repulsive, 407
  unexpanded, 407
Coriolis force, 431
Cosmic ray, 195
  multiple collisions of, 411
  residual, 209
Cosmic ray event, 276
Cosmological constant, 187,
    477, 478

Cosmological model, 141, 369.
    See also Cosmology;
    Universe
  with curved space-time, 359
  and perfect absorption, 363
Cosmology, 83, 279, 332, 349,
    459. See also Universe
  Big Bang, 79-108, 257, 280,
    334
  classical, 83
  pressure in, 85-88
  relativistic, 84-92
  Steady-state theory of, 92,
    93, 177, 188, 333-338
  thermodynamics and, 257
Counterjet, 8, 9, 20
CP symmetry, 389
  violation of, 389
CPT, 393
CPT conservation, 394
CPT violation, 389
  and decay of neutral kaon,
    394
  thermodynamics of, 394
CP violation, 445, 452
Crab Nebula, 114, 215
  filamentary structure and,
    129-130
  and polarization, 252
  pulsation of, 220
  spectrum of, 114-115, 116,
    246-247
  as x-ray source, 205, 207
Current, 306, 352
Cyanogen excitation, 336

Deceleration parameter, 84,
    187, 477, 478
  value of, 485-490
Density distribution, 282,
    330
  homogeneous, 282
Density fluctuation, 282,
    283, 288
  and galaxy formation, 287
Density wave method, 431
  and galactic structure, 431
Density wave theory, 39

Deuterium, 334. See also
  Hydrogen
 abundance of, 334
Deuterium molecule, 409
  fusion and fission analogy
    of, 409
D galaxy, 460
Differential rotation, 42
Digicon, 143, 157, 158
  description of, 144-145
Dirac theory, 338-339, 341
Disk model of elliptical
    galaxy, 430-431, 432
Dislocations, 524
Distance, four-dimensional,
    229, 301, 325
Distance measurement, 182
Distance modulus, 467, 469,
    475
Distance-redshift
    distribution, 178
Distance scale, 8, 214
  absolute, 80
  and timescale, 525
Doppler broadening, 512
Doppler effect, 178, 300, 303,
    455
Doppler shift, 80, 151, 308,
    455
  cosmological, 177
Double bubble, 91, 194
Double galaxy, 113, 171, 463-
    464
  orbital motion of, 113
Doubling, 67-68
Dust, 111, 502
Dust cloud, 544
Dust lane, 112, 113, 132
  of elliptical galaxy, 131
  of normal galaxy, 131

E galaxy. See Elliptical
    galaxy
Einstein theory, 312. See
    also Relativity
Ejection, 14, 16, 18-19, 22,
    28, 38, 41, 43, 48, 128-
    131, 194, 197
  of companion galaxy, 42

Ejection (Continued)
  direction of, 12, 128
  double, 36
  examples of, 33-40
  explosive, 152
  in galaxy, 21, 23
  isotropic, 39
  line of, 8
  of QSOs, 26
  quasar, 21
  rules of, 5, 9, 46
  secondary, 69
  spectrum of, 9
  and spiral structure, 19
  of stars, 197
  velocity of, 26
Ejection hypothesis, 11, 19-
    21, 25, 26, 28, 34, 35,
    39, 53, 55
Ejection processes, 42
Ejection velocity, 133
Elastic constant, 523
Electric field, 355, 356
  divergence of, 533
  magnitude of, 533
Electrodynamics, 347
  calculations for, 383
  Maxwell's, 351
  quantum formulation of, 365-
    373
  survey of classical, 347-350
  vector dominance model of,
    449
Electromagnetic action, 312
Electromagnetic field, 305,
    316, 348, 373, 377
  and harmonic oscillators,
    561
Electromagnetic interactions,
    257, 271, 313, 316, 343
Electromagnetic theory, 311
  action-at-a-distance, 311
Electromagnetic wave
    propagation, 364
Electromagnetism, 347, 526,
    533, 539
  and cosmology, 347
  and Coulomb's law, 347
  and gravitation, 529, 538

Electron, 337, 342, 343, 377
  gravitational radius of, 384
  motion of, 563
  and photon, 503
  relativistic, 375-378, 503
  self mass of, 445, 447-448
  as thermalizer, 337
Electron field, 376
Electron mass, 303
  variable, 186
Electron path, 378
  time reversal of, 378
Electron source, 236
Element, 102
  creation of, 104
  formation of, 102, 177
  heavy, 102, 103, 177, 188, 334, 525
  origin of, 334
  superheavy, 407
Elliptical galaxy, 3, 6, 7, 9, 14, 21, 59, 111, 133, 417, 419, 420, 423, 456, 459, 463, 481, 502, 503, 507, 508, 510
  axis of symmetry of, 420, 421, 428, 430
  bright, 494-495
  characteristics of, 6-7, 18
  companions of, 19, 45
  distance of, 50
  dwarf, 464
  dynamics of, 417-422, 423-430
  and ejection, 34
  giant, 464, 501, 503
  group of, 20
  hypothetical, 421-422
  mass of, 113
  peculiar, 18
  properties of, 417
  QSOs in, 61-63
  and quasars, 500
  and radio sources, 22, 23
  table, 10-11
  redshift of, 49
  rotation curve of, 430
  spectra of, 480, 491
  spherical, 113, 419

Elliptical galaxy (Continued)
  spheroidal, 37
  star formation in, 479
  structure of, 112, 417-422, 423-430
  surface brightness of, 417
Emission, 417. See also Bremsstrahlung; Radiation
Energy, 264, 266
Energy density, 259, 280, 537, 541
  and curvature, 534
  fluctuations of, 279
  initial, 282
Energy fluctuation, 280, 281-282, 284, 563
Energy state, 375
  negative, 374
  positive, 374, 378
Entropy production, 402
Equations of motion, 301, 329, 361
  classical, 301
  nonrelativistic, 357
  particle, 316, 521
  relativistic, 351
Ergosphere, 557, 559
Exchange-degeneracy, 408
Expansion rate, 88
Expansion velocity, 69
  of Crab Nebula, 115-116
Explosive phenomena, 203
Extragalactic objects, 460
  space densities of, 460-465
Extrinsic curvature, 534, 537, 540, 564

Faraday rotation, 31
Fermi model, 269, 270
Fermion, 272-273
Filament, 33-35, 42, 43, 46, 49, 53, 69, 130, 156, 308
  curved, 308
  high-redshift, 129
  high-velocity, 69
  luminous, 46
  of nebula, 115
  tidal, 53, 55
  velocity of, 123

Fine structure, 180
Fine structure constant, 302, 525
Fireball, 266, 267, 268, 273
  ideal, 272
  mass of, 269, 273–274
  and resonance, 274
  temperature of, 277
  velocity of, 266
Fluctuation cell, 281
Fluid, 82
  dynamical equations for, 82–83
  expanding, 83
  particles as, 317
  perfect, 86
  self-gravitating, 82
  velocity of, 105
Flux level, 20, 23, 53, 93, 129, 177
  f-meson, 444, 445, 449, 450, 451
  and graviton, 444
  and hadron, 449
Forbidden lines, 138, 146, 150, 514
  and electron density, 146–147
  of oxygen, 3, 20
  in Seyfert galaxy, 124
Frautschi's statistical bootstrap calculation, 281, 290–292
Free-particle thermodynamics, 289

Galactic collisions, 248
  and spiral arms, 430
Galactic nucleus. See Galaxy, nucleus of
Galaxy, 5, 90, 175, 310, 405
  bright, 479, 484
  Centaurus chain of, 6
  chain of, 9, 11, 13–15, 20, 22, 36, 65–67, 508
    observed by Seyfert–Table, 511
  cluster of, 47–50, 71, 79–81, 89, 90, 95, 114, 123, 194

Galaxy, cluster of (Continued)
    234, 308, 351, 460, 462, 464, 484–486, 489–492, 501–511
    distant, 335
    giant, 477
    mass of, 503–504
    properties of, 455
    quasars in, 491–494
    redshift and, 455, 479, 493
  collision of, 130
  coma-shaped, 66
  content of, 111
  cores of, 65, 196
  deceleration of, 478
  diameter of, 465
  disk, 457
  distance of, 80, 183
  distant, 334
  double, 133, 171, 463–464
  dust in, 112–113
  dwarf, 471
  dynamics of, 417–422
  ejection from, 3–15, 130, 194
  emission lines in, 45
  evolution of, 194, 334
  evolutionary sequence of, 72
  expansion of, 48
  formation of, 98, 100, 175, 176, 197, 280, 287, 288, 406
  forms of, 113
  gravitational interaction between, 118, 509
  halo of, 154
  high-redshift, 19–21, 52
  hot, 436
  kinetic energy of, 117
  local group of, 16, 476
  lower-luminosity, 61
  low-redshift, 19–21, 46, 56
  luminosity class of, 471, 472
  magnitude of, 464, 471
    absolute, 471
    vs. diameter, 472
  luminosity of, 323
    intrinsic, 324
  mass of, 111, 113, 117, 247
  Milky Way, 94, 171

Galaxy (Continued)
  Cepheids in, 469
  motion of, 476
  supernovae in, 468
  morphological type of, 72, 456-460
  multiple interacting (table), 57
  nearby, 50-57
  neighborhood of, 63-69
  nonequilibrium form table, 64-65
  nucleus of, 176, 247, 462
    coalescence of stars in, 527
    emission lines in, 504
    explosions of, 430
    properties of, 417
    as QSO, 184, 196, 198
    as quasar, 495
    radiation from, 187, 234, 239, 241
    size of, 456, 458
  number densities of, 461
    table, 461-462
  optical, 91
  organization of, 113-114
  origin of, 104
  Our. See Milky Way
  properties of, 455-456
  quasars and normal, 58-63
  QSOs and bright, 22-29, 182, 183, 184
  radio arms of, 19
  as ratio emitter, 112
  recession of, 333
  redshift of, 114-116, 491, 508, 514, 515
    and magnitudes for Arp 330 (table), 508
    in VV 282 - Table, 507
  rotation of, 12, 46, 111-113, 116, 128, 130-132, 430
    typical period of, 128
    velocities of, 154
  second generation, 9
  spectral lines in, 512-515
  spectral type, 417
  spectrum of, 116-117, 246, 479

Galaxy (Continued)
  absorption, 512-513, 514
  emission, 513-514
  major absorption lines in (table), 512
  strongest emission lines in (table), 513
  structural peculiarities of, 118, 122
  structure of, 111-113, 417-422, 436
  supercluster of, 8
    local, 8, 28
  supernormal, 113, 131
  surface brightness of, 456, 464
  synchrotron emission in - Table, 119-120
  visible, 96, 102, 477
  x-ray emission in, 234
$\gamma$-ray astronomy, 204-205
Gas, 111, 129, 185, 194, 195, 215, 432, 438, 502
  clouds of, 125, 214
  cold, 193
  diffuse, 188
  ejection of, 132-133
  in galaxies, 417
    distribution of, 436
  hot, 131, 133, 135, 145, 192-193
    absorption lines in, 124
    emission lines in, 123
  from hot stars, 151, 160, 162
  intergalactic, 192, 193, 194, 248-249
  interstellar, 204, 205
  ionized, 171
  in QSOs, 161-165
  spectrum of, 154
Geodesic deviation, 523, 535
Geometry, 523, 543, 544
  Bianchi identity and, 536
  curved space, 523, 541
  dispersion relation for, 564
  dynamics of, 527, 568
  four-dimensional, 532, 534, 538-539, 541, 542
  Riemannian, 305, 535

Geometry (Continued)
  Schwarzschild, 535, 541, 542, 544, 555
  spherically symmetric, 540
  three-dimensional, 541, 542
  time rate of change of, 527, 564
Globular cluster, 420, 432, 467. See also Cluster; Star, cluster of
  magnitude of, 467
Globular cluster mass, 105
Grain, 283
  decay of, 283, 285
  drift of, 284
  mass of, 284
Grain structure, 285
Gravitation, 347–452, 526, 534, 538
  and electromagnetism, 529
  and elementary particles, 443
  Newton's law of, 347, 545
  and space, 523
  and strong interactions, 443–452
Gravitational collapse, 405, 524, 529, 540, 541, 543–548, 561, 567
  and black hole physics, 521–528
  final state of, 565
  paradox of, 525
  of star, 560
  of universe, 560
Gravitational energy, 196
Gravitational field, 316
Gravitational interaction, 509
Gravitational radiation, 528, 550
  from black hole, 551
Gravitational wave, 443, 550
Gravitational well, 185
Gravitation theory, 311, 322
  conformally invariant theory of, 311–315, 322, 337
Graviton, 443, 448, 449, 450
  and f-meson, 444
  and lepton, 449
Gravity, 443, 524, 563

Gravity (Continued)
  f-, 452
  fluctuations of, 563–564
  quantum nature of, 443, 444

Hadron, 257, 259, 260, 267, 282, 411, 445
  decay of, 282
  and f-meson, 449
  high-energy scattering of, 262
  interactions between, 271
  kinds of, 258, 272
  mass spectrum of, 275
Hadronic era, 280, 286, 287, 288
Harmonic oscillators, 561–563
H$\beta$ line, 3
Helium, 177
  abundance of, 334
  and deuterium molecule, 409
  formation of, 177
  and hydrogen, 337
  ignition of, 232
  primordial, 177
Hercules cluster, 34, 114
High Energy Astronomical Observatory, 252
H-theorem, 394
H II region, 51, 52, 90, 417, 468, 470, 471, 513
  diameter of, 466, 468, 471, 472, 476
  emission lines and, 137, 513
  gas in, 145
  luminosity of, 476
Hubble classification of galaxies, 111, 417, 418, 456, 457
Hubble constant, 187, 318, 331, 333, 360, 371, 455, 456, 459, 460, 464, 466, 468, 475, 476, 477, 489
  observed value of, 334
Hubble curve. See Hubble line
Hubble diagram, 36, 71, 72, 118, 126, 474, 488
  dispersion in, 117
  magnitude scatter in, 476

Hubble expansion parameter, 83, 88
Hubble law, 79, 80, 95, 455, 467, 492
Hubble line, 61, 71, 335
Hubble relation, 155, 310, 324
Hyades cluster, 469
Hydrogen, 188, 189, 193. See also Deuterium
 atomic, 189
 clouds of, 470
 cold, 188, 191
 and helium, 337
 intergalactic, 190-191
 interstellar, 190
 ionized, 371
 and neutron decay, 192
 opacity of, 471
Hyperon, 410
Hypersurface, 534, 535, 540

Image Dissector Super Scanner, 143
 description of, 143-144
Index of refraction, 355, 361
Infrared emission, 125, 239
Infrared source, 132, 503
 discrete, 236
Initial neutrino flux, 103
Interacting system, 63
Interaction, 266, 270, 349, 377, 378, 380, 382, 503
 advanced, 348
 interparticle, 348
 of isolated charges, 347
 of particles, 332, 352, 368, 382
 retarded, 370
Interaction tail, 53
Intergalactic medium, 111, 188, 194
 density of, 248
Internucleon force, 407
Intrinsic curvature, 534, 537, 541
Inverse Compton effect, 205, 235, 503-504
Inverse Compton scattering, 238, 239

Irregular galaxy, 111, 417, 513
 defined, 456-457
 type I, 456-457
 type II, 457
Isotropic background, 250
Isotropy, 79
 of universe, 101, 337

Jet, 3, 4, 8, 9, 20, 130, 131, 170

$K^{\circ}$. See Kaon
Kaon, 389, 390
 decay of, 389, 394, 395
 decay products of, 394
 neutral, 389
Kaon beam, 389, 390, 392, 393
 decay of, 391
 decay products of, 391
 neutral, 389
Kaon system, 394
 neutral, 394, 395
Klein-Kaluza theory, 445, 452
K-meson. See Kaon

Large Magellanic Cloud. See Magellanic Cloud
Least action principle, 522
Leidenfrost phenomenon, 287
Lepton, 342
 decay of, 392
 and graviton, 449
 model of, 342
 and photon, 449
Lepton-hadron attraction, 450
Lepton number, 546, 547, 566
 conservation of, 546, 567
Light, 323
Light cone, 348
Light distribution, 462, 465
Lindblad resonance, 434
Line locking, 164, 165
Local field theory, 374
Local Group, 16, 30. See also Galaxy
Log N-log S curve, 91, 93, 94, 177, 251, 335, 499
 for QSOs, 336

Log N-log S curve (Continued)
  for radio galaxy, 336
Luminosity, 61, 204
  apparent, 507
  of Cepheids, 466
  infrared, 126
  intrinsic, 28, 458
    dispersion in, 491
    of quasar, 495
  of night sky, 81
  radio, 126
  of stars, 53
  x-ray, 166, 236, 238, 240, 251
Luminosity class, 458, 470, 473, 474
  calibration of (table), 460
  Van den Bergy, 458-459
Luminosity distribution, 134, 212
Luminosity-volume test, 491
  for quasars, 494-500
  and radio sources, 496-497
Lyman transition, 189, 193
Lyman limit, 138, 157, 165, 471, 496

Magellanic Cloud, 171
  distance modulus of, 171
  Large, 211, 214, 235, 237
  Small, 233
Magnetic field, 177
  Earth's, 208
  divergence of, 533
  primordial, 177
Magnetic monopole, 407
Magnitude, 23, 464
  absolute, 31, 171, 460, 470, 510
  apparent, 24, 27, 460
  bolometric, 89
  of bright stars, 466
  and color, 469
  corrections in, 479-480
  dispersion in, 117, 467, 474, 476, 482, 483
  distribution of, 489
  intrinsic, 24, 458
    dispersion in, 458

Magnitude (Continued)
  limiting, 498
  photographic, 24, 473
  x-ray, 251
Magnitude-redshift relation, 89, 478
  for QSO, 141-142
Main sequence, 469, 480
  absolute magnitude-color, 469
  apparent magnitude-color, 469
  zero-age, 469
Many body problem, 375, 409, 419, 423
Mass, 338
  nature of, 338-343
Mass distribution, 288, 462
  spherically symmetrical, 546
Mass fluctuation, 280, 286
Mass-to-light ratio, 464, 501, 502, 504, 507, 508, 510
  of clusters - Table, 506
  of elliptical galaxies, 510
  of Seyfert's sextet, 510
Mass transfer, 228
  and radiation pressure, 233
Matter, 100, 311, 406, 543
  and antimatter, 392, 393, 567
  conservation of, 88
  creation of, 333, 363
  distribution of, 405
  formation of, 177
  gravitational collapse of, 405
  hadronic, 266, 267, 450, 451
  hypercollapsed nuclear, 407-411
  initial state of, 198
  intergalactic, 150, 175
  leptonic, 451
  mean density of, 187
  nonrelativistic, 100
  rest energy of, 100
  structure of, 521
Matter-antimatter separation, 259, 287
Matter wave, 553
Maxwell field, 373
Maxwell theory, 364

Meson, 261. See also f-meson;
    Kaon; Muon, Pion
  creation of, 410
  K, 389
  P, 449
  W, 407
  π, 267
Meson trajectory, 407
Microwave radiation, 101, 102,
    196
  background, 79, 175, 176,
    192, 195, 334, 336-337
  isotropic, 336
Missing mass problem, 188,
    464, 502, 510
Momentum, 263
  longitudinal, 263
  transverse, 264, 266, 268,
    276
Momentum distribution, 263,
    264, 267, 272
Morgan classification of
    galaxies, 417, 435
Muon, 342, 343
  decay of, 343
m-z plot, 90, 155, 485
  scatter in, 481

Nebulae, 137. See also Crab
    Nebula
  gaseous, 137, 138, 145
Neutrino, 195, 267, 342
  zero mass of, 343
Neutrino density, 196, 260
Neutrino gas, 267
Neutron, 192
Neutron star, 185, 196, 215,
    217, 259, 407, 521, 527-528,
    548-549, 552
  and evolution, 232-233
  large, 450
  mass loss of, 549
  rotating, 197, 215, 552
  rotational energy of, 215
  size of, 411
Neutron-to-proton ratio, 102
N galaxy, 21, 36, 59, 67, 455,
    459, 481, 489, 491, 495
  and Seyfert galaxies, 460

N galaxy (Continued)
  spectra of, 489, 514
Night sky, 81
  darkness of, 81
  emission line of, 513
Noncircular velocity, 118, 122
Normal galaxy, 111, 112, 118,
    234, 250, 308, 417, 462
  absorption lines in, 124-125
  classification of, 456-460
  criteria for, 113-117
  gravitational field of, 113
  properties of, 456-465
  redshifts of, 175, 185
  rotation of, 113
  structure of, 113
  x-rays in, 251
Novae, 152, 171, 220, 460
  optical, 216
  x-ray, 215
Nuclear density, 257
Nuclear isomer, 407, 409-410,
    411
Nuclear mass, 277
Nuclear reactions, 102
Nuclear structure, 261
Nucleon, 280, 407
  binding energy of, 410
  cascade emission of, 411
Nucleosynthesis, 96-97, 98
  explosive, 196
  primordial, 198
Nucleus, 257
  collapsed, 405, 409-410
  decomposition of, 261
  dense, 405
  hypercollapsed, 407, 409, 411
  lifetime of stable, 409-410

Olbers paradox, 81
Omnes' theory, 287
Onsager relations, 394
Oort value, 97
Our galaxy, 112. See also
    Galaxy
  infrared radiation in, 235
  luminosity output of, 213
  nucleus of, 235
  radio emission from, 118

Our galaxy (Continued)
 radio waves in, 235
 x-ray flux from, 237, 251
 x-ray source in, 235

Pair annihilation, 377, 382
Pair creation, 382
Pair production, 445
 electron-positron, 445
Parkes flux scale, 335
Parkes source, 21, 22
Particle, 257, 271, 273, 314, 353
 absorber, 354, 355, 357, 370
  field of, 356
 acceleration of, 347, 355
 advanced field of, 354, 356, 369
 charged, 306, 529
  acceleration of, 529
  charge of, 300, 302, 351
  and Coulomb's law, 347
  nonzero mass of, 343
  system of, 306
  and velocity of light, 348
  world line of, 306, 307, 317, 348, 349, 351, 352, 359, 370, 530-531
 cloud of, 329
 collision of, 265, 268, 270, 272, 276
  experiments with, 261, 273
  high-energy, 267, 272, 277
 cosmological, 323
 creation of, 262-263, 334
 decay of, 288
 density of, 318, 362, 371
 distance between, 303
 dynamics of, 548, 560
 elementary, 258, 407
  defined, 261-262
  and gravitation, 443-452
  and pregeometry, 569
  ratio of electrical to gravitational forces for, 347
 free, 350, 366
  action of, 350, 366, 368
  Newton's law for, 351

Particle (Continued)
 heavy, 278, 286
  decay of, 279, 286
 lifetime of, 265, 266, 279
 Lorentz contracted, 262
 mass of, 276, 278, 300, 303, 307, 312, 321, 328, 350
 mass field of, 319, 341
 massless, 341
 mechanics of, 300
 noninteracting, 257
 number of, 332
 orbit of, 316, 328, 330
  circular, 558
  Kepler, 556
 path of, 301-302, 313, 316, 329, 338-339, 341, 351, 365, 370, 375, 377
  amplitude for, 384
  conjugate, 381, 382
  continuum of, 365
  reversals in, 376
 production of, 268, 270
  rate of, 268, 277
 range of, 267
 relative motion of, 529-532
 relativistic, 271, 338
 retarded field of, 354, 356, 357
 source of, 354-357
 spectrum of, 264
 strongly interacting, 257
 in universe, 318
 unstable, 264
 velocity of, 329, 347
Particle number density, 257, 259, 328
Particle velocity, 300
Path integration, 365-368
Peculiar galaxy, 19, 20, 21, 25, 41-44, 65, 136, 171, 417, 457, 514-515
 chain of, 56, 67-68
 double, 69
 elliptical, 22-23
 redshift of, 65-66
 spectral lines in - Table, 514
 spiral, 22, 23, 33

Period-luminosity law, 469, 470
Perseus cluster, 12, 13, 129, 131, 241, 244, 502-504
Photon, 364, 373, 445, 446, 451
 and H II region, 471
 and lepton, 449
 low-energy, 503
 and P-meson, 449
 source of, 373
 ultraviolet, 471
Photon/baryon ratio, 97, 103
Pion mass, 258, 284
Plasma, 135, 194, 205
 interstellar, 147
 neutral, 259
Point charge, 302
Polarization, 133, 167, 170
 linear, 170
 measurement of, 252
Population, 212. See also Star
 cool, 432, 436
 early, 21
Positron, 375, 377
Precession, 230
Precession mechanism, 230
Pregeometry, 524, 569, 571
Propagator, 325, 339, 359, 374, 376
 chain of, 374
 electron, 445
 equation for, 340
 Feynman, 339, 340, 377
 photon, 445
 for Schrödinger equation, 367
 time-symmetric, 339-340
Proportional counter, 205, 207
Protogalaxy, 286, 405
 mass of, 286
Proton, 332
 and P-meson, 449
Proton mass, 289
Pulsar, 197, 211, 215
 Crab Nebula as, 233
Pulsation, 216, 218-219, 227, 229, 231
 period of, 221

Pulsation (Continued)
 regions of, 219
 of rotating star, 222
Pulse, 227, 230
Pulses, 219
 random noise, 219
 train of, 219

QSO, 3-76, 91, 92, 111, 121, 136, 142, 145, 150, 158, 166, 178, 190, 193, 251, 252, 335, 491, 493, 497, 498, 500
 absorption lines in, 152, 157, 164-165
 multiple, 161
 absorption lines in 3C 191 (table), 151
 absorption lines identified in (table), 154
 blueshifted, 170
 bright, 30, 491
 and bright galaxies, 22-29
 carbon in, 158
 characteristics of, 145
 clusters of, 182
 core of, 163
 distance of, 175
 and elliptical galaxies, 500
 emission lines in, 153, 163
 emission lines in - Tables, 139, 140-141
 and galaxies, 182-184, 196, 198
 gas in, 161, 165
 gravitational field of, 185
 high-redshift, 138, 158
 luminosity of, 61, 180
 luminosity-volume test for, 177
 mass of, 164, 165
 mass loss in, 167
 model of, 146, 181
 radiation from, 147, 185, 187
  continuum, 147
  synchrotron, 147
 radiation pressure in, 164, 167
 radio-quiet, 137, 155, 495

QSO (Continued)
  radio-source, 148
  redshift of, 139, 141, 143,
    149, 152-156, 161, 165,
    177, 455
  spectral properties of, 145
  spectrum of, 139-140, 149,
    158, 512
    absorption lines in, 149,
      150, 153
    emission-line, 143
  variability in, 149
  and white dwarfs, 137
QSR, 3, 21-25, 27, 28, 30, 40,
    183, 310
  bright, 29
  chain of, 15
  distance of, 70-71
  redshift distribution of, 495
Quadrupole moment, 408, 550
Quantum fluctuations, 521
Quantum mechanics, 301, 317,
    341, 368
  nonrelativistic, 301
  relativistic, 338
Quantum principle, 522, 526,
    527
Quantum theory, 313, 367
  conformally invariant, 341
  relativistic, 379
Quark, 186, 198, 268, 407,
    411, 452
Quark-antiquark combination,
    408
Quark-atom, 186
Quark model, 407, 410
Quasar, 6, 8, 9, 21, 23, 25,
    27, 47, 167, 178, 237,
    493, 491, 492, 495, 497
  absorption lines of, 493
  apparent magnitude of, 27-28
  and bright galaxies, 69
  core of, 405
  emission lines in, 136
  high-redshift, 46, 136, 496
  low-redshift, 61
  luminosity-volume test for,
    494-500
  number densities of, 494,
    497-498, 499-500

Quasar (Continued)
  radio-quiet, 46, 497
  redshift of, 24, 492, 493,
    495, 497, 499
  relation to normal galaxies,
    58-63
Quasar core, 63
Quasar count, 405
  and lagging core model, 405-
    406
Quasistellar object. See QSO

Radiation, 113. See also
    Bremsstrahlung; Emission
  excess of, 213
  infrared, 205
  nonthermal, 117-118, 122,
    130, 145, 171
  QSO, 181
  thermal, 113, 215
  types of (table), 206
  in universe, 195
Radiation reaction, 356, 357
Radiation temperature, 205
Radiative damping, 354, 357,
    361, 363
Radio background, 195
Radio brightness, 24
Radio emission, 15, 118, 121,
    122, 130, 133, 134, 239,
    244, 503
  in Seyfert galaxies, 124
  and x-ray emission, 219, 238
Radio flux, 27, 91, 122, 136,
    239, 497, 498
Radio galaxy, 7, 8, 18, 21,
    93, 122, 130, 143, 147,
    183, 192, 250, 478, 481,
    488, 490, 491, 514
  forbidden lines in, 145
  physics of, 122
  representative, 123
  spectrum of, 144
Radio lobe, 6, 10-11, 18, 132,
    134, 238
  size of, 135
  x-ray emission from, 238
Radio noise, 8
Radio source, 6, 8, 11, 14,
    40, 46, 47, 53, 55, 91-93,

Radio source (Continued)
  128, 129, 131-134, 136,
  142, 183, 194, 196, 459,
  496, 497
 bright, 19, 238
 diffuse, 130
 discrete, 236
 distribution of, 335
 double, 18-21, 23, 135, 167
 double bubble, 91, 194
 elliptical galaxies as, 10-11
 and galaxies, 18-19, 22, 39
  associations of, 23, 27
 luminosity distribution of,
  93
 and Mayall's object, 69
 position of, 137
Radio spectrum, 15, 18, 22,
  167
 and x-ray spectrum, 192
Radio tail, 12, 503
Rayleigh-Jeans regime, 100,
  336
Recession velocity, 80, 123,
  125, 463, 467
Redshift, 5, 8, 11, 13, 20,
  24, 27, 28, 30, 31, 36, 44,
  47, 49, 51, 68, 81, 82,
  116, 324, 350, 372, 482,
  489, 496, 514
 absorption, 150, 155, 156,
  159
 and angular size, 90
 anomalous, 16-18, 23, 45, 46,
  49, 52
 average, 32
 and companion galaxies, 42-43
 cosmological, 31, 37, 59, 61,
  146, 150, 159, 180-182,
  455, 493
 and distance, 95, 116, 455
 distribution of, 48, 503
 Doppler, 114, 116, 184
 Einstein gravitational, 16,
  17, 444
 emission, 155, 159
 of ejecta, 23
 excess (table), 74-75
 extragalactic, 58

Redshift (Continued)
 galactic, 175, 455, 474
 gravitational, 146, 161, 164,
  185
 intrinsic, 37, 72, 185, 509
 nature of, 456
 nonvelocity, 16
 Pound-Rebka, 444
 QSO, 32, 86, 91
  distribution of, 178-180
 radiation pressure in, 161
 and recession velocity, 80
 residual, 16
 and rotation, 37
Redshift distance, 52
Redshift-distance relation,
  491
Redshift-magnitude diagram,
  49, 71, 116, 323, 458, 474,
  481-485, 487, 491
 quasar distribution and, 494
 for supernovae, 467
Redshift problem, 3-76, 197
Relativity, 307, 541
 general theory of, 307, 314,
  321, 443, 521-523, 525-529,
  533, 534, 537-539, 544,
  556, 564, 565
 special theory of, 348
Repulsive core, 407, 451. See
  also Attractive heart
Residual background, 209
Resonance, 265, 266, 268, 270,
  271, 273
 defined, 265
 and fireball, 274
 as free particle, 271
 range of, 267
Resonance lines, 150, 167
Roche lobe, 228, 229, 231-233
Rotation curve, 63, 462, 463
Rotation measure, 31, 32
Rotation velocity, 462, 512

Scattering, 262, 377, 383
 backward, 407
 elastic, 452
 of electrons, 378
 experiments in, 265

Scattering (Continued)
  forward, 407
  gravitational, 359
  high-energy, 262
  of particles, 378
  phase shift, 265
Scattering cross section, 265
  electron-proton, 337
Schrödinger's equation, 524
Schwarzschild radius, 260, 285, 318, 544
  of electron, 446
Second law of thermodynamics. See H-theorem
Self-interaction, 378-380, 384
Serpukov accelerator, 277
Seyfert galaxy, 3, 4, 14, 44, 145, 237, 250, 459, 514
  defined, 123
  infrared emission of, 125
  infrared flux of (table), 126
  mass loss of, 167
  nonthermal emission in, 125
  peculiar, 308
  physics of, 122
  properties of (table), 124
  radiation pressure in, 167
  radio, 503
  redshift of, 308
  rotation curve of, 166
Seyfert nucleus, 4, 133
  ejection by, 159
  gas in, 514
  and QSOs, 166
  spectrum of, 123, 124, 514
Seyfert spectrum, 3
  emission lines in, 124
Seyfert's sextet of galaxies, 509-510
  redshifts and magnitudes for (table), 510
S galaxy. See Spiral galaxy
Shapley-Ames galaxy, 21, 24, 25, 27
Shell, 18, 20, 185
  expanding of, 151, 215
  mass of, 18
Shred, 38
Simple harmonic motion, 530-531

Size density fluctuation, 280
Sky subtraction, 143
Small Magellanic Cloud. See Magellanic Cloud
Smoke ring, 20, 69
Solar neighborhood, composition of, 146
Space, 308. See also Space-time
  curvature of, 535
  curved, 359, 361
  de Sitter, 326
  Euclidean, 335
  flat, 308, 315, 322-326, 338, 359-363
  geometry of, 521, 565
  isospin, 342
  Minkowsky, 328-332
    electromagnetic coupling for, 343
    flat, 322
  nonEuclidean, 531
  Robertson-Walker, 315, 322, 324, 326, 328, 331, 332
  spin, 342
  stratified, 523
  structure of, 523
Space-time, 299, 526, 534
  curved, 359
  flat, 299, 305, 306
  Riemannian curved, 303
Spectral index, 24, 27
Special-relativistic time coordinates, 299
Spectrum, 21, 522
  absorption lines in, 21, 36, 116, 125
  in normal galaxy, 124
  in Seyfert galaxy, 166
  widths of, 462
  black body, 100
  electromagnetic, 203
  emission lines in, 21, 59, 66, 116, 125, 148, 462
  hadronic, 273
  lineless, 167, 171
  mass, 258, 267, 270, 275, 282, 284, 289
  low-energy, 281
  radio, 15, 18, 462

Spectrum (Continued)
  transition lines in, 138
Spherically symmetric body, 318
  problem of, 318-321
Spiral arms, 37, 41, 111, 112, 211, 212, 436, 458
  barred, 456
  corotation and, 434
  defined, 430
  and ejection, 40, 42
  emission lines in, 514
  formation of, 430-431, 432-439
  and rotation, 430
  rotational speed of, 434-435
  shape of, 433
  structure of, 213, 214, 433-434, 456
Spiral galaxy, 3, 6, 14, 21, 23, 39, 40, 46, 56, 63, 111, 417, 436, 438, 457, 459, 464, 507, 510
  barred, 41, 43, 456
    formation of, 438
  characteristics of, 7
  classification of, 456
  companions of, 19, 38
  disruption of, 43
  distance of, 50
  and ejection, 33, 34
  and QSOs, 26-27, 183
  and quasar, 46
  redshift of, 50
  young, 69
Spiral structure, 39, 430-431
  of arms, 213, 214
  ejection origin of, 39
Star, 113, 175, 542-544
  age of, 334
  atmosphere of, 225, 226, 228
  B0-type, 217
  B-V of, 480
  cluster of, 113, 469
  collapse of, 527, 545-546
  color of, 480
  cool, 438, 512
  corona of, 236
  double, 113, 148, 552

Star (Continued)
  and element formation, 188
  evolution of, 548
  formation of, 436
  hot, 138, 159-160, 470
    and QSOs, 162
    radiation pressure in, 160
    redshift of, 151
  hydrogen-burning, 469, 479
  intergalactic, 502
  luminosity of, 525
  mass loss in, 548
  orbit of, 434
    circular, 435
    elliptical, 435
    epicyclical, 435
    rosette, 435
  polytropic model for, 421
  population I, 430
  population II, 430
  pulsating, 220
  shell, 152
  spectral lines of, 136-137
  types of, 436
  white dwarf core of, 548
  x-ray, 215, 220, 233
Star sensor, 208
Steady state theory. See Cosmology; Universe
Stebbins-Whitford effect, 334-335
Stellar collisions, 134
Stellar evolution, 196, 231, 324, 478, 479
  and heavy elements, 334
  and massive stars, 231
Stephan's quintet, 50-53, 55, 56,
Strong interactions, 268, 270-272, 278, 332
  features of, 261
  and gravitation, 443-452
  statistical thermodynamics of, 257-296
  thermodynamics of, 410-411
Sun, 204, 237
  gravity of, 443
  x-ray luminosity of, 204

Superbaryon, 287-288, 411
 and cosmology, 410
Supercluster, 30. See also
   Cluster
Supercluster model, 31
Supernovae, 52, 153, 171, 196,
   197, 220, 468
 of 1054 A.D., 114
 chain-reaction, 197
 and evolution, 232
 magnitude of, 467
 remnants of, 215
 type I, 52, 467
 type II, 152
Superspace, 527, 553, 560,
   561, 564, 565
Synchrotron emission, 118, 121
 in galaxies - Table, 119-120
Synchrotron radiation, 9, 15,
   122, 149, 170, 205
 continuum, 130
 optical, 133
 and plasma, 135
 QSO, 181

Temperature, universal
   maximum, 258
Thermal bath, 394, 395
Thermal equilibrium, 280-285,
   287, 288, 394
Thermonuclear reaction, 196
Tidal interaction, 510
Time, 524, 526. See also
   Space-time
Time-reversal asymmetry, 389
Time-reversal symmetry
   violation, 389
 and oscillating universe,
   389-393
Timescale, 50, 134, 148, 156,
   197, 300, 358
 dispersion, 502, 504
 and distance scale, 525
 dynamical, 420
 evolution, 26
 expansion, 97
 of variable star, 217-218,
   220, 229, 230
Time symmetric surface, 540

Transformation, 305
 conformal, 305-310, 312, 314,
   360
Transformation law, 306
Transition, 364, 369, 370
 amplitude for, 365
 atomic, 374
 induced, 369
 Lyman, 189, 193
 spontaneous, 364, 369, 371,
   373, 380
Transition probability, 368,
   373
Tuning fork diagram, 456. See
   also Hubble classification
   of galaxies
21-cm line. See Hydrogen

Uhuru satellite, 203, 205,
   207-209, 211, 216, 219,
   238, 241, 249
Ultraviolet radiation, 471
Ultraviolet source, 138
Uncertainty principle, 526-
   527, 570
Unitary symmetry, 410
Universe, 82, 564
 absorbing, 356, 357, 370, 371
 age of, 88, 334, 476, 502
 antimatter in, 560
 average density of, 98
 Big Bang, 96, 102, 104, 105,
   175, 188, 196, 259, 567
 closed, 477, 521
 geometry of, 544
 with uniform curvature, 534
 collapse of, 445, 451, 545,
   569
 contracting, 357
 time-inverted, 392, 393
 cosmological models of, 95,
   106
 deSitter model of, 88, 326
 Einstein-deSitter model of,
   106, 322, 329, 331, 360-363
 as perfect absorber, 337
 Einstein model of, 86
 energy density of, 100
 Euclidean, 177

Universe (Continued)
  expanding, 357, 370, 411, 456, 526
  cold, 370, 389
  expanding cloud model of, 331
  as expanding fluid, 83
  expansion of, 82, 101, 117, 175, 178, 188, 197, 286, 455, 477
    deceleration of, 477
    rate of, 283
  Friedmann models of, 86, 175, 177, 187, 322–327, 328–332, 337, 478, 540
  global structure of, 260
  gravitational collapse of, 560
  homogeneous, 104, 333
  isotropy of, 101–103
  Lemaitre, 86, 180
  mass in, 320
  matter in, 96, 566
  nonexpanding, 456
  number of particles in, 319
  origin of, 325, 326
  origin of structure in, 104
  oscillating models of, 86
  oscillation of, 391
    period of, 391
    phase of, 392, 393
  properties of matter and radiation in, 187–188
  radiation in, 96
  relativistic cosmological models of, 91
  response of, 350, 359–364
    quantum, 369
  size of, 526
  static Euclidean, 91
  static Newtonian, 85, 104
  steady state, 86, 88, 175–177, 187–188, 192, 360–363, 371
  structure of, 570
  time-symmetric oscillating model of, 389, 391, 392

Van Allen radiation belt, 207

Variable star, 149, 224. See also Binary star; Star
  Cepheid, 466, 470, 471
  galactic cluster, 469
  irregular, 220
  light curve of, 225, 228, 229, 231, 466
  period of, 466, 470
Velocity dispersion, 247–248, 479, 508
  in galaxies, 455
  stellar, 462
Velocity field, 476
  isotropic, 476
  shear in, 476
Velocity of light, 299
Veneziano-type dual models, 274–275
Violent relaxation, 423
  and elliptical galaxies, 423–425, 427
Virgo cluster, 9, 18, 24, 30, 47, 49, 131, 241, 466, 467, 472–474, 476, 481
  motion of, 483

Wave, 360, 370
  advanced, 362
    blueshifted, 362
  high-frequency, 362
  retarded, 360, 361
  redshifted, 360
Weak field problem, 314, 315
Weak interaction, 343, 443
  and cosmology, 387–414
Wheeler-Feynman theory, 350, 357–360, 364, 373
  absorber theory, 354–357
  classical, 379
  quantum aspects of, 374–384
  relativistic quantized, 379
  self interaction in, 357–359
  time symmetry of, 357
White dwarf, 137, 196, 215, 225, 411, 552
  pulsation period of, 226
  and QSOs, 137
White hole, 405–406, 407, 566–567

Wigner-Weisskopf approximation, 391
W meson, 407
World line, 379, 561. See also Particle, charged
 classical, 554
 interaction between, 379, 382
 shape of, 384

X-ray, 192, 195, 215, 225, 239, 503, 528
 background, 192
 and black hole, 552
 diffuse background of, 234
 and neutron star, 552
 and radiation pressure, 233
X-ray background, 195, 209, 249
X-ray emission, 203, 214, 239, 248, 477, 503
 extragalactic, 234
 and optical emission, 219
 of quasar, 237
 and radio emission, 219
 from radio lobes, 238
 soft, 236
 as thermal bremsstrahlung, 504
X-ray flux, 235, 237, 249
 from Centaurus A - Table, 239
X-ray source, 131, 132, 203-206, 225, 226, 231, 240, 241, 244
 binary-star, 216, 233
 constant-luminosity, 212
 diffuse, 243, 250, 503
 discrete, 234
 extended, 243, 246
 extragalactic, 251
  sizes of (table), 245
 galactic, 211, 220, 251
  distribution of, 211
 intensity distribution of, 212
 mass of, 228
 origin of, 210
 point, 242, 243
 variable, 219, 237
X-ray spectrum, 192
 and radio spectrum, 192

Yukawa force, 261

Zero-velocity surface, 231
z-m relation, 49, 117, 149, 178
 and galaxies, 178